Metric Rigidity Theorems on Hermitian Locally Symmetric Manifolds

SERIES IN PURE MATHEMATICS

Part I. Monographs and Textbooks

Volume 1: Total Mean Curvature and Submanifolds on Finite Type
B Y Chen

Volume 3: Structures on Manifolds
K Yano & M Kon

Volume 4: Goldbach Conjecture
Wang Yuan (editor)

Volume 10: Compact Riemann Surfaces and Algebraic Curves
Kichoon Yang

Part II. Lecture Notes

Volume 2: A Survey of Trace Forms of Algebraic Number Fields
P E Conner & R Perlis

Volume 5: Measures on Infinite Dimensional Spaces
Y Yamasaki

Volume 8: Class Number Parity
P E Conner & J Hurrelbrink

Volume 11: Topics in Mathematical Analysis
Th M Rassias (editor)

Series in Pure Mathematics — Volume 6

Metric Rigidity Theorems on Hermitian Locally Symmetric Manifolds

Ngaiming Mok
Mathematics Department
Columbia University
USA

World Scientific
Singapore • New Jersey • London • Hong Kong

Published by

World Scientific Publishing Co. Pte. Ltd.,
P O Box 128, Farrer Road, Singapore 9128
USA office: 687 Hartwell Street, Teaneck, NJ 07666
UK office: 73 Lynton Mead, Totteridge, London N20 8DH

Library of Congress Cataloging-in-Publication Data

Mok, Ngaiming.
Metric rigidity theorems on Hermitian locally symmetric manifolds / Ngaiming Mok.
p. cm.
ISBN 9971508001
1. Complex manifolds. 2. Hermitian structures. 3. Symmetric spaces, Hermitian. I. Title.
QA614.M64 1989
514'.3--dc 19 88-30499
CIP

Printed in Singapore by Utopia Press.

To Julia, Ting-ting and
to the memory of Jia-qing

PREFACE

This monograph studies the problem of characterizing canonical metrics on Hermitian locally symmetric manifolds X of non–compact/compact type in terms of curvature conditions. We call such characterizations *metric rigidity theorems.* In case X is of non–compact type we impose the condition that it is of finite volume with respect to a canonical metric. We use throughout the notion of positivity/negativity of Hermitian holomorphic vector bundles of Griffiths. This notion is of particular importance in our context because of the curvature–decreasing property on Hermitian holomorphic subbundles. It is consequently applicable to the study of holomorphic mappings between complex manifolds. In this monograph *metric rigidity theorems* and their proofs are applied to the study of holomorphic mappings between Hermitian locally symmetric manifolds X of the same type, yielding various rigidity theorems on holomorphic mappings. For example, we prove (more generally) that if X is a quotient of finite volume of an irreducible bounded symmetric domain of rank ≥ 2 and Y is an arbitrary Kähler manifold of seminegative curvature in the sense of Griffiths, then any non–trivial holomorphic mapping $f: X \longrightarrow Y$ is necessarily a totally geodesic isometric immersion up to a scaling constant. Results in the rank–1 case and in the dual case of Hermitian symmetric manifolds of compact type are also obtained under additional geometric conditions. Other applications are made to the study of bundle homomorphisms on locally homogeneous Hermitian holomorphic vector bundles and to Hermitian metrics of negative curvature on bounded symmetric domains. Some open problems are formulated in the text and in the Appendix. In particular, we formulate a dual version of the Generalized Frankel Conjecture on characterizing compact Kähler manifolds of seminegative curvature in part motivated by considerations in this monograph. In short the subject matter lands upon a fertile meeting place of Differential Geometry, Several Complex Variables and Algebraic Geometry.

TABLE OF CONTENTS

PART II *FURTHER DEVELOPMENT*

APPENDIX

I. SEMISIMPLE LIE ALGEBRAS AND THEIR REPRESENTATIONS

II. SOME THEOREMS IN RIEMANNIAN GEOMETRY

III. CHARACTERISTIC PROJECTIVE SUBVARIETIES ASSOCIATED TO HERMITIAN SYMMETRIC MANIFOLDS

IV. A DUAL GENERALIZED FRANKEL CONJECTURE FOR COMPACT KÄHLER MANIFOLDS OF SEMINEGATIVE BISECTIONAL CURVATURE

Metric Rigidity Theorems on Hermitian Locally Symmetric Manifolds

INTRODUCTION

The subject of Hermitian locally symmetric manifolds is a very classical one in the domain of Complex Differential Geometry. Beyond compact Riemann surfaces of genus ≥ 2, the study of compact quotients of bounded symmetric domains has played a very central role. Of particular importance is the phenomenon of rigidity of complex structures of such manifolds. We refer the reader to Chap.1 for a historical account of rigidity theorems on compact quotients of bounded symmetric domains.

Following this tradition the author studied the phenomenon of Hermitian metric rigidity on compact quotients of bounded symmetric domains ([MOK3,1986–87]) and proved in particular the following theorem: Let X be a compact quotient of an irreducible bounded symmetric domain of rank ≥ 2 and h be a Hermitian metric of seminegative curvature in the sense of GRIFFITHS [GRI1], then h is necessarily a constant multiple of the Kähler–Einstein metric g. Such and similar theorems will be called *Hermitian metric rigidity theorems in the seminegative case.* Similar theorems were formulated for irreducible compact quotients of polydiscs. Very recently, TO [TO] generalized this to the case of quotients of finite volume. Our result in the compact case, combined with the Strong Rigidity Theorem of SIU [SIU2,3], yields the following generalization of a special case of Mostow's Strong Rigidity Theorem: Let (N,g) be a compact quotient of an irreducible bounded symmetric domain Ω of rank ≥ 2. Let (X,h) be a compact Kähler manifold of seminegative holomorphic bisectional curvature homotopic to N. Then, (X,h) is biholomorphic or conjugate–biholomorphic to (N,g). From the proof of Hermitian metric rigidity theorems one deduces that a non–trivial holomorphic mapping between quotients of bounded symmetric domains is necessarily a totally geodesic isometric immersion up to normalizing constants when the domain manifold is locally irreducible, of rank ≥ 2 and of finite volume. This theorem is motivated by the Super–rigidity Theorem of MARGULIS [MAR,1977] (cf. also ZIMMER [ZIM]), which already implies in our situation that the holomorphic mapping is homotopic to an isometric immersion up to normalizing constants.

The application of Hermitian metric rigidity theorems to holomorphic mappings was in fact a major motivation for formulating the former theorem.

Another motivation was to study the problem of classifying compact Kähler manifolds of semipositive holomorphic bisectional curvature. In fact, we proved ([MOK4,1987]) the dual and more difficult theorem that on an irreducible compact Hermitian symmetric manifold X of compact type of rank ≥ 2 every Kähler metric g of semipositive holomorphic bisectional curvature defines necessarily a Hermitian symmetric structure on X. Contrary to the dual situation of seminegative curvature, g is not unique up to normalizing constants and the theorem is false for Hermitian metrics. We call this the *Kähler metric rigidity theorem in the semipositive case.* Shortly afterwards the author [MOK5] resolved the Generalized Frankel Conjecture in a way which supercedes [MOK4] by a completely different method. However, the method of proof of [MOK4] is more elementary and can be used to study holomorphic mappings between irreducible Hermitian symmetric manifolds of compact type. In this direction TSAI [TSA] proved that a non–trivial holomorphic mapping between irreducible Hermitian symmetric manifolds of compact type of the same dimension is necessarily a biholomorphism if the domain manifold X is of rank ≥ 2.

In this monograph we present an expanded version of the results of [MOK3,4], incorporating new developments since then. In particular, we present To's generalization of the Hermitian metric rigidity theorem in the seminegative case to the situation of quotients of finite volume. Our Hermitian metric rigidity theorem in the seminegative case can be formulated for irreducible locally homogeneous holomorphic vector bundles on quotients of finite volume of bounded symmetric domains, as soon as the canonical metric carries seminegative and not strictly negative curvature in the sense of GRIFFITHS [GRI1]. This applies also to the case of complex hyperbolic space forms (i.e., quotients of the unit ball) where our previous rigidity theorem for holomorphic mappings does not apply. In [MOK7] we proved a Hermitian metric rigidity theorem on complex hyperbolic space forms to show that any local biholomorphism between complex hyperbolic space forms is necessarily an isometry when the domain manifold is of finite volume. The result of [MOK7] motivated CAO–MOK [CM] to study holomorphic immersions f between complex hyperbolic space forms when the domain manifold X is compact. We proved that if the complex codimension of the immersion f is smaller than the complex dimension of X, then f is necessarily an isometric immersion up to a normalizing constant. We do not know if the restriction on dimensions is necessary. On the other hand, it is unknown if the result of [CM] persists when the domain manifold X is only assumed to be of finite volume.

The monograph is divided into two parts. In Part I we present the background and basic results of [MOK3,4]. We present in a coherent fashion the necessary background for the study of Complex Differential Geometry on Hermitian locally symmetric manifolds. For the purpose of illustration we will give quite detailed and explicit discussion/calculations on the classical bounded symmetric domains. For the understanding of this portion of the monograph we assume that the reader has some background in Riemannian Geometry and some rudimentary knowledge of complex manifolds. Basic facts and principles in Hermitian and Kähler Geometries will be developed essentially from scratch. We will need a significant amount of background materials in the theory of Riemannian/Hermitian symmetric manifolds and compact semisimple Lie groups/algebras and their representations. We adopt the policy of stating facts from general theory and giving standard references, while giving proofs only when the statement and/or its proof is very directly related to the proof of our basic results. Part II of the monograph presents more recent developments. For the understanding of this portion considerably more background beyond the general theory of Hermitian symmetric manifolds will be needed. At the beginning of each chapter in Part II we will present the necessary prerequisites for the particular chapter. Most background results will be stated without proof but references will be given where needed.

The proof of Hermitan metric rigidity theorems in the seminegative case relies essentially on an integral formula involving first Chern forms on certain sub-manifolds of the projectivized tangent bundle. Such an integral formula and the subsequent proof of the Hermitian metric rigidity theorem for compact manifolds can be regarded as applications of very basic principles of Hermitian Differential Geometry. However, in order to verify the integral formula in a uniform way for all compact quotients of irreducible bounded symmetric domains of rank ≥ 2, we will need the general theory of realization of Hermitian symmetric manifolds of non–compact type as bounded symmetric domains, due to HARISH–CHANDRA [HA], and some elementary theory of representations of compact semisimple Lie groups. For the sake of coherence we will present the background materials on semisimple Lie groups/algebras and the representation theory of compact semisimple Lie groups/algebras in the Appendix. The proof of the Kähler metric rigidity theorem in the semipositive case (for irreducible Hermitian symmetric manifolds X of compact type and of rank ≥ 2) is based first of all on a dual integral formula on certain submanifolds of the projectivized cotangent bundle. The key point, however, is to exploit the Kähler condition by studying a space of Riemann spheres totally geodesic with respect to any choice of Hermitian symmetric structure on X.

In Chap.1 we give an introduction and a summary of results. Background knowledge on Hermitian and Kähler manifolds will be presented in Chap.2. In Chap.3 we present the general background on Riemannian/Hermitian symmetric manifolds necessary for our purpose. The duality between Hermitian symmetric manifolds of non–compact and compact types will be emphasized. In particular, we give a proof of the Borel Embedding Theorem. In Chap.4 we study some classical bounded symmetric domains explicitly so as to make available examples for the ensuing general discussion. In Chap.5 we present the general theory of bounded symmetric domains. In particular, we give a proof of the Harish–Chandra Embedding Theorem. In Chap.6 we give a proof of the Hermitian metric rigidity theorem in the seminegative case for compact quotients. We do this by using an integral formula on a holomorphic fibre bundle $\mathcal{S}$ contained in the projectivized tangent bundle, constructed in the locally irreducible case Ω/Γ, $\Omega = G/K$, from a dominant weight vector of the isotropy representation of K. $\mathcal{S}$ will be called the characteristic bundle. For the deduction of this theorem in the compact case from the integral formula we will give both the original algebraic proof in [MOK3] and a more geometric argument based on Moore's Ergodicity Theorem. The latter proof applies also to the case of irreducible quotients of the polydisc. In Chap.7 a principal object will be the characteristic bundle $\mathcal{S}$ over an irreducible Hermitian symmetric manifold of compact type, defined as in the dual seminegative case but corresponding at the same time geometrically to the set of all tangent directions to minimal rational curves. The crux of our argument is to show that minimal rational curves are totally geodesic with respect to any Kähler metric h of semipositive bisectional curvature.

In Part II of the monograph, Chaps.8–11, we present recent developement beyond the basic results. In Chap.8 we present To's extension of the Hermitian metric rigidity theorem to the case of finite volume. The proof relies on the use of minimal compactifications of such manifolds X of SATAKE–BAILY–BOREL ([SA1]

& BAILY–BOREL [BB]) and toroidal compactifications of [AMRT]; and some precise estimates of the canonical metric near the compactifying divisors of toroidal compactifications, as given in [MUM1]. To's theorem imposes absolutely no growth condition on the Hermitian metric h of seminegative curvature on X. One might expect that the growth condition needed is automatic due to the fact that the minimal compactification X_{min} is obtained by adding a subvariety of codimension ≥ 2, or just from the pseudoconcavity of X, as a consequence of extension theorems. This is however not the case. Nonetheless, in case h is Kähler one can deduce To's Theorem by using extension theorems of closed positive currents. We present such an alternative elementary proof of To's theorem in the Kähler case, which is sufficient for proving a rigidity theorem for holomorphic mappings. In Chap.9 we prove a rigidity theorem for holomorphic immersions between equi-dimensional complex hyperbolic space forms ([MOK7]). The basic tool is a Hermitian metric rigidity theorem on some locally homogeneous vector bundle over such manifolds. We also prove a rigidity theorem for holomorphic immersion of compact hyperbolic space forms in case of low codimensions (CAO–MOK [CM]). While the proof does not follow from metric rigidity theorems it is motivated by such theorems. In Chap.10 we generalize the Hermitian metric rigidity theorem to certain locally homogeneous bundles (including the one encountered in Chap.9) and use it to study homomorphisms between such bundles. In the last chapter, Chap.11, we present the proof of TSAI [TSA] that any holomorphic map between equi-dimensional irreducible Hermitian symmetric manifolds of compact type is necessarily a biholomorphism, provided that the rank of the domain manifold is at least 2. The proof is based on a localization of the argument of [MOK4] to show that preimages of minimal rational curves are totally geodesic with respect to Kähler–Einstein metrics on the domain manifold.

It is hoped that the monograph will be of value to graduate students with some background in Riemannian Geometry and Several Complex Variables. A substantial portion of Part II can be regarded as applications of methods of Algebraic Geometry (compactifications of arithmetic varieties, splitting of vector bundles over rational curves) and Several Complex Variables (maximum principle for plurisubharmonic functions, extension theorems for closed positive currents) to some classical problems in Complex Differential Geometry. For readers not familiar with these methods hopefully the monograph will provide some motivation for supplementary reading as well.

In App.I we collect the necessary background materials on semisimple Lie groups/algebras and the representation of compact semisimple Lie algebras. In App.II we state some theorems in Riemannian Geometry which we used in the discussion on Riemannian symmetric manifolds. In App.III we give a description of the typical fibre S_0 of S. It will be shown that such S_0, which we call characteristic projective subvarieties, are in one–to–one correspondence with Kähler submanifolds of the projective space with parallel second fundamental form. In App.IV we formulate a dual version of the Generalized Frankel Conjecture for compact Kähler manifolds of seminegative holomorphic bisectional curvature. The formulation is motivated by the basic results of Part I and their proofs.

I would like to thank Prof. Kuranishi for some valuable suggestions on the first draft of the monograph. Thanks are due to H.–D. Cao, E. Falbel, I.–H. Tsai and S.–K. Yeung for reading portions of the preliminary manuscript. I am most grateful to W.–K. To, whose very thorough and careful proof–reading has immensely simplified the task of editing. The late Prof. Jia–qing Zhong read the first plan of the book. His criticisms have been invaluable for the overall organization of the book. The memory of his unfailing friendship and enthusiasm has been a driving force in difficult times.

PART I

BACKGROUND AND FIRST RESULTS

CHAPTER 1 HISTORICAL BACKGROUND AND SUMMARY OF RESULTS

§1 Historical Background

(1.1) The subject of metric rigidity theorems on Riemannian locally symmetric manifolds of seminegative sectional curvature is a very classical one. The first definitive result in this direction is the Strong Rigidity Theorem of Mostow:

THEOREM 1 (MOSTOW [MOS,1973])
Let (X,g) and (Y,h) be compact Riemannian locally symmetric manifolds of negative Ricci curvature. Suppose $\pi_1(X) \cong \pi_1(Y)$ and (X,g) has no closed two–dimensional totally geodesic submanifolds which are local direct factors. Then, (X,g) and (Y,h) are isometric up to normalizing constants.

Here we say that (N,s) is a local direct factor if and only if it is isomorphic to a factor in the de Rham decomposition of some finite covering of (X,g). The assumption that (X,g) and (Y,h) are of negative Ricci curvature means that the universal covering manifold $(\tilde{X},\tilde{g})$ is a Riemannian symmetric manifold of non–compact type, i.e., $\tilde{X} = G/K$, where G is a non–compact semisimple Lie group and $K \subset G$ is a maximal compact subgroup. In the de Rham decomposition of $(\tilde{X},\tilde{g})$, we have $(\tilde{X},\tilde{g}) \cong (\tilde{X}_1,\tilde{g}_1) \times \dots \times (\tilde{X}_r,\tilde{g}_r)$, where each $(\tilde{X}_i,\tilde{g}_i)$ is irreducible and of non–compact type. The normalizing constants arise from the choice of normalizing constants in the irreducible factors. The assumption that (X,g) has no closed two–dimensional totally geodesic submanifolds which are local direct factors excludes the case of compact Riemann surfaces S of genus ≥ 2 and fibre bundles arising from them.

Mostow's Strong Rigidity Theorem was later generalized to the situation of Riemannian locally symmetric manifolds of finite volume by PRASAD [PRA,1973] in case of rank–1 and by MARGULIS [MAR,1977] in case of rank ≥ 2.

In the works of Mostow, Prasad and Margulis, the semisimple Lie groups G play a fundamental role. In the general category of Riemannian manifolds, we have the following more recent theorem of GROMOV [GROM].

THEOREM 2 (GROMOV [GROM,1981])
Let (X,g) be a compact irreducible Riemannian locally symmetric manifold of negative Ricci curvature and of rank ≥ 2. Suppose (Y,h) is a compact Riemannian manifold of seminegative sectional curvature such that $\pi_1(X) \cong \pi_1(Y)$. Then, (Y,h) is isometric to (X,g) up to normalizing constants.

By an irreducible Riemannian locally symmetric manifold (X,g) we mean one for which no finite covering can be split isometrically into global factors of positive dimensions. (X,g) can be irreducible while the universal covering manifold is not. In this case we say that (X,g) is irreducible and locally reducible. The hypothesis that (X,g) is irreducible can easily be relaxed to cover the reducible case by a result of EBERLEIN [EBE,1983]. More precisely, up to a finite covering (X,g) splits isometrically into a product of irreducible Riemannian locally symmetric manifolds (X_i,g_i) and Gromov's Theorem remains valid as long as each (X_i,g_i) is of rank ≥ 2. Recently, Gromov's Theorem was generalized by BALLMANN–EBERLEIN [BE,1986] to the case of finite volume given that (Y,h) is of finite volume and of bounded seminegative sectional curvature. Obviously Gromov's Theorem would fail in the rank–1 situation since the Riemannian locally symmetric manifold (X,g) would be of negative sectional curvature and can thus be perturbed. The proof of Gromov's Theorem and its subsequent generalizations rely in part on the statement and proofs of Mostow's Strong Rigidity Theorem.

(1.2) In this monograph our interest lies on Hermitian locally symmetric manifolds. We will first of all deal with Hermitian locally symmetric manifolds of negative Ricci curvature and of finite volume. By the Harish–Chandra Embedding Theorem ([HA,1956]) of Hermitian symmetric manifolds of non–compact type onto bounded symmetric domains Ω, we are dealing equivalently with quotients of bounded symmetric domains by torsion–free discrete groups of holomorphic isometries such that the quotient manifold is of finite volume with respect to the Bergman metric (cf. §4.1). Henceforth by automorphisms of Ω, denoted by $\text{Aut}(\Omega)$ we will mean biholomorphic self–mappings on Ω; by a quotient of Ω we will mean a quotient of Ω by a torsion–free discrete group of automorphisms (which are necessarily isometries with respect to the Bergman metric). For such manifolds there was first of all the local rigidity theorem of CALABI–VESENTINI [CV] and of BOREL [BO4] on the complex structures:

THEOREM 1 (CALABI–VESENTINI [CV,1960] & BOREL [BO4,1960])
Let X be a compact quotient of an irreducible bounded symmetric domain of complex dimension ≥ 2. Then, $H^1(X,T_X) = 0$ for the holomorphic tangent bundle T_X. In particular, X is locally rigid as a complex manifold.

As in the case of Mostow's Strong Rigidity Theorem, the case of compact Riemann surfaces S of genus ≥ 2 is excluded. In fact, S admits a moduli space as a compact complex manifold parametrized effectively by (3g–3) complex variables, where g = genus(S). The case of irreducible but locally reducible compact quotients is not dealt with in these results. Later, MATSUSHIMA–SHIMURA [MA–S] studied irreducible compact quotients of polydiscs and proved in particular

THEOREM 2 (MATSUSHIMA–SHIMURA [MA–S,1963])
Let X be an irreducible compact quotient of the polydisc Δ^n, $n \geq 2$. Then, $H^1(X,T_X) = 0$ for the holomorphic tangent bundle T_X. In particular, X is locally rigid as a complex manifold.

In the results of Calabi–Vesentini, Borel and Matsushima–Shimura, much more precise vanishing theorems (also for higher cohomology groups) were obtained. The purpose of [MS] was actually to compute the dimension of certain spaces of automorphic forms attached to X. Thm.2 above was obtained as a by–product.

In 1978, SIU [SIU2,3] began the study of strong rigidity of compact quotients of bounded symmetric domains as Kähler manifolds. He proved

THEOREM 3 ([SIU2,3;1978])
Let N be a compact quotient of an irreducible bounded symmetric domain of complex dimension ≥ 2. Suppose X is a compact Kähler manifold homotopic to N. Then, X is either biholomorphic or conjugate–biholomorphic to N.

The situation where X is not necessarily irreducible and locally reducible were later studied in JOST–YAU [JY1] and MOK [MOK1]. The most pertinent case is that of (globally) irreducible compact quotients of the polydisc Δ^n, $n \geq 2$. They proved in this case

THEOREM 4 (JOST–YAU [JY1,1985] for $n = 2$, MOK [MOK1,1985])
Let N be an irreducible compact quotient of the polydisc Δ^n, $n \geq 2$. Suppose X is a compact Kähler manifold homotopic to N. Then, there exists a diffeomorphism $f: X \rightarrow N$ such that for the lifting $F: \tilde{X} \rightarrow \tilde{N} = \Delta^n$ to the universal covering spaces $\tilde{X}$ and $\tilde{N} = \Delta^n$, $F = (F_1,\ldots,F_n)$, each F_i is either holomorphic or anti–holomorphic.

Recently, the Strong Rigidity Theorems for Kähler manifolds stated above were generalized to the case of quotients of finite volume with additional assumptions on the domain manifold (SIU [SIU4] and JOST–YAU [JY2] in case of rank–1, MOK [MOK6] in case of irreducible quotients of the polydisc and Jost–Yau [JY3] for the higher rank situation).

While the formulation of Siu's Strong Rigidity Theorem and its subsequent generalizations do not involve additional properties of the Kähler metric, the stated theorems and easy generalizations contain the Hermitian case of Mostow's Strong Rigidity Theorem, since any biholomorphism between two Hermitian symmetric manifolds of non–compact type is necessarily an isometry up to normalizing constants, and since any two compact $K(\pi,1)$'s (i.e., toplogical spaces with vanishing homotopy groups π_k for $k \geq 2$, e.g., complete Riemannian manifolds of seminegative sectional curvature) with isomorphic fundamental groups are necessarily homotopic.

(1.3) Concerning continuous mappings between Riemannian locally symmetric manifolds of negative Ricci curvature and of finite volume the Super–rigidity Theorem of MARGULIS [MAR] (cf. also ZIMMER [ZIM]) implies the following

THEOREM 1 (consequence of MARGULIS [MAR,1977])
Let (X,g) and (X',g') be irreducible Riemannian locally symmetric manifolds of negative Ricci curvature and of finite volume. Assume that (X,g) is of rank ≥ 2. Suppose $f_0: X \rightarrow X'$ is a continuous map. Then, f_0 is homotopic to some $f: X \rightarrow X'$, which is an isometric immersion up to normalizing constants.

The Super–rigidity Theorem of Margulis is formulated in terms of semisimple Lie groups G and G' which are the identity components of the groups of isometries of the universal covering spaces $(\tilde{X},\tilde{g})$ and $(\tilde{X}',\tilde{g}')$. Let Γ and Γ' be the fundamental groups of X and X' resp. The Super–rigidity Theorem of Margulis asserts that any group homomorphism from Γ to Γ' is induced by a (smooth) group homomorphism from G to G'. The translation into the more geometric statement above is immediate from the way that the canonical metrics are defined using Killing forms.

§2 Statement of Results

(2.1) We study in this monograph Hermitian/Kähler metrics on Hermitian locally symmetric manifolds of compact or non–compact type. There are two most basic results in this monograph. The first result is on compact quotients of bounded symmetric domains. The second basic result is on Hermitian symmetric manifolds of compact type. For the precise meaning of the terminology used in the statements and explanations of the results, we refer the reader to Chaps.2 & 3.

THEOREM 1 (MOK [MOK3,1987])

Let X be a compact quotient of an irreducible bounded symmetric domain Ω of rank ≥ 2. Let h be a Hermitian metric of seminegative curvature in the sense of GRIFFITHS [GRI1]. Then, h is necessarily a constant multiple of the canonical metric g.

Here the canonical metric means the Riemannian metric arising from the Killing form of the Lie algebra $\mathfrak{g}$ of infinitesimal isometries. It is equivalently, up to a normalizing constant, to the Bergman metric or the Kähler–Einstein metric (cf. [Ch.2, §1 & Ch.5, (1.2), Prop.2]). Thm.1 can also be formulated for irreducible quotients of the polydisc (cf. [Ch.6, §4]). We will call Thm.1 and its generalizations *Hermitian metric rigidity theorems in the seminegative case.*

THEOREM 2 (MOK [MOK4,1987])

Let (X_c,g) be an irreducible Hermitian symmetric manifold of compact type and of rank ≥ 2. Suppose h is a Kähler metric of semipositive holomorphic bisectional curvature on X_c. Then, there exists a biholomorphism Φ of X and a constant c such that $h = c\Phi^*g$.

The strict Riemannian analogue of Thm. 2 is false. For instance, CHEEGER

[CHE] constructed non–symmetric Riemannian metrics of semipositive sectional curvature on symmetric manifolds of compact type. Shortly after proving Thm.2, the author [Mok5] was able to resolve the Generalized Frankel Conjecture on classifying compact Kähler manifolds of semipositive holomorphic bisectional curvature in a way that supercedes Thm.2 (cf. [App.(IV.2)] for the exact statement). The proof of the latter conjecture uses the non–linear technique of evolution equations. The proof of Thm.2 presented here is more transparent and has other applications, notably to the study of rigidity phenomena for holomorphic mappings between Hermitian symmetric manifolds of compact type. We will call Thm.2 the *Kähler metric rigidity theorem in the semipositive case.*

We state here two consequences of Thm.1 which were part of the original motivation for formulating the latter theorem. Thm.3 can be regarded as a Kähler analogue of Gromov's Theorem [(1.1), Thm.2].

THEOREM 3 (MOK [MOK3])

Let (X,h) be a compact Kähler manifold of seminegative holomorphic bisectional curvature homotopic to a Hermitian locally symmetric manifold (N,g) uniformized by an irreducible bounded symmetric domain Ω of rank ≥ 2. Then, (X,h) is biholomorphically or conjugate–biholomorphically isometric to (N,g).

THEOREM 4

Let (X,g) be a Hermitian locally symmetric manifold of finite volume uniformized by an irreducible bounded symmetric domain Ω of rank ≥ 2. Suppose (N,h) is any Hermitian locally symmetric manifold of non–compact type. Then, any holomorphic mapping $f: X \rightarrow N$ is necessarily a totally geodesic isometric immersion up to a normalizing constant.

A slight modification of Thm.4 can also be formulated for the irreducible and locally reducible case. Furthermore, Thm.4 is a special case of a rigidity theorem for holomorphic mappings for which the target manifold is only assumed to be a Kähler manifold of seminegative holomorphic bisectional curvature. We will call Thm.4 and its generalizations rigidity theorems for holomorphic mappings in the seminegative case.

Thm.4 for quotients of finite volume can be deduced from a slight modification of Thm.3 and the Ahlfors–Schwarz lemma. This will be given in [Ch.4, (1.2), Prop.4]. Full generalizations of Thms.1–4 to quotients of finite volume and to locally homogeneous holomorphic vector bundles will be given in Chaps.8–10.

Thm.3 is an immediate consequence of Thm.1 and the Strong Rigidity Theorem for Kähler manifolds of SIU [(1.2), Thm.3]. In (1.3) we will deduce an essential part of Thm.4 from Thm.1.

§3 Deduction of Some Results from the Hermitian Metric Rigidity Theorem in the Seminegative Case

(3.1) In this section we are going to deduce an essential part of the rigidity theorem for holomorphic mappings in the seminegative case [(2.1), Thm.4] from the Hermitian metric rigidity theorem [(2.1), Thm.1]. The purpose is to illustrate a basic principle in Hermitian Differential Geometry. The same principle will be used in the proof of [(2.1), Thm.1]. For this purpose, we state

PROPOSITION 1 (cf. [Ch.2, (3.2), Prop.1])

Let X be a complex manifold and (V,h) be a Hermitian holomorphic vector bundle on X of seminegative curvature. Let V' be a holomorphic vector subbundle of V. Then, $(V',h|_{V'})$ is of seminegative curvature.

Here the notion of negativity on a Hermitian holomorphic vector bundle is always understood to be in the sense of GRIFFITHS [GRI1]. (For details of this and various other notions of negativity, cf. [Ch.2, §3]). We will say that (X,h) is of seminegative curvature if for the holomorphic tangent bundle T_X with the induced Hermitian metric, denoted also by h, (T_X,h) is of seminegative curvature. An immediate consequence is the following

PROPOSITION 2 (cf. [Ch.2, (3.2), Prop.2])

Let X be a complex manifold and g, h be two Hermitian metrics of seminegative curvature on X. Then, $(X,g+h)$ is also of seminegative curvature.

Prop.2 follows from Prop.1 and the fact that the diagonal embedding $\delta: X \longrightarrow X \times X$ induces an isometric immersion $(X,g+h) \longrightarrow (X,g) \times (X,h)$.

Let (X,g) be a compact Hermitian locally symmetric manifold uniformized by an irreducible bounded symmetric domain Ω of rank ≥ 2. Let (N,h) be a Hermitian locally symmetric manifold of non–compact type. Both (X,g) and (N,h) are of seminegative curvature in the sense of GRIFFITHS [GRI1] (cf. [Ch.2, (3.3), Prop.1 & Ch.3, (1.3), Prop.2]). Let $f: X \longrightarrow N$ be a non–constant holomorphic mapping. We are going to deduce that f is an isometry up to a scaling constant from the Hermitian metric rigidity theorem [(2.1), Thm.1]. Consider the holomorphic immersion $\tau: X \longrightarrow X \times N$ defined by the assignment $x \longrightarrow (x,f(x))$. τ realizes $(X,g+f^*h)$ isometrically as a submanifold of $(X,g) \times (N,h)$, so that by Prop.1, $(X,g+f^*h)$ is of seminegative curvature. From the Hermitian metric rigidity theorem [(2.1), Thm.1] we deduce that $g + f^*h = cg$ for some constant c. In other words, $f^*h = (c-1)g$, so that f is an isometric immersion up to a normalizing constant.

It will be seen in [Ch.6, §5] that f is in fact a totally geodesic isometric immersion. The proof of the total geodesy of f will follow rather easily from the proof of the Hermitian metric rigidity theorem.

CHAPTER 2 FUNDAMENTALS OF HERMITIAN AND KÄHLER GEOMETRIES

§1 Hermitian and Kähler Metrics

(1.1) Let X be an n–dimensional complex manifold. The complex structure of X gives rise to a J–operator. Let (z_j), $1 \le j \le n$, be a system of local holomorphic coordinates on X. Write as usual $z_j = x_j + \sqrt{-1}y_j$. Then, in terms of the usual basis $\{\partial/\partial x_j, \partial/\partial y_j: 1 \le j \le n\}$, the natural J–operator is defined by $J(\partial/\partial x_j) = \partial/\partial y_j$; $J(\partial/\partial y_j) = -\partial/\partial x_j$. J is an endomorphism of the real tangent bundle such that $J^2 = -\mathrm{id}$. Extend J by complex linearity into the complexified tangent bundle $T_X^{\mathbf{C}}$. Then, at $x \in X$ $T_x^{\mathbf{C}}$ splits into the direct sum of the $(\pm\sqrt{-1})$–eigenspaces of J. Write $T_x^{1,0}$ and $T_x^{0,1}$ resp. for the $(+\sqrt{-1})$ and $(-\sqrt{-1})$–eigenspaces and call elements of these spaces (complexified) tangent vectors of type (1,0) and (0,1) resp. Write $T_X^{1,0} \subset T_X^{\mathbf{C}}$ for the complex vector subbundle $\bigcup_{x\in X} T_x^{1,0}$, etc. We have the bundle decomposition $T_X^{\mathbf{C}} = T_X^{1,0} \oplus T_X^{0,1}$. In terms of the local holomorphic coordinates $(z_j)_{1\le j\le n}$, a basis of $T_x^{1,0}$ at x is given by $\{\partial/\partial z_j: 1 \le j \le n\}$, where $\frac{\partial}{\partial z_j} = \frac{1}{2}(\frac{\partial}{\partial x_j} - \sqrt{-1}\frac{\partial}{\partial y_j})$. Similarly a basis of $T_x^{0,1}$ is given by $\{\partial/\partial \bar{z}_j: 1 \le j \le n\}$, where $\frac{\partial}{\partial \bar{z}_j} = \frac{1}{2}(\frac{\partial}{\partial x_j} + \sqrt{-1}\frac{\partial}{\partial y_j})$. From now on by a tangent vector we will mean a complexified tangent vector unless specified otherwise.

DEFINITION 1

A Hermitian metric g on a complex manifold X is a J–invariant Riemannian metric on the underlying smooth manifold X, i.e., g satisfies $g(Ju,Jv) = g(u,v)$ for real tangent vectors u and v.

Extend g by complex–bilinearity to $T_X^{\mathbf{C}}$ and denote the the extended complex symmetric bilinear form by $g(\cdot,\cdot)$. The condition that g is J–invariant is equivalent to the condition that $g(u,v) = 0$ for u and v of the same type and that in terms of local holomorphic coordinates $(z_j)_{1\le j\le n}$, the n–by–n matrix G with (i,j)–th entry $g_{i\bar{j}}$ defined by $g_{i\bar{j}} = g(\partial/\partial z_i, \partial/\partial \bar{z}_j)$ is Hermitian symmetric (and automatically positive definite). We write $(.,.)$ for $g(.,\bar{.})$, $\bar{\ }$ denoting conjugation in $T_X^{\mathbf{C}}$. Thus, at each point $x \in X$, $(.,.)$ defines Hermitian inner products on $T_x^{1,0}$ and $T_x^{0,1}$. In terms of (z_j) the Hermitian metric g is given by $g = 2Re(\Sigma\, g_{i\bar{j}}\, dz^i \otimes d\bar{z}^j)$. Also associated to the Hermitian metric g is the real tensor $A = Im(\Sigma\, g_{i\bar{j}}\, dz^i \otimes d\bar{z}^j)$. It follows from the Hermitian property of $(g_{i\bar{j}})$ that A is skew–symmetric. It can be identified with $\omega = \sqrt{-1}\Sigma\, g_{i\bar{j}}\, dz^i \wedge d\bar{z}^j$, an alternating (1,1)–form. We call ω the Hermitian form of (X,g). By partition of unity any complex manifold can be endowed with Hermitian metrics.

(1.2) Of special interest among Hermitian metrics is the class of Kähler metrics. We give the following geometric definition of Kähler manifolds. Recall that a Hermitian manifold (X,g) is by definition also a Riemannian manifold. It makes sense therefore to talk about parallel transport on (X,g).

DEFINITION 2

A Hermitian manifold (X,g) is said to be Kähler if and only if the types of complexified tangent vectors are preserved under parallel transport.

The preceding definition, while most geometric, is not the easiest to use. We are going to give and prove equivalent definitions for the Kähler property. Denote in this section by ∇ the Riemannian connection of (X,g).

PROPOSITION 1

Let (X,g) be a Hermitian manifold such that g is given by $2Re(\Sigma g_{i\bar{j}}\, dz^i \otimes d\bar{z}^j)$ in local holomorphic coordinates (z_j). Then, (X,g) is Kähler if and only if one of the following equivalent conditions is satisfied:

(1) types of complexified tangent vectors are preserved under parallel transport;

(2) for any parallel (real) vector field η along a smooth curve γ, $J\eta$ is also parallel;

(3) $\nabla J \equiv 0$, i.e., the almost complex structure J is parallel;

(4) $\nabla\omega \equiv 0$, i.e., the Hermitian form ω is parallel;

(5) $d\omega \equiv 0$, i.e., the Hermitian form ω is closed;

(6) locally there exists a potential function φ such that $g_{i\bar{j}} = \dfrac{\partial^2 \varphi}{\partial z_i \partial \bar{z}_j}$;

(7) at every point $P \in X$ there exists complex geodesic coordinates (z_i) in the sense that the Hermitian metric g is represented by the Hermitian matrix $(g_{i\bar{j}})$ satisfying $g_{i\bar{j}}(P) = \delta_{ij}$ and $dg_{i\bar{j}}(P) = 0$.

Proof:
(1) is the geometric definition for Kähler manifolds we adopted. We prove

(1) $\Longrightarrow$ (2) Suppose η is a parallel (real) vector field along a smooth curve $\gamma := \{\gamma(t): -\epsilon < t < \epsilon\}$. We are going to show that $J\eta$ is also a parallel vector field. Let $\eta = \eta_{1,0} + \eta_{0,1}$ be the unique decomposition of the vector field η into components of type (1,0) and (0,1) resp. Let ψ and ψ' be the parallel transport of $\eta_{1,0}(o)$ and $\eta_{0,1}(o)$ along γ resp. By hypothesis (1) ψ and ψ' are vector fields of types (1,0) and (0,1) resp. By the uniqueness of the decomposition of tangent vectors into components of types (1,0) and (0,1) it follows that $\psi \equiv \eta_{1,0}$ and $\psi' \equiv \eta_{0,1}$ along γ. In other words, $\eta_{1,0}$ and $\eta_{0,1}$ are both parallel along γ. Since $J\eta = \sqrt{-1}\eta_{1,0} - \sqrt{-1}\eta_{0,1}$, it follows that $J\eta$ is parallel along γ.

(2) $\Longrightarrow$ (1) Let $\eta_{1,0}(o)$ be a tangent vector of type (1,0) at $\gamma(o)$ and write ψ for its parallel transport along γ. Let $\psi = \psi_{1,0} + \psi_{0,1}$ be the decomposition into components of types (1,0) and (0,1). By hypothesis (2) $J\psi = \sqrt{-1}\psi_{1,0} - \sqrt{-1}\psi_{0,1}$ is also parallel. It follows that both $\psi_{1,0}$ and $\psi_{0,1}$ are parallel along γ. Since $\psi_{0,1}(o) = 0$ we conclude that $\psi_{0,1}$ vanishes identically on γ, so that the parallel transport of $\eta_{1,0}(o)$ remains to be of type (1,0) along γ, proving (2) $\Longrightarrow$ (1).

(2) $\Longrightarrow$ (3) Let $P \in X$ and v be any real tangent vector at P. Choose the curve γ such that $\gamma(o) = P$ and v is tangent to γ. Let η be a parallel vector field along γ. The hypothesis (2) implies that $J\eta$ is also parallel. It follows from $\nabla_v(\eta) = 0$ and $\nabla_v(J\eta) = 0$ that $\nabla_v(J)(\eta) = 0$. Since v and $\eta(o)$ are arbitrary at P it follows that in fact $\nabla J \equiv 0$.

(3) $\Longrightarrow$ (2) In the notations of the preceding paragraph, if $\nabla_v(\eta) = 0$ and $\nabla_v J = 0$ it is immediate that $\nabla_v(J\eta) = 0$. Hence, if η is parallel along γ, so is $J\eta$.

(3) ⟹ (4) Consider the 2–tensor $G = \Sigma g_{i\bar{j}}\, dz^i \otimes d\bar{z}^j$. For any two real vectors $2Re\xi$ and $2Re\eta$, $\xi, \eta \in T_X^{1,0}$, we have $G(2Re\xi, 2Re\eta) = G(\xi+\bar{\xi}, \eta+\bar{\eta}) = 2Re G(\xi,\eta)$. From the Hermitian property of the matrix $(g_{i\bar{j}})$ it follows that $S = Re(G)$ is symmetric and that $A = Im(G)$ is skew–symmetric. Recall that 2S gives the Riemannian metric tensor. Clearly if we identify A with the corresponding real alternating 2–form, then A gives the Hermitian form ω. We are going to show that $\nabla A \equiv 0$, i.e., $\nabla\omega \equiv 0$. Since ∇ is the Riemannian connection on (X,g) we have $\nabla S \equiv 0$. First we relate A to S and J. We claim that for any two real tangent vectors $u = 2Re\xi$ and $v = 2Re\eta$, $S(u,v) = A(Ju,v)$. To see this we note first that $G(Ju,v) = G(\sqrt{-1}\xi - \sqrt{-1}\bar{\xi}, \eta+\bar{\eta}) = G(\sqrt{-1}\xi, \bar{\eta}) = \sqrt{-1}G(\xi,\bar{\eta}) = \sqrt{-1}G(u,v)$. Since $G = S + \sqrt{-1}A$ we have $(S+\sqrt{-1}A)(Ju,v) = \sqrt{-1}(S+\sqrt{-1}A)(u,v)$. Equating the imaginary parts we get immediately $S(u,v) = A(Ju,v)$. We are going to deduce that $\nabla A \equiv 0$. Fix a point $P \in X$ and let γ be any smooth curve passing through P. Let u and v be 2 parallel real vector fields along γ and w be a vector tangent to γ at P. Then, applying ∇_w to both sides of $S(u,v) = A(Ju,v)$ we obtain from $\nabla S \equiv 0$ and $\nabla J \equiv 0$ (by hypothesis) that $\nabla_w A(Ju,v) = 0$. Since u, v, w are arbitrary at P, it follows that $\nabla A \equiv 0$, proving (3) ⟹ (4).

(4) ⟹ (3) The same argument as in the preceding paragraph shows that $\nabla S \equiv 0$ (always true) and $\nabla A \equiv 0$ (by hypothesis) imply $\nabla J \equiv 0$.

(4) ⟹ (5) We write the Riemannian metric as $\Sigma h_{pq}\, dx^p \otimes dx^q$. Choosing normal geodesic coordinates at an arbitrary point $P \in X$ the hypothesis (4) implies that $dh_{pq}(P) = 0$. It follows that $d\omega = 0$.

(5) ⟹ (6) It suffices to solve the equation $\sqrt{-1}\partial\bar{\partial}\varphi = \omega$. Fix an open set U on X biholomorphic to a Euclidean polydisc. Since ω is closed by hypothesis it follows from the Poincaré lemma that there exists a real 1–form η on U such that $d\eta = \omega$. Decompose η into components of type (1,0) and (0,1) and write $\eta = \eta^{1,0} + \eta^{0,1}$, $\eta^{0,1} = \overline{\eta^{1,0}}$. From $d\eta = \omega$ and comparing components of the same type on both sides we obtain $\partial\eta^{0,1} = 0$; $\bar{\partial}\eta^{0,1} = 0$ and $\bar{\partial}\eta^{1,0} + \partial\eta^{0,1} = \omega$. From the Dolbeault–Grothendieck lemma there exists on U a smooth function ψ such

that $\bar\partial\psi = \eta^{0,1}$. It follows from $\bar\partial\eta^{1,0} + \partial\eta^{0,1} = \omega$ that $\partial\bar\partial(\psi - \bar\psi) = \omega$. Hence, the real function $\varphi = -\sqrt{-1}(\psi - \bar\psi)$ solves $\sqrt{-1}\partial\bar\partial\varphi = \omega$, i.e., $\dfrac{\partial^2\varphi}{\partial z_i \partial \bar z_j} = g_{i\bar j}$.

(6) $\Longrightarrow$ (5) If $\omega = \sqrt{-1}\partial\bar\partial\varphi$ it is immendiate that $d\omega = \sqrt{-1}d(\partial\bar\partial\varphi) = 0$ (since $d = \partial + \bar\partial$ and $\partial^2 = \bar\partial^2 = 0$).

(5), (6) $\Longrightarrow$ (7) Let $P \in X$ be an arbitrary point. We are going to construct complex geodesic coordinates at P. Obviously by making a unitary change of coordinates we may assume that holomorphic local coordinates (w_i) have been chosen such that, writing $\omega = \sqrt{-1}(\Sigma\, h_{i\bar j}\, dw^i \wedge d\bar w^j)$, we have $h_{i\bar j}(P) = \delta_{ij}$. We are going to make a holomorphic change of coordinates from (w_α) to (z_i) of the form $w_i = z_i + \Sigma c^i_{jk} z_j z_k$ such that $c^i_{jk} = c^i_{kj}$. Expanding ω in terms of (z_i) we obtain $\omega = \sqrt{-1}(\Sigma\, g_{i\bar j}\, dz^i \wedge d\bar z^j)$ with $g_{i\bar j}(P) = \delta_{ij}$ and

$$\partial_k g_{i\bar j}(P) = \partial_k h_{i\bar j}(P) + c^j_{ki} + c^j_{ik} = \partial_k h_{i\bar j}(P) + 2c^j_{ki},$$

where $\partial_k \equiv \partial/\partial z_k$. To get complex geodesic coordinates at P it suffices to set

$$2c^j_{ki} = -\partial_k h_{i\bar j}(P).$$

Since $c^j_{ki} = c^j_{ik}$, the latter set of equations is consistent if $\partial_k h_{i\bar j}(P) = \partial_i h_{k\bar j}(P)$, which is precisely the condition $d\omega(P) = 0$ guaranteed by hypothesis (5)/(6).

(7) $\Longrightarrow$ (4) With respect to the complex geodesic coordinates (z_i) at $P \in X$ the connection form Γ of the underlying Riemannian manifold (X,g) vanishes at P. Writing $\omega = \sqrt{-1}(\Sigma\, g_{i\bar j}\, dw^i \wedge d\bar w^j)$ it follows immediately that $\nabla\omega \equiv 0$.

The proof of Prop.1 is completed. ▌

REMARKS

The geometric definition (1) of Kähler metrics can be rephrased by saying that the holonomy group of (X,g) is reduced to (contained in) $U(n) \subset O(2n)$ (given by the embedding $T^{1,0}_P \subset T^{\mathbb{C}}_P$ at any $P \in X$).

§2 The Hermitian Connection and its Curvature

(2.1) Let X be a complex manifold and V be a holomorphic vector bundle over X. Let h be a Hermitian metric on V, i.e., a collection of Hermitian inner products on the fibres V_x, $x \in X$, varying smoothly with x. We denote the Hermitian inner products by $(.,.)$. We are going to define a method of differentiating smooth sections of V compatible with the Hermitian metric h. First, a connection D on (V,h) is a consistent way of differentiating smooth sections s of V over open sets U: for any tangent vector field ξ on U, $D_\xi s$ is also a smooth section of V over U, with the properties that D is complex–linear in both ξ and s and that D satisfies the product rule $D_\xi(fs) = \xi(f)s + fD_\xi s$ for smooth function f over U. A connection D on V is said to be a complex connection if and only if for any local holomorphic section σ and any tangent vector η of type (0,1) in the domain of definition of s, $D_\eta \sigma = 0$. D is said to be a metric connection if and only if it is compatible with the Hermitian metric h: i.e., for any open set U, any *real* tangent vector v on U, and for any two smooth sections s and t over U, $v(t,s) = (D_v t,s) + (t,D_v s)$. For a complexified tangent vector ξ it follows that $\xi(t,s) = (D_\xi t,s) + (t,D_{\bar\xi} s)$.

We are going to define a complex metric connection on (V,h). First of all, we remark that the requirement that D be complex is consistent with the product rule since the transition functions for V are holomorphic. Let U be a coordinate open set on X with holomorphic local coordinates (z_i) such that V is holomorphically trivial over U. Let $\{e_\alpha\}$ be a holomorphic basis of $V|_U$ and write $s = s^\alpha e_\alpha$ for a smooth section of V over U. Here and henceforth we adopt the Einstein summation of summing over indexes that appear once as superscipts and once as subscripts. Let $\eta = \eta^i \frac{\partial}{\partial z_i}$ be a smooth vector field of type (1,0) over U. Clearly by the product rule to define a complex connection D of V it suffices to define $D_i e_\alpha = \Gamma^\gamma_{i\alpha} e_\gamma$, where from now on $D_i \equiv D_{\partial/\partial z_i}$, etc. To define the Riemann–Christoffel symbols $(\Gamma^\gamma_{i\alpha})$ we impose the additional condition that D be metric, i.e., we require

$$\partial_i(e_\alpha,e_\beta) = (D_i e_\alpha,e_\beta) + (e_\alpha,D_{\bar i} e_\beta). \qquad (1)$$

Write $(e_\alpha,e_\beta) = h_{\alpha\bar\beta}$. The matrix H with (α,β)–th entry $h_{\alpha\bar\beta}$ is Hermitian symmetric and positive definite. Since the left hand side of (1) is $\partial_i h_{\alpha\bar\beta}$ and since $D_{\bar{i}}e_\beta = 0$ by the assumption that D is a complex connection, the requirement that D be metric determines $(\Gamma^\gamma_{i\alpha})$ uniquely by the equations $\partial_i(e_\alpha, e_\beta) = (D_i e_\alpha, e_\beta)$, giving

$$h_{\gamma\bar{\beta}}\Gamma^\gamma_{i\alpha} = \partial_i h_{\alpha\bar{\beta}}, \text{ i.e.,}$$

$$\Gamma^\gamma_{i\alpha} = h^{\gamma\bar{\beta}}.\partial_i h_{\alpha\bar{\beta}}, \tag{2}$$

where $(h^{\alpha\bar{\beta}})$ stands for the conjugate inverse of the matrix $(h_{\alpha\bar{\beta}})$, i.e., $h_{\alpha\bar{\beta}}h^{\gamma\bar{\beta}} = \delta^\gamma_\alpha$ for the Kronecker delta (δ^γ_α). It is clear that the Riemann–Christoffel symbols $(\Gamma^\gamma_{i\alpha})$ defined by (2) give rise to a unique complex metric connection D on (V,h). We call D the Hermitian connection of (V,h).

There is the following interpretation of the Hermitian connection D. Let (V*,h*) be the dual bundle of (V,h) and $\{e^\alpha\}$ be a holomorphic basis of V* over U dual to the basis $\{e_\alpha\}$. The Hermitian inner product can be written as $H = \Sigma h_{\alpha\bar{\beta}} e^\alpha \otimes \bar{e}^\beta$. (By abuse of notation we use H to denote both the tensor and the matrix $(h_{\alpha\bar{\beta}})$.) A connection D on V induces by the compatibility with the dual pairing between V and V* (defined independent of Hermitian metrics) a connection on V*. When D is a complex connection so is the dual connection on V*. By conjugation the connection extends to the conjugate bundles of V and V*. By requiring the product rule on tensor products D extends to the tensor algebra obtained from V, V* and their conjugate bundles. Denote by the same symbol D such an extension. The requirement that D be a metric connection on V is equivalent to the requirement $DH = 0$. Let s be a smooth section of V over U. There is a way of lifting s to a section $\bar{s}^*$ of $\bar{V}^*$ by defining $\bar{s}^* = s_{\bar{\beta}}\bar{e}^\beta$, where $s_{\bar{\beta}} = h_{\alpha\bar{\beta}}s^\alpha$. $\bar{s}^*$ is the contraction of H with s. Write Φ for this lifting operation so that $\bar{s}^* = \Phi(s)$. Let η be a tangent vector field of type (1,0) on U. If D is a complex metric connection on (V,h), then $D_\eta\bar{s}^* = \Phi(D_\eta s)$. We have $D_\eta\bar{s}^* = \overline{D_{\bar{\eta}}s^*}$, s^* being the conjugate of $\bar{s}^*$. Since $\bar{\eta}$ is a tangent vector field of type (0,1) and D is a complex connection $D_{\bar{\eta}}s^*$ and hence $D_\eta\bar{s}^*$ is well–defined, so that $D_\eta s$ can be defined from $D_\eta\bar{s}^* = \Phi(D_\eta s)$.

(2.2) Let (X,g) be a Hermitian manifold. The restriction of the Hermitian metric $(.,.) = g(.,\bar{.})$ to $T_X^{1,0}$ defines a Hermitian metric on $T_X^{1,0}$ (which can be identified with the holomorphic tangent bundle T_X). By conjugation D extends to a connection on $T_X^{\mathbf{C}} = T_X^{1,0} \oplus T_X^{0,1}$. On the other hand since by definition g is a Riemannian metric on the underlying smooth manifold X, there is also a Riemannian connection ∇ on (X,g), which extends to the complexified tangent bundle $T_X^{\mathbf{C}}$. Comparing the two connections D and ∇ we have

PROPOSITION 1

The Hermitian connection D agrees with the Riemannian connection ∇ if and only if (X,g) is Kähler. In other words, a Hermitian manifold (X,g) is Kähler if and only if the Hermitian connection D is torsion–free.

Proof:

Write $G = \Sigma g_{i\bar{j}}\, dz^i \otimes d\bar{z}^j = S + \sqrt{-1}A$, where S and A are real tensors. $2S = G + \overline{G}$ is the underlying Riemannian metric of the smooth manifold X. Since for any (complexified) tangent vector η, $D_\eta \overline{G} = \overline{D_{\bar{\eta}} G}$, $DG \equiv 0$ implies $DS \equiv 0$. In other words, D is compatible with the Riemannian metric $2Re(\Sigma\, g_{i\bar{j}}\, dz^i \otimes d\bar{z}^j)$. By the uniqueness of the Riemannian connection it follows that $D \equiv \nabla$ if and only if D is torsion–free. We compute the torsion tensor T of (X,g) in terms of holomorphic coordinates $\{z_i\}$. Since D is a complex connection $T(\eta,\xi) = 0$ if η and ξ are of opposite type. On the other hand, we have

$$\begin{aligned} T\left(\frac{\partial}{\partial z_i},\frac{\partial}{\partial z_j}\right) &= D_i\left(\frac{\partial}{\partial z_j}\right) - D_j\left(\frac{\partial}{\partial z_i}\right) \\ &= \sum_k \left(\Gamma^k_{ij} - \Gamma^k_{ji}\right)\frac{\partial}{\partial z_k}. \end{aligned}$$

By taking conjugates we get the formula for $T(\partial/\partial\bar{z}_i,\partial/\partial\bar{z}_j)$. It follows that the Hermitian connection D is torsion–free if and only if $\Gamma^k_{ij} = \Gamma^k_{ji}$ for all i, j and k. For a point $P \in X$, choosing holomorphic coordinates (z_i) such that $g_{i\bar{j}}(P) = \delta_{ij}$, we have $\Gamma^k_{ij} = \partial_i g_{j\bar{k}}$. Thus, the Hermitian connection D is torsion–free if and only if $\partial_i g_{j\bar{k}}(P) = \partial_j g_{i\bar{k}}(P)$, i.e., $d\omega(P) = 0$ for arbitrary points $P \in X$. In other words, D is the Riemannian connection if and only if (X,g) is Kähler.

(2.3) We define now the curvature Θ of the Hermitian holomorphic vector bundle (V,h) of rank r with respect to the Hermitian connection D. With respect to a system of local holomorphic coordinates $(z_i)_{1\le i\le n}$ over an open set U and a local holomorphic basis $\{e_\alpha\}_{1\le\alpha\le r}$ of $V|_U$ over U, we express the Hermitian connection in the form $D_ie_\alpha = \Gamma^\gamma_{i\alpha}e_\gamma$, in terms of the Riemann–Christoffel symbols $(\Gamma^\gamma_{i\alpha})$. Let Γ be the End(V)–valued one form defined by $\Gamma = \Gamma^\gamma_{i\alpha}\, e_\gamma \otimes e^\alpha \otimes dz^i$. We call Γ the connection 1–form of (V,h) over U. As for Riemannian connections Γ (not just the components) depends on the choice of the local holomorphic coordinates (z_i). Γ is however independent of the choice of the holomorphic basis $\{e_\alpha\}$ of $V|_U$. We can also consider Γ as given by a row vector of rank r whose γ–th entry is $\Gamma^\gamma_{i\alpha}\, e^\alpha \otimes dz^i$. By abuse of notation we also denote this row vector of V–valued 1–forms by Γ and call it the Hermitian connection. Let e be the column vector of rank r consisting of the basis $\{e_\alpha\}$. We can write symbolically $De = \Gamma \otimes e$. Let f be a row vector of rank r smooth functions on U. Then, by the product rule $D(fe) = df \otimes e + f\, De$.

As in Riemannian Geometry the curvature Θ of (V,h) measures the deviation from commutativity of two covariant derivatives. Let ψ be a V–valued p–form on U. Denote by $D\psi$ the covariant exterior derivative of ψ using the Hermitian connection D in the same way as one defines exterior differentiation on p–forms. For a V –valued 1–form $\psi \otimes e$ we have $D(\psi \otimes e) = d\psi \otimes e - \psi \wedge De$. For the flat connection d we have $d^2(fe) = 0$. For (V,h) we have

$$\begin{aligned} D^2(fe) &= D(df \otimes e + f\,De) \\ &= d^2f \otimes e - (df \wedge \Gamma) \otimes e + (df \wedge \Gamma) \otimes e + f\,D^2e \\ &= f\,D^2e = f\,d\Gamma \otimes e - f\,(\Gamma \wedge \Gamma) \otimes e. \end{aligned}$$

We define the curvature of (V,h) to be the End(V)–valued 2–form $\Theta = \sqrt{-1}(d\Gamma - \Gamma \wedge \Gamma)$. (For the meaning of $\sqrt{-1}$ cf. §5 on first Chern forms). By abuse of notations we also denote by Θ the r–by–r matrix $(\Theta_\alpha{}^\beta)$ of 2–forms, so that we have $D^2(fe) = -\sqrt{-1}\, f\, \Theta \otimes e$. Since $D^2(fe)$ is well–defined it follows that Θ is independent of the choice of the local holomorphic coordinates (z_i). For the purpose of local computations we will use

LEMMA 1

Fix an arbitrary point $P \in X$ and let (z_i) be some holomorphic coordinate system on a neighborhood U of x, over which $V|_U$ is holomorphically trivial. Then, there exists a choice of holomorphic basis $\{e_\alpha\}$ of $V|_U$ such that for the Hermitian metric $h = \Sigma h_{\alpha\bar\beta} e^\alpha \otimes \bar{e}^\beta$ we have $h_{\alpha\bar\beta}(P) = \delta_{\alpha\beta}$ and $dh_{\alpha\bar\beta}(P) = 0$.

Proof:

The proof is similar to that of proving the existence of complex geodesic coordinates on Kähler manifolds (cf. [(1.2), Prop.1]). Choose the holomorphic local coordinates (z^i) such that $z^i(P) = 0$. Let $\{e'_\alpha\}$ be a holomorphic basis of $V|_U$ (with dual basis $\{e'^\alpha\}$) such that the matrix $(h'_{\alpha\bar\beta})$ representing h in terms of $\{e'_\alpha\}$ satisfies $h'_{\alpha\bar\beta}(P) = \delta_{\alpha\beta}$. We introduce a new choice of holomorphic basis $\{e_\alpha\}$ such that for the dual bases $\{e^\alpha\}$ we have $e^\alpha = e'^\alpha + \Sigma_{\beta,i} c^\alpha_{\beta i} z^i e^\beta$. Then, we have $h_{\alpha\bar\beta}(P) = \delta_{\alpha\beta}$ and

$$\partial_i h_{\alpha\bar\beta}(P) = \partial_i h'_{\alpha\bar\beta}(P) + c^\beta_{\alpha i}.$$

It suffices now to set $c^\beta_{\alpha i} = -\partial_i h'_{\alpha\bar\beta}$.

We call the holomorphic basis $\{e_\alpha\}$ of $V|_U$ a special holomorphic basis at P. For a vector $v \in V$ over U we write $v = \Sigma v^\alpha e_\alpha$ and call (v^α) special fibre coordinates (adapted to h at P).

REMARKS

Here the choice of holomorphic basis is independent of the choice of the holomorphic coordinate system (z^i). Thus, if (X,g) is a Hermitian manifold and (T_X,g) is the associated Hermitian holomorphic tangent bundle, one can still find special fibre coordinate systems even if (X,g) is not Kähler.

PROPOSITION 1

The End(V)–valued curvature 2–form Θ is of type $(1,1)$.

Proof:

Let $P \in X$ be an arbitrary point. Choose special local fibre coordinates (v^α)

adapted to h at P. Then, we have $\Theta = \sqrt{-1}d\Gamma$. Recall that $\Gamma^{\gamma}_{i\alpha} = \partial_i h_{\alpha\bar\beta} h^{\gamma\bar\beta}$ so that $\Gamma = (\partial H)H^{-1}$. We have in terms of (v^{α})

$$\Theta = \sqrt{-1}\, d\Gamma = \sqrt{-1}\, d((\partial H)H^{-1}) = \sqrt{-1}\, \bar\partial\partial H.$$

In particular, this shows that the curvature Θ is of type (1,1). We write

$$\Theta = \sqrt{-1}\, \Theta_{\alpha}{}^{\beta} e^{\alpha} \otimes e_{\beta} = \sqrt{-1}\, \Theta_{\alpha}{}^{\beta}{}_{i\bar j} e^{\alpha} \otimes e_{\beta}\, dz^i \wedge d\bar z^j.$$

In general for any choice of local holomorphic base and fiber coordinates (z^i) and (v^{α}) resp. we have from the formulas $\Theta = \sqrt{-1}\,(d\Gamma - \Gamma \wedge \Gamma)$ and $\Gamma = (\partial H)H^{-1}$

$$\Theta_{\alpha}{}^{\beta}{}_{i\bar j} = -h^{\beta\bar\mu}\, \partial_i\partial_{\bar j} h_{\alpha\bar\mu} + h^{\beta\bar\gamma}\, h^{\mu\bar\nu}\, \partial_i h_{\alpha\bar\nu}\, \partial_{\bar j} h_{\mu\bar\gamma}.$$

We can also identify V^* with $\bar V$ using the Hermitian metric h. This way

$$\Theta = \sqrt{-1}\, \Theta_{\alpha\bar\beta}\, e^{\alpha} \otimes \overline{e^{\beta}} = \sqrt{-1}\, \Theta_{\alpha\bar\beta i\bar j}\, e^{\alpha} \otimes \overline{e^{\beta}}\, dz^i \wedge d\bar z^j,$$

$$\Theta_{\alpha\bar\beta i\bar j} = -\partial_i\partial_{\bar j} h_{\alpha\bar\beta} + h^{\mu\bar\nu}\, \partial_i h_{\alpha\bar\nu}\, \partial_{\bar j} h_{\mu\bar\beta}.$$

We will use the same notation Θ for the two ways of writing the curvature.

§3 Different Notions of Positivity/Negativity of Curvature

(3.1) Let X be a complex manifold and (V,h) be a Hermitian holomorphic vector bundle, with curvature Θ. For any $x \in X$ we define a Hermitian bilinear form P on $V_x \otimes T_x(X)$ by defining

$$P(v \otimes \eta, v' \otimes \eta') = \Theta_{v\bar v' \eta\bar\eta'}$$

and extending to $V_x \otimes T_x(X)$ by Hermitian bilinearity. We have

DEFINITION 1

We say that (V,h) is positive (resp. semipositive) in the sense of Nakano at $x \in X$ if and only if P is a positive definite (resp. positive semidefinite) on $V_x \otimes T_x(X)$.

The notion of negativity and seminegativity in the sense of Nakano is similarly defined.

Associated to Θ is another Hermitian bilinear form Q on $V_x \otimes T_x(X)$ defined by the formula

$$Q(v \otimes \eta', v' \otimes \eta) = \Theta_{v\bar{v}'\eta\bar{\eta}'}$$

and extended by Hermitian bilinearity. We have

DEFINITION 2

We say that (V,h) positive (resp. semipositive) in the dual sense of Nakano at $x \in X$ if and only if Q is positive definite (resp. positive semidefinite) on $V_x \otimes T_x(X)$.

Let (V^*,h^*) be the Hermitian dual bundle of (V,h). Let Θ^* be the curvature of (V^*,h^*). Fix $x \in X$ and choose holomorphic fiber coordinates such that $h_{\alpha\bar{\beta}}(x) = \delta_{\alpha\beta}$. Then in terms of such coordinates we have from the formula for curvatures $\Theta_{\alpha\bar{\beta}i\bar{j}} = -\Theta^*_{\beta\bar{\alpha}i\bar{j}}$. It follows readily that (V,h) is positive (resp. negative) in the sense of Nakano if and only if (V^*,h^*) is negative (resp. positive) in the dual sense of Nakano.

The notion of positivity in the sense of Griffiths (as given in [GRI1]) is the most commonly used notion of positivity, which is defined by

DEFINITION 3

We say that (V,h) is positive (resp. semipositive) in the sense of Griffiths at $x \in X$ if for any non–zero vector $v \in V_x$ and any non–zero tangent vector $\eta \in T_x(X)$ we have $\Theta_{v\bar{v}\eta\bar{\eta}} > 0$ (resp. $\Theta_{v\bar{v}\eta\bar{\eta}} \geq 0$).

We say that (the curvature of) the Hermitian holomorphic vector bundle (V,h) over X is positive in the sense of Nakano if it is true at every point of X, etc. It is clear from the definitions that

LEMMA 1

If (V,h) is positive (resp. semipositive) in the sense of Nakano or in the dual sense of Nakano, then (V,h) is positive (resp. semipositive) in the sense of Griffiths.

Proof:

By definition we have $\Theta_{v\bar{v}\eta\bar{\eta}} = P(v \otimes \eta , v \otimes \eta) = Q(v \otimes \bar{\eta} , v \otimes \bar{\eta})$. ▌

(3.2) Consider now a holomorpohic vector subbundle V' of V. We are going to compare the curvature Θ' of $(V',h|_{V'})$ with the curvature Θ of (V,h). Suppose V' is of rank r'. Fix $x \in X$ and let (z^i) be local holomorphic coordinates and $\{e_\alpha\}_{1\leq\alpha\leq r'}$ be a special holomorphic basis of $(V',h|_{V'})$ over some open neighborhood U of x. In terms of these coordinates we have

$$\Theta'_{\alpha\bar{\beta}i\bar{j}} = -\partial_i\partial_{\bar{j}}h_{\alpha\bar{\beta}}$$

for $1 \leq i,j \leq n$ and $1 \leq \alpha,\beta \leq r'$. Extend $\{e_\alpha\}_{1\leq\alpha\leq r'}$ to a holomorphic basis $\{e_\alpha\}_{1\leq\alpha\leq r}$ of V over U (possibly after shrinking U). Then, for $1 \leq i,j \leq n$ and $1 \leq \alpha,\beta \leq r'$ we have

$$\Theta_{\alpha\bar{\beta}i\bar{j}} = -\partial_i\partial_{\bar{j}}h_{\alpha\bar{\beta}} + \sum_{r'<\mu\leq r} \partial_i h_{\alpha\bar{\mu}}\, \partial_{\bar{j}} h_{\mu\bar{\beta}}.$$

In particular, for $1 \leq \alpha,\beta \leq r'$, we can write

$$\Theta'_{\alpha\bar{\beta}i\bar{j}} = \Theta_{\alpha\bar{\beta}i\bar{j}} - (\sigma_{\alpha i},\sigma_{\beta j}), \text{ where}$$

$$\sigma_{\alpha i} = \sum_{r'<\mu\leq r} \partial_i h_{\alpha\bar{\mu}}\, e_\mu$$

is a vector at x orthogonal to V'_x and $(.,.)$ denotes the Hermitian inner product given by h. Denote by N_x the orthogonal complement of V'_x in V_x. We obtain from $(\sigma_{\alpha i})$ by extension a complex bilinear map $\sigma\colon V_x \otimes T_x(X) \to N_x$, which is the second fundamental form of the isometric embedding $(V',h|_{V'}) \hookrightarrow (V,h)$. In particular, we have the Gauss–Codazzi equation

$$\Theta'_{\alpha\bar{\alpha}i\bar{i}} = \Theta_{\alpha\bar{\alpha}i\bar{i}} - \|\sigma_{\alpha i}\|^2,$$

where $\|.\|$ denotes the length measured in terms of h. Thus, we have

PROPOSITION 1

Let (V,h) be a Hermitian holomorphic vector bundle of seminegative curvature in the sense of Griffiths. Then, any Hermitian holomorphic vector subbundle $(V',h|_{V'})$ is also of seminegative curvature.

Henceforth when we say that (V,h) is of seminegative curvature, we will always mean that (V,h) is seminegative in the sense of Griffiths. Prop.1 suggests that it is easier to work with Hermitian holomorphic vector bundles of seminegative curvature. We have in fact

PROPOSITION 2

Let V be a holomorphic vector bundle and h_1, h_2 be two Hermitian metrics of seminegative curvature on V. Then, (V,h_1+h_2) is also of seminegative curvature.

Proof:

We have an isometric embedding $(V,h_1+h_2) \hookrightarrow (V,h_1) \oplus (V,h_2)$. It is clear that the direct sum $(V,h_1) \oplus (V,h_2)$ is of seminegative curvature. From Prop.1 it follows that (V,h_1+h_2) is also of seminegative curvature. As a consequence, the space of Hermitian metrics on V of seminegative curvature, if non-empty, forms a convex set in the space of Hermitian metrics on V.

(3.3) Consider now a Hermitian manifold (X,g). We have an associated Hermitian holomorphic tangent bundle (T_X,g). We will say that (X,g) is of seminegative curvature if and only if (T_X,g) is. Let now (X,g) be Kähler. Since the Riemannian connection ∇ on (X,g) agrees with the Hermitian connection D, there are additional notions of positivity/negativity coming from Riemannian Geometry. Let R be the curvature tensor of the Riemannian manifold (X,g) and u, v be two linearly independent real tangent vectors at some point $x \in X$. The sectional curvature $K(u,v)$ is given by

$$K(u,v) = \frac{R(u,v;v,u)}{\|u \wedge v\|^2}.$$

In terms of coordinates the Riemannian manifold (X,g) is of positive sectional curvature if and only if for $u \wedge v \neq 0$ with $u = u^i \frac{\partial}{\partial x_i}$, $v = v^i \frac{\partial}{\partial x_i}$, we have

$$R_{ijkl}\,(u^iv^j - v^iu^j)\,(u^kv^l - v^ku^l) > 0$$

With J denoting the almost complex structure of (X,g), we call K(u,Ju) the holomorphic sectional curvature of the J–invariant real 2–plane generated by u. We extend the curvature tensor R by complex linearity in the 4 variables to $T_X^{\mathbf{C}}$. Using the decomposition $T_X^{\mathbf{C}} = T_X^{1,0} \oplus T_X^{0,1}$ we express the curvature tensor in terms of the basis $\{\partial/\partial z_i, \partial/\partial \bar{z}_i\}_{1\le i\le n}$ of $T_X^{\mathbf{C}}$ for $z_i = x_i + \sqrt{-1}y_i$; $y_i = x_{n+i}$. Here and henceforth we will use this basis in writing the curvature components unless specified otherwise. Since the Riemannian connection ∇ of the Kähler manifold (X,g) agrees with the Hermitian connection D and the curvature Θ of D is an $\mathrm{End}(T_X)$–valued (1,1)–form it follows readily that the only possible non–vanishing terms of the curvature components are of the form $R_{i\bar{j}k\bar{l}}$ and those obtained from the universal symmetries of the curvature tensor, e.g., $R_{\bar{j}ik\bar{l}}$. We say that the curvature tensor R is of type (2,2). The curvature tensor R of Kähler manifolds (X,g) has the additional symmetry property $R_{i\bar{j}k\bar{l}} = R_{i\bar{l}k\bar{j}}$ as a consequence of the Bianchi identity and the fact that R is of type (2,2). In fact, we have

$$R_{i\bar{j}k\bar{l}} = -R_{ik\bar{l}\bar{j}} - R_{i\bar{l}\bar{j}k} = -R_{i\bar{l}\bar{j}k} = R_{i\bar{l}k\bar{j}}.$$

It is covenient to identify the real tangent bundle $T_X^{1,0}$ with the real tangent bundle $T_X^{\mathbf{R}}$ via $\xi \mapsto 2Re\xi$. Writing $\xi = \xi^i \frac{\partial}{\partial z_i}$ and $u = 2Re\xi = \xi + \bar{\xi}$, we sometimes call $\dfrac{R(u,Ju;Ju,u)}{\|u\|^4}$ the holomorphic sectional curvature in the direction of ξ. We have

$$\begin{aligned} R(u,Ju;Ju,u) &= R(\xi+\bar{\xi}, \sqrt{-1}\xi-\sqrt{-1}\,\bar{\xi}; \sqrt{-1}\xi-\sqrt{-1}\,\bar{\xi}, \xi+\bar{\xi}) \\ &= 4\,R(\xi,\bar{\xi};\xi,\bar{\xi}) = 4\,R_{i\bar{j}k\bar{l}}\,\xi^i\bar{\xi}^j\xi^k\bar{\xi}^l. \end{aligned}$$

When $u = 2Re\xi$ is of unit length ξ is of length $1/\sqrt{2}$. Thus, for ξ of unit length the holomorphic sectional curvature in the direction of ξ is given by $R_{\xi\bar{\xi}\xi\bar{\xi}}$.

We define the notion of holomorphic bisectional curvature.

DEFINITION 1

Let (X,g) be a Kähler manifold, $x \in X$ an arbitrary point and $\xi, \eta \in T_x^{1,0}$. Write $u = 2Re\xi$ and $v = 2Re\eta$. We define the holomorphic bisectional curvature in the directions (ξ,η) to be $\frac{R(u,Ju;Jv,v)}{\|u\|^2\|v\|^2}$.

Thus, holomorphic bisectional curvatures are generalizations of holomorphic sectional curvatures and for ξ, η of unit length the holomorphic bisectional curvature in the directions (ξ,η) is given by $R_{\xi\bar{\xi}\eta\bar{\eta}}$. Concerning the relation between Riemannian sectional curvatures and holomorphic bisectional curvatures we have

PROPOSITION 1

Suppose the Kähler manifold (X,g) is of positive (resp. negative) Riemannian sectional curvature. Then, (X,g) is of positive (resp. negative) holomorphic bisectional curvature.

<u>Proof</u>:

It suffices to express a holomorphic bisectional curvature as a sum of two Riemannian sectional curvatures. Let $\xi, \eta \in T_x^{1,0}$ Write $u = 2Re\xi$ and $v = 2Re\eta$. We have by the first Bianchi identity

$$R(u,Ju;Jv,v) = -R(u,Jv;v,Ju) - R(u,v;Ju,Jv) \qquad (1)$$

On the other hand, since (X,g) is Kähler we have for tangent vectors A, B, C and D at x

$$R(A,B;JC,JD) = g(J(R(A,B)C),JD) = R(A,B;C,D) \qquad (2)$$

From (1) and (2)

$$\begin{aligned} R(u,Ju;Jv,v) &= R(u,Jv;Jv,u) - R(u,v;u,v) \\ &= R(u,Jv;Jv,u) + R(u,v;v,u), \end{aligned}$$

expressing a holomorphic bisectional curvature as a sum of two Riemannian sectional curvatures, proving Prop.1. ∎

Let $S \subset X$ be a local complex submanifold of X. $(S,g|_S)$ is Kähler since the restriction of a the closed Kähler form ω on X to S is closed. To compute the holomorphic bisectional curvatures of $(S,g|_S)$ it is equivalent to compute the curvature of the Hermitian holomorphic vector subbundle $(T_S,g|_S) \hookrightarrow (T_X,g)$. Denote by R^S the curvature tensor of the Kähler submanifold $(S,g|_S)$. We have the Gauss–Codazzi equation for holomorphic bisectional curvatures

$$R^S_{\xi\bar{\xi}\eta\bar{\eta}} = R_{\xi\bar{\xi}\eta\bar{\eta}} - \|\sigma(\xi,\eta)\|^2, \tag{3}$$

Here σ denotes the second fundamental form of the isometric embedding $(T_S,g|_S) \hookrightarrow (T,g)$. Regarding (X,g) and $(S,g|_S)$ as Riemannian manifolds there is a second fundamental form σ_o which is pointwise a symmetric bilinear map $T^{\mathbf{R}}_x(S) \times T^{\mathbf{R}}_x(S) \longrightarrow N^{\mathbf{R}}_x$, where $N^{\mathbf{R}}_x$ denotes the normal space at x of the submanifold S in X. σ_o is given by $\sigma_o(u,v)(x) = pr_N \nabla_u v(x)$ for any extension of u and v to a smooth vector field in a neighborhood and for pr_N denoting the orthogonal projection into $N^{\mathbf{R}}_x$. Extending by complex bilinearity we get σ_o: $T^{\mathbf{C}}_x(S) \times T^{\mathbf{C}}_x(S) \longrightarrow N^{\mathbf{C}}_x$. From the fact that $\nabla \equiv D$ is a complex connection it follows that $\sigma_o(A,B) = 0$ if A and B are of opposite type and that furthermore the restriction $\sigma_{1,0}$ of σ_o to $T^{1,0}_x(X) \times T^{1,0}_x(X)$ takes values in $N^{1,0}_x$. Here we use the decomposition $N^{\mathbf{C}} = N^{1,0} \oplus N^{0,1}$ into types. The second fundamental form σ: $T_x(X) \times T_x(X) \longrightarrow N_x$ can be identified with $\sigma_{1,0}$ using the identification of T_X with $T^{1,0}_X$. Both σ_o and σ will both be referred to as the second fundamental form of S in X, depending on the context.

Denote components of the Ricci curvature tensor of the Kähler manifold (X,g) by (R_{ab}); $a, b \in \{1, \dots n; \bar{1}, \dots, \bar{n}\}$. Since R is of type (2,2), we have

$$R_{i\bar{j}} = g^{ab} R_{iab\bar{j}} = g^{\bar{l}k} R_{i\bar{l}k\bar{j}} = g^{k\bar{l}} R_{i\bar{j}k\bar{l}},$$

while $R_{ij} = R_{\bar{i}\bar{j}} = 0$. We define the Ricci curvature form of the Kähler manifold (X,g) to be $\mathrm{Ric} = \sqrt{-1}\, R_{i\bar{j}}\, dz^i \wedge d\bar{z}^j$. (X,g) is of positive (resp. negative) Ricci curvature if and only if Ric is a positive (resp. negative) (1,1)–form.

It is clear that the Kähler manifold (X,g) is of positive (resp. negative) bisectional curvature if and only if the associated Hermitian tangent bundle (T_X,g) is of positive (resp. negative) curvature in the sense of Griffiths. Furthermore, in this case (X,g) is of positive (resp. negative) Ricci curvature.

For Kähler manifolds the notion of positivity/negativity in terms of holomorphic bisectional curvatures plays an essential role in characterizing Kähler manifolds with curvature properties. Because of the Gauss–Codazzi equation for bisectional curvatures we have the curvature–decreasing (meaning non–increasing) property for Kähler submanifolds. The curvature–decreasing property is not true for Riemannian sectional curvatures. Other than verifying this by pointwise computations there is the following global reason: Suppose (X,g) is a simply–connected Riemannian manifold with non–positive sectional curvature, then by the theorem of Cartan–Hadamard (cf. CHEEGER–EBIN [CE]) X is diffeomorphic to a Euclidean space. If the curvature–decreasing property on Kähler submanifolds were true for Riemannian sectional curvatures every simply–connected closed complex submanifold of the Euclidean space $\mathbf{C}^n$ (endowed with the flat metric) would be contractible. This contradicts with the fact that any Stein manifold S can be embedded as a closed complex submanifold $S \hookrightarrow \mathbf{C}^n$ (BISHOP [BI] and NARASIMHAN [NA1]) and that given any finite real n–dimensional CW–complex K, there exists a complex n–dimensional Stein manifold S homotopic to K (NARASIMHAN [NA2]). This argument also suggests that the assumption that a complete Kähler manifold is of seminegative bisectional curvature does not impose strong topological restrictions on the manifold. For problems of strong rigidity of Kähler manifolds (cf. SIU [SIU2]), the notion of seminegativity of curvature in the dual sense of Nakano will be important. This is not surprising since the notion of positivity/negativity of Hermitian holomorphic vector bundles in the sense/dual sense of Nakano was defined for formulating vanishing theorems for cohomology groups of such bundles and strong rigidity can be regarded as a non–linear version of vanishing theorems.

§4 Projectivization of Hermitian Holomorphic Line Bundles

(4.1) We start with recalling the definition of first Chern classes of holomorphic line bundles. Let X be a complex manifold and L be a holomorphic line bundle over X. Let $\mathcal{U} = \{U_\alpha\}$ be a covering of X by open sets such that $L|U_\alpha$ is holomorphically trivial with a holomorphic basis $e^{(\alpha)}$. Over U_α we write a smooth section as $s = s^{(\alpha)}e^{(\alpha)}$. Over $U_{\alpha\beta} := U_\alpha \cap U_\beta$ we have $s^{(\alpha)} = \varphi_{\alpha\beta}s^{(\beta)}$. The holomorphic transition functions $\{\varphi_{\alpha\beta}\}$ constitutes a Čech 1–cocycle in $\mathcal{Z}^1(\mathcal{U},\mathcal{O}^*)$ where $\mathcal{O}^*$ denotes the sheaf of germs of non–vanishing holomorphic functions. The cohomology class thus defined in $H^1(X,\mathcal{O}^*)$ is independent of the choice of the covering $\mathcal{U}$ The abelian group of isomorphism classes of holomorphic line bundles over X is called the Picard group over X, to be denoted by $\text{Pic}(X)$. The assignment $L \longrightarrow \{\varphi_{\alpha\beta}\} \in H^1(X,\mathcal{O}^*)$ establishes an isomorphism $\text{Pic}(X) \cong H^1(X,\mathcal{O}^*)$. Over X there is the exponential short exact sequence

$$0 \longrightarrow \mathbf{Z} \longrightarrow \mathcal{O} \xrightarrow{e} \mathcal{O}^* \longrightarrow 0, \tag{1}$$

where $\mathbf{Z}$ is the constant sheaf of integers and e is defined by $e(f) = e^{2\pi i f}$. The long exact sequence associated to (1) gives a connecting homomorphism

$$\delta\colon H^1(X,\mathcal{O}^*) \longrightarrow H^2(X,\mathbf{Z}).$$

For $L \in \text{Pic}(X)$, $\delta(L) \in H^2(X,\mathbf{Z})$ is called the first Chern class of L, denoted by $c_1(L)$. By abuse of language one also calls the image of $c_1(L)$ in $H^2(X,\mathbf{C})$ given by the canonical map $H^2(X,\mathbf{Z}) \longrightarrow H^2(X,\mathbf{C})$ the first Chern class of L.

Without loss of generality we may always assume that each $U_\alpha \in \mathcal{U}$ is simply–connected. On $U_{\alpha\beta\gamma} := U_\alpha \cap U_\beta \cap U_\gamma$ one can then define $c_{\alpha\beta\gamma} = \frac{-\sqrt{-1}}{2\pi}\{\log(\varphi_{\alpha\beta}) + \log(\varphi_{\beta\gamma}) + \log(\varphi_{\gamma\alpha})\}$, which an integer by the cocycle condition $\varphi_{\alpha\beta}\varphi_{\beta\gamma}\varphi_{\gamma\alpha} = 1$. To represent the first Chern class $c_1(L) \in H^2(X,\mathbf{C})$ using differential forms one uses the Leray isomorphism between the Čech cohomology group $H^2(X,\mathbf{C})$ and the de Rham cohomology group $H^2_{dR}(X,\mathbf{C})$ using a double sequence and standard diagram–chasing arguments. This gives

PROPOSITION 1

Let h be a Hermitian metric on the holomorphic line bundle L over X such that in terms of holomorphic bases $\{e^{(\alpha)}\}$ over $\mathscr{U} = \{U_\alpha\}$, we have $\|e^{(\alpha)}\|^2 = h^{(\alpha)}$ ($\|\cdot\|$ denoting norms measured by h). Then the first Chern class $c_1(L) \in H^2(X,\mathbb{R})$ is represented by the real closed (1,1)–form $c_1(L,h) = \frac{-\sqrt{-1}}{2\pi}\, \partial\overline{\partial} \log h^{(\alpha)}$.

For a proof cf. e.g. GRIFFITHS–HARRIS [GH, Ch.1, p.139ff.]. We remark that the expression $\partial\overline{\partial} \log h^{(\alpha)}$ is well–defined independent of α since the $\{h^{(\alpha)}\}$ are related by $h^{(\beta)} = |\varphi_{\alpha\beta}|^2 h^{(\alpha)}$, so that

$$\partial\overline{\partial} \log h^{(\beta)} = \partial\overline{\partial} \log h^{(\alpha)} + \partial\overline{\partial} \log |\varphi_{\alpha\beta}|^2 = \partial\overline{\partial} \log h^{(\alpha)}$$

since for a nowhere vanishing holomorphic function f, $\log |f|$ is pluriharmonic. In the proof of the metric rigidity theorem we will make use of the topological invariance of the de Rham class $[c_1(L,h)]$. This is of course a consequence of Prop.1. Alternatively, we can take the definition of $c_1(L,h)$ as given in Prop.1 and see that it is independent of the Hermitian metric. In fact, any two Hermitian metrics h and h' on L are related by $h' = sh$, where s is a global smooth function on X. We have then the equality of de Rham cohomology classes

$$[c_1(L,h')] = [c_1(L,h)] + [\frac{-\sqrt{-1}}{2\pi}\, \partial\overline{\partial} \log s] = [c_1(L,h)]$$

since $\partial\overline{\partial} \log s = d(\overline{\partial} \log s)$.

(4.2) Let (V,h) be a Hermitian holomorphic vector bundle over X. We are going to associate to (V,h) a Hermitian holomorphic line bundle L by projectivization. We start with the trivial vector bundle V_o given by $\mathbf{C}^r \to \{o\}$ over a single point equipped with the Euclidean metric. As a holomorphic vector bundle V_o can be projectivized by the Hopf process using "polar coordinates". We have the holomorphic fibration $\pi: \mathbf{C}^r - \{o\} \to \mathbb{P}^{r-1}$ defined by assigning v to the complex line $[v]: \mathbf{C}v \in \mathbb{P}^{r-1}$. We assert that π realizes $\mathbf{C}^r - \{o\}$ as a $\mathbf{C}^*$–bundle. This can be seen by explicit trivializations over affine open subsets of $\mathbb{P}^{r-1}$. Let $(w_1,\dots,w_r)$ be Euclidean coordinates of $\mathbf{C}^r$ and denote by $[w_1,\dots,w_r]$

the corresponding homogeneous coordinates on $\mathbb{P}^{r-1}$. For $1 \leq k \leq r$ let U_k be the affine open subset $\{[w_1,\dots,w_r]: w_k \neq 0\}$. Define $\varphi_k: \pi^{-1}(U_k) \longrightarrow U_k \times \mathbf{C}^*$ by

$$\varphi_k(w_1,\dots,w_r) = ([w_1,\dots,w_r];w_k).$$

φ_k is a trivialization of π over U_k. On $U_{kh} = U_k \cap U_h$ define φ_{kh} as $\varphi_k \circ \varphi_h^{-1}$. We have $\varphi_{kh}([w_1,\dots,w_r];w_h) = ([w_1,\dots,w_r];w_k)$, so that for $\lambda_h \in \mathbf{C}^*$

$$\varphi_{kh}([w_1,\dots,w_r];\lambda_h) = \left([w_1,\dots,w_r]\,;\,\left[\frac{w_k}{w_h}\right]\lambda_h\right).$$

$\{\varphi_{kh}\}_{0\leq k,h\leq r}$ constitutes a 1–cocylce which defines on $\pi: \mathbf{C}^r - \{0\} \longrightarrow \mathbb{P}^{r+1}$ the structure of a holomorphic $\mathbf{C}^*$–bundle. The same cocycle $\{\varphi_{kh}\}$ defines a holomorphic line bundle L which is obtained by adjoining the zero section $\cong \mathbb{P}^{r-1}$ to L. In other words L is obtained from $\mathbf{C}^r$ by blowing up the origin. Using the Euclidean inner product on $\mathbf{C}^r$ the isomorphism L – {zero section} $\cong \mathbf{C}^r - \{0\}$ induces a Hermitian metric θ on L. We compute its curvature using the isomorphism φ_k: $L|U_k \cong U_k \times \mathbf{C}$. The inhomogeneous coordinates for $[w_1,\dots,w_r] \in U_k$ are given by $\left(\frac{w_1}{w_k},\dots,\frac{w_{k-1}}{w_k},\frac{w_{k+1}}{w_k},\dots,\frac{w_r}{w_k}\right) = w^{(k)}$. The point $(w_1,\dots,w_r) \in \mathbf{C}^r$ is identified with $(w^{(k)};w_k)$ so that $([w_1,\dots,w_r]\,;\,\lambda_k) \in U_k \times \mathbf{C}$ is identified with $\lambda_k\left(\frac{w_1}{w_k},\dots,\frac{w_{k-1}}{w_k}, 1, \frac{w_{k+1}}{w_k},\dots,\frac{w_r}{w_k}\right) \in \mathbf{C}^r$. From this it follows that

$$\|([w_1,\dots,w_r];\lambda_k)\|^2 = \sum_{1\leq i\leq r} \left|\frac{w_i}{w_k}\right|^2 . |\lambda_k|^2.$$

The curvature form of (L,θ) over $\mathbb{P}^{r-1}$ is therefore given by

$$\Theta(L,\theta) = -\sqrt{-1}\,\partial\bar\partial \log \sum_{1\leq i\leq r} \left|\frac{w_i}{w_k}\right|^2$$

$$= -\sqrt{-1}\,\partial\bar\partial \log\,(1 + \|w^{(k)}\|^2),$$

where $\|w^{(k)}\|$ refers to the length of $w^{(k)} \in U_k \cong \mathbf{C}^{r-1}$ in terms of the Euclidean metric on $\mathbf{C}^{r-1}$. Let φ denote the Kähler form of the Fubini–Study metric on

$\mathbb{P}^{r-1}$ such that the pull–back $\pi^*\varphi$ is given by $\sqrt{-1}\,\partial\bar\partial \log \|w\|^2$, then (L,θ) is a Hermitian holomorphic line bundle of constant negative curvature with respect to $(\mathbb{P}^{r-1},\varphi)$. As a holomorphic line bundle L is dual to the hyperplane section line bundle on $\mathbb{P}^{r-1}$, i.e., $L \cong \mathcal{O}(-1)$.

Consider now a Hermitian holomorphic vector bundle (V,h) of rank r over an n–dimensional complex manifold X. The Hopf process described above can be carried out fiber by fiber to yield the projectivization $\mathbb{P}(V)$ of V and a Hermitian holomorphic line bundle $(L,\hat{h})$ over $\mathbb{P}(V)$. Restricted to each fiber V_x over $x \in X$ this gives a Hermitian holomorphic line bundle $(L_x,\hat{h}_x) \to \mathbb{P}(V_x)$ isomorphic to $(L,\theta) \to \mathbb{P}^{r-1}$ of the previous paragraph. We are going to compute the curvature of $(L,\hat{h})$.

Let $x \in X$ and $(z^i)_{1\le i\le n}$ be a system of local holomorphic coordinates at x. Let $(v^\alpha)_{1\le\alpha\le r}$ be special holomorphic fiber coordinates for V at x. Denote by $\{e_\alpha\}_{1\le\alpha\le r}$ the corresponding holomorphic basis of V over a neighborhood of x. Consider now the point $[\mu] := (x,[e_r]) \in \mathbb{P}(V_x)$. In a neighborhood of $[\mu]$ in $\mathbb{P}(V)$ we use holomorphic coordinates $(z^1,\dots,z^n\,;\,u^1,\dots,u^{r-1})$, where $u^k := v^k/v^r$ (and $v^r \neq 0$). For the Hermitian holomorphic line bundle $(L,\hat{h})$ over $\mathbb{P}(V)$ we use the holomorphic fiber coordinate $\lambda = v^r$ in a neighborhood of $[\mu]$. We assert

PROPOSITION 1

In terms of the coordinates given above λ is a special holomorphic fiber coordinates for L adapted to $\hat{h}$ at $[\mu]$ and, denoting the curvature form of (V,h) by Θ, we have

$$\Theta(L,\hat{h})([\mu]) = -\sqrt{-1}\sum_{1\le\alpha\le r-1} du^\alpha \wedge d\bar{u}^\alpha + \sqrt{-1}\sum_{1\le i,j\le n} \Theta_{r\bar{r}i\bar{j}}\, dz^i \wedge d\bar{z}^j.$$

As a consequence $(L,\hat{h})$ is of seminegative curvature if and only if (V,h) is of seminegative curvature.

Proof:

Write Ψ: $V - X \cong L - (\text{zero section})$ for the canonical isomorphism. Write e for the holomorphic basis of L in a neighborhood of $[\mu]$ in $\mathbb{P}(V)$ defined by

$$\Psi^{-1}e([v]) = v.$$

Denote by e^* for the dual basis of e and $h = \hat{h}_o\, e^* \otimes \bar{e}^*$ on a neighborhood of $[\mu]$. We have

$$\hat{h}_o = \|e\|^2 = \sum_{1\le\alpha,\beta\le r-1} h_{\alpha\bar{\beta}}(z)\, u^\alpha \bar{u}^\beta + 2Re \sum_{1\le\alpha\le r-1} h_{\alpha\bar{r}}(z)\, u^\alpha + h_{r\bar{r}}(z)$$

and $\|e([\mu])\| = 1$. The point $[\mu]$ has coordinates $(z^1(x),\dots,z^n(x);0,\dots,0)$. Consequently,

$$d\hat{h}_o([\mu]) = 0$$

since $\frac{\partial}{\partial z_i} h_{\alpha\bar{\beta}}(x) = 0$ and, for $1 \le \alpha \le r-1$, $\frac{\partial}{\partial u_\alpha} \hat{h}_o = \Sigma h_{\alpha\bar{\beta}}\bar{u}^\beta + 2Re(\Sigma h_{\alpha\bar{r}}) = 0$ at $[\mu]$ with $u^\alpha([\mu]) = 0$ and $h_{\alpha\bar{r}} = 0$. Thus,

$$\begin{aligned}\Theta(L,\hat{h})[\mu] &= -\sqrt{-1}\,\partial\bar{\partial}\log\|e\|^2([\mu])\\ &= -\sqrt{-1}\left(\sum_{1\le\alpha\le r-1} du^\alpha \wedge d\bar{u}^\alpha + \sum_{1\le i,j\le n} \frac{\partial^2 h_{r\bar{r}}}{\partial z_i \partial\bar{z}_j} dz^i \wedge d\bar{z}^j\right)\\ &= -\sqrt{-1}\sum_{1\le\alpha\le r-1} du^\alpha \wedge d\bar{u}^\alpha + \sqrt{-1}\sum_{1\le i,j\le n} \Theta_{r\bar{r}i\bar{j}}\, dz^i \wedge d\bar{z}^j,\end{aligned}$$

proving Prop. 1. ▌

For any (V,h) and any $x \in X$ the restriction of $\Theta(L,\hat{h})$ to $\mathbb{P}(V_x)$ is negative definite. Given (V,h) of positive curvature one can obtain an associated Hermitian holomorphic line bundle of positive curvature as follows: The dual bundle (V^*,h^*) is of negative curvature. The tautological line bundle (Λ,θ) of (V^*,h^*) is of negative curvature by Prop.1. Its dual Hermitian holomorphic line bundle (Λ^*,θ^*) is then of positive curvature.

CHAPTER 3 RIEMANNIAN AND HERMITIAN SYMMETRIC MANIFOLDS

§1 Definition and Basic Properties of Riemannian Symmetric Manifolds

(1.1) For a general reference on materials covered by this chapter we refer the reader to HELGASON [HEL]. For basic facts and references on semisimple Lie algebras cf. [App.I]. Let (X,g) be a Riemannian manifold. A non–trivial isometry σ of (X,g) is said to be an involution if and only if $\sigma^2 \equiv \mathrm{id}$. We have

DEFINITION 1

A Riemannian manifold (X,g) is said to be Riemannian symmetric if and only if at each point $x \in X$ there exists an involution σ_x such that x is an isolated fixed point of σ_x.

Suppose (X,g) is a Riemannian symmetric manifold. Let $x \in X$ and $\gamma = \{\gamma(t)\colon -\epsilon < t < \epsilon\}$ be a geodesic emanating from x such that $\gamma(o) = x$. Denote by σ_t the involution at $\gamma(t)$. σ_t is necessarily unique. In fact, if $\exp_x$ denotes the exponential map at $x \in X$, then we have $\sigma_x(\exp_x(v)) = \exp_x(-v))$ in a neighborhood of x. Denote the composition $\sigma_o \circ \sigma_{-t/2}$ by τ_t. The isometry τ_t is called a transvection. Along the geodesic γ it is given by $\tau_t(\gamma(s)) = \gamma(t+s)$. From this it follows readily that a Riemannian symmetric manifold (X,g) is necessarily geodesically complete. The group Aut(X,g) of isometries of the complete Riemannian manifold (X,g) is a real Lie transformation group by MYERS–STEENROD [MS]. Let $G = \mathrm{Aut}_o(X,g)$ be the identity component of Aut(X,g). Using the transvections $\{\tau_t\}$ one deduces that G acts transitively on (X,g), making X into a homogeneous space G/K, where $K \subset G$ denotes the isotropy subgroup at x. K is necessarily compact.

E. Cartan, who defined the notion of Riemannian symmetric manifolds, classified them by a reduction to the classification of simple Lie algebras (cf. [HEL, Chaps. V & X]). Let (X,g) be a Riemannian symmetric manifold. Write X = G/K as a homogeneous space as above. Denote the identity coset eK by o. Write $\mathfrak{g}$ and $\mathfrak{k}$ for the Lie algebras of G and K resp. Identify the tangent space $T_o(X)$ with the Lie algebra $\mathfrak{g}/\mathfrak{k}$. The involution σ_o at o induces a tangent map $d\sigma_o\colon \mathfrak{g} \to \mathfrak{g}$. We write $\theta = d\sigma_o$. θ is an involution on $\mathfrak{g}$, i.e., θ is a non-trivial automorphism of the Lie algebra $\mathfrak{g}$ such that $\theta^2 = \mathrm{id}$. We have a canonical decomposition of $\mathfrak{g}$ into $\mathfrak{k} + \mathfrak{m}$, where $\mathfrak{m}$ is the (-1)–eigenspace of θ and the Lie algebra $\mathfrak{k}$ is the $(+1)$–eigenspace of θ. $\mathfrak{m}$ can be identified with $T_o(X) \cong \mathfrak{g}/\mathfrak{k}$.

(1.2) Let now (X,g) be a simply–connected Riemannian symmetric manifold. (X,g) can be decomposed isometrically using de Rham's Theorem ([App.II.1, Thm.1]). With respect to the canonical decomposition $\mathfrak{g} = \mathfrak{k} + \mathfrak{m}$ the isotropy subgroup $K \subset G = \mathrm{Aut}_0(X,g)$ acts effectively on $\mathfrak{m}$ by adjoint action. Let $\mathfrak{m} = \Sigma\mathfrak{m}_i$ be the decomposition of the K–representation space $\mathfrak{m}$ into irreducible K–representation spaces. Since each $\mathfrak{m}_i$ is K–invariant, by transporting $\mathfrak{m}_i$ to points $gK \in G/K \cong X$ this gives rise to a decomposition of the tangent bundle T_X into a direct sum of vector subbundles $T_X = \Sigma T_i$. By [App.II.2, Prop.1] each T_i is invariant under parallel transport on (M,g), so that by de Rham's Theorem we have an isometric decomposition $(X,g) \cong (X_0,g_0) \times (X_1,g_1) \times \ldots \times (X_r,g_r)$, where (X_0,g_0) is flat and for each $i \geq 1$, (X_i,g_i) is a Riemannian symmetric manifold $\cong G_i/K_i$ such that in the canonical decomposition $\mathfrak{g}_i = \mathfrak{k}_i + \mathfrak{m}_i$ the isotropy group K_i acts irreducibly on $\mathfrak{m}_i \cong T_0(X_i)$.

We now discuss the notion of irreducible Riemannian symmetric manifolds of compact and non–compact types. By E. Cartan's criterion a Lie algebra $\mathfrak{g}$ is semisimple if and only if the Killing form $B_{\mathfrak{g}}(x,y) = \mathrm{Tr}_{\mathfrak{g}}(\mathrm{ad}(x)\mathrm{ad}(y))$ is non–degenerate. We say that a semisimple Lie algebra $\mathfrak{g}$ is compact if and only if the Killing form is negative definite. It is known that a semisimple Lie group G is compact if and only if the associated Lie algebra $\mathfrak{g}$ is compact (cf. [HEL, Ch.II, §2, Prop.(6.6), p.132]). Let now (X,g) be a simply–connected Riemannian symmetric manifold with $X \cong G/K$, $G = \mathrm{Aut}_0(X,g)$ such that the compact group K acts irreducibly on $\mathfrak{m} = T_0(X)$. We call X an irreducible symmetric manifold. Let $\mathfrak{g} = \mathfrak{k} + \mathfrak{m}$ be the canonical decomposition of the (± 1)–eigenspaces with respect to the involution $\theta = d\sigma_0$. From the fact that θ is an automorphism it follows that

$$[\mathfrak{k},\mathfrak{k}] \subset \mathfrak{k}, \quad [\mathfrak{m},\mathfrak{m}] \subset \mathfrak{k}, \quad [\mathfrak{k},\mathfrak{m}] \subset \mathfrak{m}. \tag{1}$$

From (1) it is clear that $\mathfrak{k}$ and $\mathfrak{m}$ are orthogonal with respect to the Killing form

$B_{\mathfrak{g}}$. $B_{\mathfrak{g}}|_{\mathfrak{k}} \equiv B_{\mathfrak{k}}$ is negative definite. Since the Killing form $B_{\mathfrak{g}}$ is invariant under K and K acts isometrically on $T_o(X)$ it follows that $B_{\mathfrak{g}}|_{\mathfrak{m}}$ is λ times the inner product g on $\mathfrak{m}$. If $\lambda = 0$, then (X,g) is flat. If $\lambda < 0$, then the Killing form $B_{\mathfrak{g}}$ is negative definite on $\mathfrak{g}$, so that $\mathfrak{g}$ is compact and semisimple. If $\lambda > 0$, then (X,g) is non–compact and semisimple. In the latter two cases we say that (X,g) is an irreducible Riemannian symmetric manifold of compact type and non–compact type resp. In general when G is semisimple we say that (X,g) is of semisimple type.

Let $\mathfrak{g}$ be a semisimple Lie algebra, $\mathfrak{k}$ be a compact subalgebra and θ be an involution on $\mathfrak{g}$ such that $\mathfrak{k}$ is the fixed point set of θ. We call $(\mathfrak{g},\mathfrak{k},\theta)$ a semisimple orthogonal symmetric Lie algebra. As described above, given any simply-connected Riemannian symmetric manifold (X,g) without flat factors there corresponds such a pair $(\mathfrak{g},\mathfrak{k},\theta)$. Conversely, given such a pair $(\mathfrak{g},\mathfrak{k},\theta)$ one can reconstruct a simply–connected Riemannian symmetric manifold (X,g) by taking G to be a simply–connected Lie group with Lie algebra $\mathfrak{g}$, $K \subset G$ to be a connected subgroup corresponding to the Lie subalgebra $\mathfrak{k} \subset \mathfrak{g}$, and the Riemannian metric g to be induced by a right–invariant metric on G. θ induces a group isomorphism σ on G defining an involution on G and hence on $X = G/K$ at $o = eK$, making (X,g) into a Riemannian symmetric manifold. Write $\mathfrak{g} = \mathfrak{k} + \mathfrak{m}$ for the canonical decomposition of $\mathfrak{g}$ into (± 1)–eigenspaces of θ. We say that $(\mathfrak{g},\mathfrak{k},\theta)$ is of compact type (resp. of non–compact type) if and only if the Killing form $B_{\mathfrak{g}}$ is negative definite (resp. positive definite on $\mathfrak{m}$). We say that $(\mathfrak{g},\mathfrak{k},\theta)$ is irreducible if and only if K acts irreducibly on $\mathfrak{m}$. In this case it is clear that $\mathfrak{k}$ is a maximal compact subalgebra.

Note that the simply–connected group G does not necessarily act effectively on X. For example if $\mathfrak{g} = \mathfrak{sl}(2,\mathbb{R})$ we obtain $\mathfrak{g}^{\mathbb{C}} = \mathfrak{sl}(2,\mathbb{C})$ and $X \cong SL(2,\mathbb{R})/S^1$ is the upper half–plane H, equipped with the Poincaré metric. $SL(2,\mathbb{R})$ acts on H by fractional linear transformations given by $\begin{bmatrix} a & b \\ c & d \end{bmatrix}(z) = \frac{az+b}{cz+d}$, with the kernel $\{\pm 1\}$. $Aut_o(X,g)$ is then identified with $PSL(2,\mathbb{R}) = SL(2,\mathbb{R})/\{\pm 1\}$. For a semisimple orthogonal symmetric Lie algebra $(\mathfrak{g},\mathfrak{k},\theta)$ and the corresponding simply–connected Riemannian symmetric manifold (X,g) we have in general $Aut_o(X,g) \cong G/Z(G)$, where Z(G) denotes the center of G. Clearly the involution σ on G preserves the center Z(G) as a set, so that σ descends to an involution on $G/Z(G)$, which is centerless and isomorphic to the adjoint group of $\mathfrak{g}$. From now on we will use G to denote instead the adjoint group corresponding to $\mathfrak{g}$.

There is a duality between non–compact semisimple $\mathfrak{g}_o$ and compact semisimple Lie algebras $\mathfrak{g}_c$ and their involutions. For a semisimple orthogonal symmetric Lie algebra $(\mathfrak{g},\mathfrak{k},\theta)$ with canonical decomposition $\mathfrak{g} = \mathfrak{k} + \mathfrak{m}$, define $\mathfrak{g}^*$ to be the real vector space $\mathfrak{k} + \sqrt{-1}\mathfrak{m}$. Denote by θ^* the involution on $\mathfrak{g}^*$ given by $\theta^*(k + \sqrt{-1}m) = k - \sqrt{-1}m$ for $k \in \mathfrak{k}$ and $m \in \mathfrak{m}$. Denote by $\mathfrak{g}^{\mathbf{C}}$ the complexification of $\mathfrak{g}$. We have $\mathfrak{g}^* \subset \mathfrak{g}^{\mathbf{C}}$. The relations (1) then imply that $\mathfrak{g}^*$ is closed under the Lie bracket $[.,.]$, i.e., $\mathfrak{g}^* \subset \mathfrak{g}^{\mathbf{C}}$ is a real Lie subalgebra. Clearly $\mathfrak{g}^{**} = \mathfrak{g}$. If $\mathfrak{g}$ is simple and non–compact, then from the definition $B_{\mathfrak{g}^*}$ is negative definite, so that $\mathfrak{g}^*$ is compact. We say that two Riemannian symmetric manifolds of semisimple type are dual to each other if and only if their associated orthogonal symmetric pairs are dual to each other. The problem of classifying simply–connected Riemannian symmetric manifolds can be reduced to classifying simple complex Lie algebras $\mathfrak{g}^{\mathbf{C}}$ and their involutions, giving rise to various dual pairs of compact and non–compact real forms $(\mathfrak{g}_c, \mathfrak{g}_o)$ (cf. [HEL, Ch.X, p.438ff.]). It is known from classification theory that every simple complex Lie algebra $\mathfrak{g}^{\mathbf{C}}$ admits involutions (which may not be conjugate under inner automorphisms).

We summarize basic structure theorems for Riemannian symmetric manifolds as follows:

PROPOSITION 1

Let (X,g) be a simply–connected Riemannian symmetric manifold. Then, $(X,g) \cong (X_c,g_c) \times (X_o,g_o) \times (X_e,g_e)$, where (X_e,g_e) is a flat Euclidean space; (X_c,g_c) and (X_o,g_o) are resp. Riemannian symmetric manifolds of compact and non-compact type.

PROPOSITION 2

Let (X_c,g_c) be a simply–connected irreducible symmetric manifold of compact type. Then, $(X_c,g_c) \cong (X_1,g_1) \times \ldots \times (X_r,g_r)$, where each (X_i,g_i) is an irreducible Riemannian symmetric manifold. If (X,g) is irreducible then either $X \cong G_c/K$, where $G_c = \mathrm{Aut}_o(X_c,g_c)$ is a compact centerless simple Lie group and K is some maximal connected proper subgroup; or, X_c is a compact simply–connected simple Lie group equipped with a bi–invariant Riemannian metric g_c.

We say that (X_c,g_c) is of types I and II in the two cases resp.

PROPOSITION 3

A Riemannian symmetric manifold (X_o,g_o) of non–compact type is necessarily simply–connected. We have $(X_o,g_o) \cong (X_1,g_1) \times \ldots \times (X_r,g_r)$, where each (X_i,g_i) is an irreducible symmetric manifold of non–compact type. If (X,g) is irreducible then either $X \cong G_0/K$, where $G_0 = \mathrm{Aut}_o(X,g)$ is a non–compact centerless simple Lie group and $K \subset G_0$ is some maximal connected compact subgroup; or, G_0 is the underlying real Lie group of a centerless non–compact adjoint Lie group over $\mathbb{C}$ and (X_o,g_o) is dual to a compact simply–connected irreducible Riemannian symmetric manifold (X_c,g_c) of type II.

For Props. 2 and 3 cf. [HEL, Ch. VIII, p.379ff.] and [HEL, Ch. VI, p.252ff.].

(1.3) Let (X,g) be an irreducible Riemannian symmetric manifold of semi-simple type. Write $G = \mathrm{Aut}_o(X,g)$ and $K \subset G$ for the isotropy group at some point $o \in X$. We compute the curvature of $X \cong G/K$ at $o = eK$ using the Killing form. Write σ_o for the involution at o and $\theta = d\sigma_o$ for the corresponding involution on the Lie algebra $\mathfrak{g}$. Recall that the curvature tensor is given on (X,g) by the formula

$$R(A,B)C = \nabla_A\nabla_B C - \nabla_B\nabla_A C - \nabla_{[A,B]}C,$$

where A, B, C are smooth extensions of the vectors A, B, C at o to a neighborhood. Here $[A,B]$ stands for the Poisson bracket on vector fields. Recall the identification $T_o(X) \cong \mathfrak{g}/\mathfrak{k}$. Using the canonical decomposition $\mathfrak{g} = \mathfrak{k} + \mathfrak{m}$ with respect to the involution $\theta = d\sigma_o$ we may further identify $T_o(X)$ with $\mathfrak{m} \subset \mathfrak{g}$. On the other hand $\mathfrak{g}$ can be identified with the Lie algebra $\mathfrak{g}'$ of Killing vector fields (i.e., infinitesimal isometries). The Lie bracket on $\mathfrak{g}'$ is given by the

Poisson bracket. The identification of $\mathfrak{g}$ with $\mathfrak{g}'$ gives an extension of vectors A, B, $C \in T_0(X) \cong \mathfrak{m} \subset \mathfrak{g}$ to Killing vector fields on X, given by infinitesimal transvections $(d\tau_t/dt)|_{t=0}$. Recall from [App.I.1, Prop.1] that the transvections $\{\tau_t\}$ induce parallel transport along their integral curves (which are geodesics) emanating from o. We start with (cf. e.g. WOLF [WOL1, Thm.(8.4.1), p.245–246])

PROPOSITION 1

At $o = eK$ we have $R(A,B)C = -[[A,B],C]$.

Proof:

Given a one–parameter family of diffeomorphisms $\{\varphi_t\}$ with $V = d\varphi_t/dt$ and a smooth tensor field T denote by $\mathcal{L}_V T = \lim_{t\to 0} \frac{1}{t}(\varphi_t^* T - T)$ the Lie derivative of T against V. The Lie derivative for vector fields A and B gives $\mathcal{L}_A B = [A,B]$. The Jacobi identity for the Poisson bracket on vector fields A, B and C can be expressed as $(\mathcal{L}_A \mathcal{L}_B - \mathcal{L}_B \mathcal{L}_A)C = \mathcal{L}_{[A,B]}C$. When A arises from a one-parameter family of isometries $\{\varphi_t\}$ it follows from $\varphi_t^*(\nabla_B C) = \nabla_{\varphi_t^* B}(\varphi_t^* C)$ that we have $\mathcal{L}_A \nabla_B C = \nabla_{[A,B]} C + \nabla_B \mathcal{L}_A C$ Finally, if $\{\varphi_t\}$ induces parallel transport along an integral curve $\gamma = \{\gamma(t): -\epsilon < t < \epsilon\}$ of A with $\gamma(0) = o$, then we have $\mathcal{L}_A T(o) = \nabla_A T(o)$. In fact, if $T = T' + tW$ with T' parallel along γ, then we have $\mathcal{L}_A T(o) = \nabla_A T(o) = W(o)$. We use the infinitesimal transvections A, B, $C \in \mathfrak{m}$ for the calculation of curvature. We note that for A, $B \in \mathfrak{m}$, $[A,B] \in [\mathfrak{m},\mathfrak{m}] \subset \mathfrak{k}$ so that $[A,B](o) = 0$ as $\mathfrak{k}$ is given by infinitesimal isotropies at 0.

$$\begin{aligned} \nabla_A \nabla_B C(o) &= \mathcal{L}_A \nabla_B C(o) \\ = \mathcal{L}_B \mathcal{L}_A C(o) + \nabla_{[A,B]} C(o) &= \mathcal{L}_B \mathcal{L}_A C(o) \end{aligned}$$

It follows now

$$\begin{aligned} R(A,B)(C)(o) &= \nabla_A \nabla_B C(o) - \nabla_B \nabla_A C(o) \\ = -(\mathcal{L}_A \mathcal{L}_B - \mathcal{L}_B \mathcal{L}_A)C(o) &= -[[A,B],C](o), \end{aligned}$$

as asserted in the proposition.

As a consequence of Prop.1 we obtain

PROPOSITION 2

Let (X,g) be a Riemannian symmetric manifold of compact (resp. non-compact) type. Then, the sectional curvature of (X,g) is semipositive (resp. seminegative). If (X,g) is furthermore irreducible then it is of constant positive (negative) Ricci curvature.

Proof:

It suffices to prove Prop.2 under the assumption that (X,g) is irreducible. In the notations of Prop.1 and denoting the inner product on (X,g) by $(.,.)$ we have

$$R(A,B;B,A) = (R(A,B)B, A) = (-[[A,B],B], A).$$

When (X,g) is irreducible and of compact type the Riemannian inner product on $T_0(X) \cong \mathfrak{m}$ is given by $-cB_{\mathfrak{g}}|_{\mathfrak{m}}$, where $B_{\mathfrak{g}}$ is the Killing form and c is a positive constant since both $(.,.)$ and $B_{\mathfrak{g}}|_{\mathfrak{m}}$ are invariant under K and K acts irreducibly on $\mathfrak{m}$. On the other hand from the invariance of $B_{\mathfrak{g}}$ under inner automorphisms we have

$$B_{\mathfrak{g}}([u,v],w) = -B_{\mathfrak{g}}(v,[u,w]).$$

It follows that for (X,g) irreducible and of compact type we have for $A, B \in \mathfrak{m}$ orthogonal and of unit length

$$K(A,B) = -cB_{\mathfrak{g}}([-[A,B],B], A) = -cB_{\mathfrak{g}}([A,B], [A,B]) \geq 0$$

since $B_{\mathfrak{g}}$ is negative definite on the compact semisimple Lie algebra $\mathfrak{g}$. If on the other hand (X,g) is irreducible and of non-compact type, then $B_{\mathfrak{g}}|_{\mathfrak{m}}$ is positive definite so that the Riemannian inner product $(.,.)$ is given by $cB_{\mathfrak{g}}|_{\mathfrak{m}}$ for some positive constant c. It follows in this case that (X,g) is of seminegative Riemannian sectional curvature. In both cases the Ricci curvature tensor defines a symmetric bilinear form ρ on $T_0(X) \cong \mathfrak{m}$ invariant under the action of K. It follows again from the irreducibility of isotropy representation of K on $\mathfrak{m}$ that ρ must be a constant multiple of the Riemannian inner product $(.,.)$. If the constant were zero it follows from the formula for sectional curvatures using the Lie bracket that the Killing form $B_{\mathfrak{g}}$ would vanish identically on $\mathfrak{m}$, contradicting with the semisimplicity of $\mathfrak{g}$. The proof of Prop.2 is completed.

§2 Hermitian Symmetric Manifolds

(2.1) We specialize now to Hermitian symmetric manifolds. First, we have

DEFINITION 1

Let (X,g) be a Riemannian symmetric manifold. (X,g) is said to be a Hermitian symmetric manifold if (X,g) is a Hermitian manifold and the involution σ_x at each point $x \in X$ is a holomorphic isometry.

We will write J for the (integrable) almost complex structure on X and denote by $\text{Aut}(X,J,g)$ the group of holomorphic isometries on (X,g), i.e., isometries preserving the almost complex structure J. We have

PROPOSITION 1

A Hermitian symmetric manifold (X,g) is Kähler. Suppose furthermore that (X,g) is simply-connected and that $(X,g) \cong (X_c,g_c) \times (X_o,g_o) \times (X_e,g_e)$ is the decomposition of the underlying Riemannian symmetric manifold into factors of compact, non-compact and Euclidean types, then each factor is itself a Hermitian symmetric manifold. Furthermore, the irreducible de Rham factors of (X_c,g_c) and (X_o,g_o) are also Hermitian symmetric manifolds.

Proof:

The almost complex structure J is a tensor of the form $J = J_\alpha^\beta\, dx^\alpha \otimes \frac{\partial}{\partial x_\beta}$ in terms of local coordinates $\{x_\alpha\}$. J is invariant under the involutions $\sigma = \sigma_x$, $x \in X$. It follows that $\nabla_{\sigma A}(\sigma J) = \sigma(\nabla_A J)$ for any vector $A \in T_x(X)$. From type considerations we have $\sigma(\nabla_A J) = \nabla_A J$ so that $\nabla_{-A} J = \nabla_A J$, proving that $\nabla J \equiv 0$. From [Ch.2, (1.2), Prop.1] it follows that (X,g) is Kähler.(*) (Cf. REMARKS.)

Write $X = G/K$ with $G = \text{Aut}_o(X,g)$ Let $T = \Sigma T_i$ be a decomposition of T_X into parallel subbundles corresponding to a decomposition at $o = eK$ of $T_o(X) \cong \mathfrak{m}$ into irreducible components of the isotropy representation of K on $\mathfrak{m}$. Since J is parallel on X for each T_i the subbundle $JT_i \subset T_X$ is also parallel. As the curvature tensor is invariant under J, JT_i corresponds to a flat submanifold if T_i does. From the uniqueness part of de Rham decompositions (cf. [App.II, Thm.1]) it follows that the Euclidean component (X_o,g_o) is invariant under J, i.e., X_e is a complex Euclidean space $\mathbf{C}^n$ equipped with the flat metric. Suppose (X_i,g_i) is an isometric factor of (X,g) of semisimple type corresponding to some T_i. For every real number λ the distribution $V_\lambda := \{v + \lambda Jv: v \in T_i\}$ gives a subbundle of T_X invariant under holonomy. This contradicts with the uniqueness part of de Rham's Theorem unless we have $JT_i = T_i$, i.e., X_i is a complex manifold and (X_i,g_i) is a Hermitian symmetric manifold, proving Prop.1.

REMARKS

(*) Let T be a tensor of contravariant degree p and covariant degree q on a Riemannian symmetric manifold (X,g) invariant under involutions σ_x, $x \in X$. The argument given shows that if $p + q$ is odd then $T \equiv 0$.

Let (X,g) be a Hermitian symmetric manifold of semisimple type. It is known that $G = \mathrm{Aut}_0(X,g)$ agrees with the identity component $\mathrm{Aut}_0(X,J,g)$ of the group of holomorphic isometries (cf. [HEL, Ch.V, Thm.(4.1), p.243]). Thus, by Prop.1, the classification problem for simply–connected Hermitian symmetric manifolds is reduced to the question of classifying irreducible simply–connected Riemannian symmetric manifolds (X,g) admitting a G–invariant integrable almost complex structure J. We have

PROPOSITION 2 (cf. [HEL, Ch.VIII, p.352ff.])

Let (X,g) be an irreducible Riemannian symmetric manifold of semisimple type. Write as usual $G = \mathrm{Aut}_0(X,g)$, $X \cong G/K$. Then, (X,g) admits an integrable almost complex structure (hence the structure of a Hermitian symmetric manifold) if and only if K is not semisimple. In this case G is simple and the center of K is isomorphic to the circle group S^1 and the complex structure is unique. Moreover, every Hermitian Riemannian symmetric manifold is simply–connected.

If (X,g) is irreducible and of compact type it follows from [(1.2), Prop.2] that either G is simple or $\mathfrak{g} = \mathfrak{h} \oplus \mathfrak{h}$, $\mathfrak{k}$ is the diagonal of $\mathfrak{h} \oplus \mathfrak{h}$ for some compact simple Lie algebra $\mathfrak{h}$. In the latter case $\mathfrak{k}$ is centerless and (X,g) does not admit a Hermitian symmetric structure. An analogous statement applies to (X,g) of non–compact type by duality. For the last statement of Prop.2 here we recall first of all that all Riemannian symmetric manifolds of non–compact type are simply–connected ([(1.2), Prop.3]). Let (X,g) be a Hermitian symmetric manifold of compact type, then by [(2.1), Prop.2] it is of positive Ricci curvature. A theorem of KOBAYASHI [KO] asserts that any compact Kähler manifold of positive Ricci curvature is simply–connected.

Let (X,g) be as in the proposition such that the center Z(K) of K is positive–dimensional. We are only going to describe how the integrable almost complex structure arises from Z(K). We work with G/K and the associated semisimple orthogonal symmetric Lie algebra $(\mathfrak{g},\mathfrak{k};\theta)$. G is centerless. Write $\mathfrak{g} = \mathfrak{k} + \mathfrak{m}$ for the canonical decomposition. Since Z(K) is positive–dimensional there exists an element $j \in Z(K)$ of order 4. Write $s = j^2$. Then, $s^2 = e$. Consider the automorphism σ on G defined by $\sigma(g) = sgs^{-1}$. Clearly $\sigma^2(g) = g$ and the fixed point set of σ is the centraliser Z(s) of s in G with Lie algebra $\mathfrak{z}(s) \subset \mathfrak{g}$. As $s \in Z(K)$ we have $K \subset Z(s)$. Since G is centerless $Z(s) \neq G$. As $\mathfrak{k} \subset \mathfrak{g}$ is a maximal proper subalgebra we have $\mathfrak{k} = \mathfrak{z}(s)$. Consider $\theta' := d\sigma = Ad(s) : \mathfrak{g} \longrightarrow \mathfrak{g}$. θ' is an involution with fixed point set $\mathfrak{k}$. Let $\mathfrak{g} = \mathfrak{k} + \mathfrak{m}'$ be the canonical decomposition with respect to θ', so that $\theta'|_{\mathfrak{m}'} = -id$. $\mathfrak{m}'$ agrees with the orthogonal complement of $\mathfrak{k}$ in $\mathfrak{g}$ with respect to the Killing form $B_{\mathfrak{g}}$ so that in fact we must have $\theta = \theta'$ and $\mathfrak{m} = \mathfrak{m}'$. Define η: $G \longrightarrow G$ by $\eta(g) = jgj^{-1}$. Define $J: \mathfrak{m} \longrightarrow \mathfrak{m}$ as $d\eta|_{\mathfrak{m}} = Ad(j)|_{\mathfrak{m}}$. Then, from $\eta^2 = \sigma$ we have $J^2 = -id$. Since J is by definition K–invariant transporting by G we get a G–invariant almost complex structure J on $X \cong G/K$. From the irreducibility of (X,g) we have $g(v,w) = cB_{\mathfrak{g}}(v,w)$ for some non–zero constant c. The Riemannian metric g on X is Hermitian with respect to J since

$$\begin{aligned} g(Jv,Jw) &= g(Ad(j)v,Ad(j)w) = c\,B_{\mathfrak{g}}(Ad(j)v,Ad(j)w) \\ &= c\,B_{\mathfrak{g}}(v,w) = g(v,w) \end{aligned}$$

by the invariance of the Killing form under the adjoint action of G.

To show that J is integrable we resort to the integrability criterion of Newlander–Nirenberg [NN]. Given an almost complex manifold (X,J) of real dimension 2n one has the decomposition $T_X^{\mathbb{C}} \cong T_X^{1,0} \oplus T_X^{0,1}$ into $(\pm\sqrt{-1})$–eignspaces (cf. [Ch.2, (1.1)]). This gives rise to the notion of forms of type (p,q). In a neighborhood U of a point $x \in X$ choose a frame field $\{v_i\}_{1\le i\le n}$ of $T_U^{1,0}$

and denote by $\{\omega^i\}_{1\le i\le n}$ the dual basis of $(T_U^{1,0})^*$. One can define the $\overline{\partial}_J$–operators on functions f by defining $\overline{\partial}_J(f) = \overline{v_i}(f)\,\overline{\omega}^i$. Using the bases $\{v_i\}$ and $\{\omega^i\}$ one extends the definition of the $\overline{\partial}_J$–operator to (p,q)–forms. A necessary condition for the almost complex structure J to come from an actual complex structure is that $\overline{\partial}_J^2 = 0$. One can verify algebraically that if $\overline{\partial}_J^2 = 0$ on functions the same is true for (p,q)–forms. The condition $\overline{\partial}_J^2 = 0$ on functions is equivalent to the more geometric condition that $T_X^{1,0}$ is closed under Lie brackets. EHRESMANN [EH] showed in the case of real–analytic (X,J) that the latter condition is actually sufficient for the integrability of (X,J). The proof uses only the integrability condition of Frobenius on distributions (cf. also KOBAYASHI-NOMIZU [KN, Vol.II, App.8, p.321–324]). The general case was proved by Nirenberg–Newlander using the condition $\overline{\partial}_J^2 = 0$ on functions. The problem is local in nature. It suffices to find local "holomorphic coordinates", i.e., smooth complex–valued functions $\{f_1,\ldots,f_n\}$ such that $\overline{\partial}_J(f_i) \equiv 0$ and such that $\{Re(f_i), Im(f_i)\}_{1\le i\le n}$ constitutes a system of real coordinates locally. It then follows immediately from the chain rule that the different choices of $\{f_1,\ldots,f_n\}$ on overlaps of coordinate patches are related by holomorphic transformations, giving X the structure of a complex manifold compatible with J. In [NN] the existence problem for holomorphic coordinates is solved by using the Cauchy kernel and the iteration method of Korn–Lichtenstein. An alternative proof can be obtained by solving the $\overline{\partial}$–Neumann problem (KOHN [KOH]) or by using $\overline{\partial}$ with L^2–estimates (HÖRMANDER [HÖR]). For the formulation of various equivalent forms of the integrability condition cf. [KN, Vol.II, Ch.IX, Thm.2.8, p.125].

In the case of the almost–complex structure J on the Riemannian symmetric manifold (X,g) one can check the integrability condition by using the Nijenhuis torsion tensor N (cf. [HEL, Ch.VIII, p.352]). Given any two vector fields A and B on (X,J) one considers the vector

$$N(A,B) = [A,B] + J[JA,B] + J[A,JB] - [JA,JB].$$

It can be proved that N(A,B)(x) depends only on the values of A and B at x so that N defines a tensor on X, contravariant of degree 1 and covariant of de–

gree 2. The assumption that $T_x^{1,0}$ is closed under Lie brackets is equivalent to the vanishing of the Nijenhuis tensor N. On the other hand since J and hence N is G–invariant and the tensor N is of odd total degree, we have $N \equiv 0$ (cf. Remarks after [(1.2), Prop.2]).

In §3 we will prove the Borel Embedding Theorem for realizing Hermitian manifolds of non–compact type (X_o,g_o). It will be seen that the proof there can actually be used to show the integrability statement of Prop.2.

§3 The Borel Embedding Theorem

(3.1) Let $\mathfrak{g}_o$ be a simple Lie algebra of non–compact type, $\mathfrak{k} \subset \mathfrak{g}_o$ a maximal compact subalgebra. Let θ_o be a Cartan involution on $\mathfrak{g}_o$, i.e., θ_o is a Lie algebra automorphism such that $(.,.) = -B_{\mathfrak{g}}(u,\theta_o v)$ is positive definite. Suppose $\mathfrak{k}$ is not semisimple. Let $\mathfrak{g}^{\mathbb{C}} = \mathfrak{k}^{\mathbb{C}} + \mathfrak{m}^{\mathbb{C}}$ be the complexification of $\mathfrak{g}$. Write $\mathfrak{g}_c = \mathfrak{k} + \sqrt{-1}\mathfrak{m}$ for the compact dual of $\mathfrak{g}_o$. Write $G^{\mathbb{C}}$ for the adjoint Lie group of $\mathfrak{g}^{\mathbb{C}}$. Let $G_0, G_c, K \subset G^{\mathbb{C}}$ be the connected real Lie subgroups corresponding to the real Lie subalgebras $\mathfrak{g}_o, \mathfrak{g}_c, \mathfrak{k}$ resp. G_0 and G_c are centerless. In fact if the involution $\theta_o: \mathfrak{g}_o \to \mathfrak{g}_o$ extends by complexification to an involution $\theta^{\mathbb{C}}: \mathfrak{g}^{\mathbb{C}} \to \mathfrak{g}^{\mathbb{C}}$ and hence by restriction an involution $\theta_c: \mathfrak{g}_c \to \mathfrak{g}_c$. $\theta^{\mathbb{C}}$ gives rise to an involution σ on $G^{\mathbb{C}}$. σ preserves G_0 and G_c. K is the identity component of the fixed point sets of $\sigma|G_0$ and $\sigma|G_c$. Since $\mathfrak{k}$ is not semisimple some element of the center Z(K) gives rise to an integrable G_0–invariant almost complex structure J on $X_o = G_0/K$ and similarly on $X_c = G_c/K$. We obtain thus from the Killing forms of $\mathfrak{g}_o$ and $\mathfrak{g}_c$ dual Hermitian symmetric manifolds (X_o,g_o) and (X_c,g_c).

From [(2.1), Prop.1] all irreducible Hermitian symmetric manifolds of compact or non–compact type arise in the way we just described. By taking direct products of (X_o,g_o) we obtain all Hermitian symmetric manifolds of non–compact type. From now on we will let $(\mathfrak{g}_o,\mathfrak{k};\theta)$ denote a direct sum of simple non–compact orthogonal Lie algebras with non–semisimple maximal compact subalgebras, etc., so that (X_o,g_o) and (X_c,g_c) are possibly reducible. Write $o = eK$ both on X_o and X_c. Let $\mathfrak{m}^{\mathbb{C}} = \mathfrak{m}^+ + \mathfrak{m}^-$ be the direct sum decomposition of $\mathfrak{m}^{\mathbb{C}}$ into the $(\pm\sqrt{-1})$–eigenspaces of the (complexification of the) almost complex structure J at o. This corresponds to the two decompositions $T^{\mathbb{C}}(X_o) = T^{1,0}(X_o) + T^{0,1}(X_o)$ and $T^{\mathbb{C}}(X_c) = T^{1,0}(X_c) + T^{0,1}(X_c)$. Since the isometries K on (X_o,g_o) and (X_c,g_c) preserves the almost complex structure J, $\mathfrak{m}^+$ and $\mathfrak{m}^-$ are invariant under the adjoint action of K, i.e., $[\mathfrak{k},\mathfrak{m}^+] \subset \mathfrak{m}^+$ and $[\mathfrak{k},\mathfrak{m}^-] \subset \mathfrak{m}^-$.

LEMMA 1

$\mathfrak{m}^+$ and $\mathfrak{m}^-$ are abelian subalgebras of $\mathfrak{g}^{\mathbb{C}}$. Moreover, the complex vector subspace $\mathfrak{p} = \mathfrak{k}^{\mathbb{C}} + \mathfrak{m}^- \subset \mathfrak{g}^{\mathbb{C}}$ is a complex Lie subalgebra of $\mathfrak{g}^{\mathbb{C}}$.

Proof:

Recall that in the construction of J in (2.1) J is the restriction to $\mathfrak{m}$ of an inner automorphism of $\mathfrak{g}_o$ (which we also denoted by J) and hence by complexification of $\mathfrak{g}^{\mathbb{C}}$. Since $[\mathfrak{m}^{\mathbb{C}},\mathfrak{m}^{\mathbb{C}}] \subset \mathfrak{k}^{\mathbb{C}}$ and J fixes $\mathfrak{k}^{\mathbb{C}}$, for $m_1, m_2 \in \mathfrak{m}^{\mathbb{C}}$ we have $[Jm_1,Jm_2] = [m_1,m_2]$. On the other hand for $m_1^+, m_2^+ \in \mathfrak{m}^+$ we have $[Jm_1^+,Jm_2^+] = [\sqrt{-1}m_1^+,\sqrt{-1}m_2^+] = -[m_1^+,m_2^+]$, so that necessarily $[\mathfrak{m}^+,\mathfrak{m}^+] = 0$. The proof of $[\mathfrak{m}^-,\mathfrak{m}^-] = 0$ is the same. $[\mathfrak{p},\mathfrak{p}] \subset [\mathfrak{k}^{\mathbb{C}},\mathfrak{k}^{\mathbb{C}}] + [\mathfrak{k}^{\mathbb{C}},\mathfrak{m}^-] \subset \mathfrak{k}^{\mathbb{C}} + \mathfrak{m}^- = \mathfrak{p}$ so that $\mathfrak{p} \subset \mathfrak{m}^{\mathbb{C}}$ is a complex Lie subalgebra.

Let $P = \exp(\mathfrak{p})$ denote the real Lie subgroup of $G^{\mathbb{C}}$ corresponding to $\mathfrak{p}$. Relating the dual pair (X_o,X_c) we prove

THEOREM 1 (BOREL EMBEDDING THEOREM, cf. WOLF [WOL2])

The embedding $G_c \hookrightarrow G^{\mathbb{C}}$ induces a biholomorphism $X_c \cong G_c/K \hookrightarrow G^{\mathbb{C}}/P$ onto the complex homogeneous manifold $G^{\mathbb{C}}/P$. Moreover, the embedding $G_0 \hookrightarrow G^{\mathbb{C}}$ induces an open embedding $X_o \cong G_0/K \hookrightarrow G^{\mathbb{C}}/P \cong X_c$, realizing X_o as an open subset of its compact dual X_c.

Proof:

Since $P \subset G$ is a complex Lie subgroup the homogeneous space $G^{\mathbb{C}}/P$ carries a natural complex structure. As $K \subset P$ the embedding $G_c \hookrightarrow G^{\mathbb{C}}$ of Lie groups induces a smooth mapping $\varphi: X_c \cong G_c/K \hookrightarrow G^{\mathbb{C}}/P$. Consider the decomposition $\mathfrak{g}^{\mathbb{C}} = \mathfrak{k}^{\mathbb{C}} + \mathfrak{m}^- + \mathfrak{m}^+ = \mathfrak{p} + \mathfrak{m}^+$. We have the identification $T_o^{\mathbb{R}}(G^{\mathbb{C}}/P) = \mathfrak{g}^{\mathbb{C}}/\mathfrak{p} \cong \mathfrak{m}^+$. Here and henceforth we will write $T_X^{\mathbb{R}}$ for the real tangent bundle over X and reserve the notation T_X for the holomorphic tangent bundle. At $o = eP$ the tangent map $d\varphi(o)$: $T_o^{\mathbb{R}}(X_c) \to T_o^{\mathbb{R}}(G^{\mathbb{C}}/P)$ is given by the map $\sqrt{-1}\mathfrak{m} \to \mathfrak{m}^+$ induced by the natural injection $\mathfrak{k} + \sqrt{-1}\mathfrak{m} \to \mathfrak{k}^{\mathbb{C}} + \mathfrak{m}^{\mathbb{C}} = \mathfrak{p} + \mathfrak{m}^+$. Since $\sqrt{-1}\mathfrak{m} \cap \mathfrak{m}^- = \{0\}$ and $\dim_{\mathbb{R}}(\sqrt{-1}\mathfrak{m}) = \dim_{\mathbb{R}}(\mathfrak{m}^+)$ it follows that $d\varphi(o)$ is an isomorphism of real vector spaces. By homogeneity it follows that the smooth map $\varphi: X_c \to G^{\mathbb{C}}/P$ is a local diffeomorphism. Since X_c is compact φ is necessarily a smooth covering map. We assert that φ is holomorphic. The J–operator on the complex homogeneous space $G^{\mathbb{C}}/P$ at $o = eP$, with the identification $T_o^{\mathbb{R}}(G^{\mathbb{C}}/P) \cong \mathfrak{m}^+$ (as a real vector space) is given by $J(\mathfrak{m}^+) = \sqrt{-1}\mathfrak{m}^+$. By the Cauchy–Riemann equations φ is holomorphic if and only if it preserves the J–operator. The decomposition of an element $v \in \sqrt{-1}\mathfrak{m}$ into elements of types (1,0) and (0,1) is given by $v = \frac{1}{2}(v - \sqrt{-1}v) + \frac{1}{2}(v + \sqrt{-1}v)$, so that for $v = \sqrt{-1}m \in \sqrt{-1}\mathfrak{m} \cong T_o^{\mathbb{R}}(X_c)$ at $o = eK \in X_c$ we have $d\varphi(o)(v) = \frac{1}{2}(v - \sqrt{-1}Jv)$. It follows that

$$d\varphi(o)(Jv) = \frac{1}{2}(Jv - \sqrt{-1}J^2v) = \frac{1}{2}(Jv + \sqrt{-1}v)$$
$$= \frac{\sqrt{-1}}{2}(v - \sqrt{-1}Jv) = \sqrt{-1}d\varphi(o)(v),$$

showing that $d\varphi(o)$ preserves the almost complex structure. We have therefore shown that $\varphi: X_c \to G^{\mathbb{C}}/P$ is a holomorphic covering map. As such $G^{\mathbb{C}}/P$ inherits a G_c–invariant metric $g_c^{\#}$. The involutions on X_c, defined via an inner automorphism by an element s of G_c, descends to involutions on $G^{\mathbb{C}}/P$, making $(G^{\mathbb{C}}/P, g_c^{\#})$ into a Hermitian symmetric manifold of compact type. By [(2.1), Prop.2] it must be simply–connected so that φ is in fact a biholomorphism identifying X_c with the complex homogeneous space $G^{\mathbb{C}}/P$.

The proof that $\psi: X_o \cong G_0/K \hookrightarrow G^{\mathbf{C}}/P \cong X_c$ is an open embedding is similar. Write $\Omega \subset G^{\mathbf{C}}/P$ for the image $\psi(G_0)$. It suffices to observe that $\Omega = G_0(o)$ and as such inherits the structure of a Hermitian symmetric manifold as above. It follows again from [(2.1), Prop.2] that in fact Ω is simply–connected and ψ is an open embedding, proving Thm.1.

REMARKS

Without knowing the integrability of the almost complex structures J on X_o and X_c the proof of the Borel Embedding Theorem shows that the identifications $\varphi: X_c \cong G^{\mathbf{C}}/P$ and $\psi: X_o \cong \Omega \subset G^{\mathbf{C}}/P$ preserves the J–operator. With these identifications it can therefore be used to show that the almost complex structures (X_o,J) and (X_c,J) are integrable, giving an alternative proof of [(2.1), Prop.1].

It is convenient to give another description of the G–invariant J–operator, as follows:

PROPOSITION 1

Let (X,g), $X \cong G/K$, be a Hermitian symmetric manifold of semisimple type and J be the G–invariant integrable almost complex structure on X. We write $\mathfrak{g} = \mathfrak{k} + \mathfrak{m}$ for the canonical decomposition of $\mathfrak{g}$. Then, there exists an element z of the center $\mathfrak{z}(\mathfrak{k})$ of $\mathfrak{k}$ such that for $v \in \mathfrak{m} \cong T_o^{\mathbf{R}}(X)$ we have $Jv = [z,v]$.

<u>Proof</u>:

It suffices to consider the case when (X,g) is irreducible. The center $\mathfrak{z}(\mathfrak{k})$ of $\mathfrak{k}$ is fixed under the adjoint action of K. Identify the center Z(K) with the rotation group $\{r_\theta: 0 \le \theta < 2\pi\}$ such that on $\mathfrak{m}$, $Jv = \mathrm{Ad}(r_{\pi/2})v$. $\mathrm{Ad}(r_\theta)|_{\mathfrak{m}^+}: \mathfrak{m}^+ \to \mathfrak{m}^+$ is a K–endomorphism (i.e., an endomorphism of $\mathfrak{m}^+$ as a K–representation space). As K acts irreducibly on $\mathfrak{m}^+$ this forces $\mathrm{Ad}(r_\theta)\xi = e^{i\theta}\xi$. Differentiating against θ we have $ad(z)\xi = \sqrt{-1}\,\xi$ for a generator z of $\mathfrak{z}(\mathfrak{k})$. As z is real by conjugation $ad(z)\xi = -\sqrt{-1}\,\zeta$ for $\zeta \in \mathfrak{m}^-$. Consequently $Jv = [z,v]$ for $v \in \mathfrak{m}$.

CHAPTER 4 BOUNDED SYMMETRIC DOMAINS — THE CLASSICAL CASES

§1 The Bergman and Carathéodory Metrics on Bounded Domains

(1.1) We are going to study Hermitian metrics on quotients of Hermitian symmetric manifolds of non–compact type X_0. For this we will need to realize such manifolds as domains. The Borel Embedding Theorem [Ch.3, (3.1), Thm.1] shows that X_0 is biholomorphic to a domain in its compact dual X_c. In Chapter 5 we will see that X_0 can actually be realized as bounded domains in some Euclidean space $\mathbf{C}^n$, by the Harish–Chandra Embedding Theorem. We will first of all introduce the notion of bounded symmetric domains and discuss the classical cases. We have

DEFINITION 1

Let $\Omega \subset\subset \mathbf{C}^n$ be a bounded domain in a Euclidean space. We say that Ω is a bounded symmetric domain if and only if at each $x \in \Omega$ there exists a biholomorphism $\sigma_x : \Omega \longrightarrow \Omega$ such that $\sigma_x^2 = \mathrm{id}$ and x is an isolated fixed point of σ_x.

We will denote by $\mathrm{Aut}(\Omega)$ the group of holomorphic self–mappings on Ω. The mapping $\sigma_x \in \mathrm{Aut}(\Omega)$ in the definition will be called a holomorphic involution. In Chapter 5 it will be shown that all Hermitian symmetric manifolds of non–compact type can be realized as bounded symmetric domains. We will first show that any bounded symmetric domain Ω can be equipped with an $\mathrm{Aut}(\Omega)$–invariant Kähler metric g making (Ω,g) into a Hermitian symmetric manifold. For this we define the notion of Bergman metrics on bounded domains.

(1.2) Let now $\Omega \subset\subset \mathbf{C}^n$ be an arbitrary bounded domain. Denote by $d\lambda$ the Lebesgue measure and by $L^2(\Omega)$ the Hilbert space of functions square–integrable with respect to $d\lambda$. We define $H^2(\Omega) \subset L^2(\Omega)$ to be the space of square–integrable holomorphic functions. Denote by $(.,.)$ the inner product on $L^2(\Omega)$ and by $\|.\|$ the L^2–norm. Let $f \in H^2(\Omega)$. From the mean–value inequality for the subharmonic function $|f|^2$ it follows that $|f(x)|^2 \le \mathrm{Const.}\,\frac{\|f\|^2}{\delta^{2n}(x,\partial\Omega)}$, where δ denotes the Euclidean distance. By Montel's Theorem $H^2(\Omega)$ is complete with respect to the L^2–norm. Let now $\{f_i\}_{1\le i<\infty}$ be an orthonormal basis of the Hilbert space $H^2(\Omega)$. We define the Bergman kernel $K(z,w)$ by

$$K(z,w) = \sum f_i(z)\,\overline{f_i(w)}.$$

PROPOSITION 1

The infinite sum $K(z,w)$ converges uniformly on compact sets to a function jointly real–analytic in the variables (z,w), holomorphic in z and anti–holomorphic in w. $K(z,w)$ is square–integrable in the variables (z,w) separately, is independent of the choice of the holomorphic basis $\{f_i\}_{1\le i<\infty}$ and possesses a reproducing property for square–integrable holomorphic functions f given by

$$f(z) = \int_\Omega K(z,w)\, f(w)\, d\lambda(w).$$

Proof:

Consider formally the infinite sum $K'(z,w) = K(z,\overline{w})$. K' is the limit of holomorphic functions on $\Omega \times \Omega'$, where Ω' denotes $\{w \in \mathbf{C}^n : \overline{w} \in \Omega\}$. By Montel's Theorem the infinite sum K converges uniformly on compact subsets of $\Omega \times \Omega$ as a continuous function if and only if K' converges uniformly on compact subsets of $\Omega \times \Omega$ to a holomorphic function. To prove the latter by the Cauchy–Schwarz inequality it suffices to show that on any compact subset M of Ω we have the estimate $\sum_i \int_M |f_i|^2 \le C_M < \infty$ for some constant C_M. By the mean–value inequality it is enough to show that for any $x \in \Omega$, and for B denoting the unit ball of the Hilbert space $H^2(\Omega)$, we have $(\Sigma|f_i(x)|^2)^{1/2} \le \sup_{f\in B}|f(x)| := \mu(x)$. (The proof actually shows that equality holds.) Let I be a finite set of positive integers. Then, for $F = \Sigma_{i\in I}\overline{f_i(x)}f_i$ we have $\|F\|^2 = \Sigma_{i\in I}|f_i(x)|^2$ and $|F(x)| = \Sigma_{i\in I}|f_i(x)|^2 = \|F\|^2$. We deduce thus from $|F(x)| \le \|F\|\mu(x)$ that in fact $\|F\|^2 \le \|F\|\mu(x)$, giving $(\Sigma_{i\in I}|f_i(x)|^2)^{1/2} = \|F\| \le \mu(x)$. As I is arbitrary this shows $(\Sigma|f_i(x)|^2)^{1/2} \le \mu(x)$, proving our assertion, so that $K(z,w)$ converges uniformly on compact subsets of $\Omega \times \Omega$. It remains to prove the reproducing property of $K(z,w)$ and that $K(z,w)$ is independent of the choice of orthonormal basis $\{f_i\}_{i\in I}$. To prove the reproducing property for $f \in H^2(\Omega)$

expanding f in terms of the orthonormal basis $\{f_i\}$ as $\sum (f,f_i)\, f_i$, we have

$$\int_\Omega f(w)\, K(z,w)\, d\lambda(w) = \int_\Omega \Big(\sum_i (f,f_i)\, f_i(w)\Big) \Big(\sum_j f_j(z)\, \overline{f_j(w)}\Big)\, d\lambda(w)$$
$$= \sum_{i,j} \delta_{ij} (f,f_i)\, f_j(z) = \sum_j (f,f_j)\, f_j(z) = f(z).$$

To prove that $K(z,w)$ is independent of the choice of the orthonormal basis $\{f_i\}$ we show more generally that if $H(z,w)$ is Hermitian symmetric in (z,w) and holomorphic in z; and possesses the same reproducing property as K, then $H \equiv K$. To see this we have from the reproducing property of both H and K the equation

$$\int_\Omega f(w)\, (K(z,w) - H(z,w))\, d\lambda(w) = 0$$

for any $f \in H^2(\Omega)$. Since both H and K are holomorphic in the first variable, for z fixed we deduce that f is orthogonal to the holomorphic function $g_z(w) := K(w,z) - H(w,z)$. As f is arbitrary this implies that $g_z(w) \equiv 0$ for any $z \in \Omega$, i.e., $H(z,w) \equiv K(z,w)$. ∎

From the Bergman kernel $K(z,w)$ we are going to derive a Kähler metric. The function $\log K(z,z)$ is plurisubharmonic because it is the logarithm of a sum of norm squares of holomorphic functions. In fact, for any finite set $\{F_i\}$ of holomorphic functions we have the identity

$$\sqrt{-1}\, \partial\overline{\partial} \log \Sigma |F_i|^2 = \frac{1}{(\Sigma |F_i|^2)^2} \sum \sqrt{-1}\, (F_i dF_j - F_j dF_i) \wedge (\overline{F_i dF_j} - \overline{F_j dF_i})$$

obtained using the Lagrange identity (cf. e.g. SKODA [SKO1]). We assert that $\log K(z,z)$ is actually strictly plurisubharmonic. Writing $\omega = \sqrt{-1}\, \partial\overline{\partial} \log K(z,z)$ we have to show that in fact $\omega(v \wedge \overline{v}) > 0$ for any non–zero tangent vector v of type $(1,0)$. Since $K(z,w)$ is independent of the choice of an orthonormal basis $\{f_i\}_{1 \le i < \infty}$ of $H^2(\Omega)$ fixing $x \in \Omega$ and $v \in T_x^{1,0}$ one may choose f_1 constant and f_2 such that $df_2(v) \neq 0$. From the formula above it follows that $\omega(v \wedge \overline{v}) > 0$.

One significance of the Bergman metric ds_B^2 is given by

PROPOSITION 2
The Bergman metric is invariant under $\mathrm{Aut}(\Omega)$.

Proof:
Instead of using the Bergman kernel we use the Bergman kernel form on $\Omega \times \Omega$, defined by $\kappa(z,w) = (\sqrt{-1})^n K(z,w)\, dz^1 \wedge \ldots \wedge dz^n \wedge d\overline{w}^1 \wedge \ldots \wedge d\overline{w}^n$. The factor $(\sqrt{-1})^n$ is put there to guarantee that $\kappa(z,z)$ is a positive (n,n)–form. The advantage of using $\kappa(z,w)$ is that there is a natural inner product $(.,.)$ on (n,0)–forms defined independent of metrics, given by $(\mu,\nu) := \int_\Omega \left[\frac{\sqrt{-1}}{2}\right]^n \mu \wedge \overline{\nu}$. Denote by K_Ω the canonical line bundle and write $H^2(\Omega,K_\Omega)$ for the Hilbert space of square–integrable (n,0)–forms on Ω with respect to $(.,.)$. If to each $f \in H^2(\Omega)$ we associate the (n,0)–form $\varphi = f(dz^1 \wedge \ldots \wedge dz^n)$ we obtain an isometry between the Hilbert spaces $H^2(\Omega)$ and $H^2(\Omega,K_\Omega)$. Let now ψ be any non–vanishing holomorphic n–form on Ω. Write $\kappa(z,z) = k_\psi(z,z)\,(\sqrt{-1})^n\, \psi \wedge \overline{\psi}$. Then, $\partial\overline{\partial} \log k_\psi(z,z)$ is independent of the choice of ψ. In fact if η is another choice for ψ we have $k_\eta(z,z) = k_\psi(z,z)\left|\frac{\psi}{\eta}\right|^2$, where $\frac{\psi}{\eta}$ is a non–vanishing holomorphic function, and hence $\log k_\eta(z,z) = \log k_\psi(z,z) + \log\left|\frac{\psi}{\eta}\right|^2$. Using the fact that $\log\left|\frac{\psi}{\eta}\right|^2$ is pluriharmonic we obtain therefore $\partial\overline{\partial} \log k_\eta(z,z) = \partial\overline{\partial} \log k_\psi(z,z)$, as asserted. Suppose now $\gamma \in \mathrm{Aut}(\Omega)$. From the definition of $H^2(\Omega,K_\Omega)$ it is immediate that γ induces an isometry on $H^2(\Omega,K_\Omega)$. Hence, $\kappa(z,z)$ is invariant under γ. Let now ψ be the holomorphic n–form $dz^1 \wedge \ldots \wedge dz^n$ and η be $\gamma^*(\psi) = D_\gamma(z)\,\psi$, where $D_\gamma(z)$ is the Jacobian determinant of γ. From $\kappa(z,z) = K(z,z)\,(\sqrt{-1})^n\, \psi \wedge \overline{\psi} = K(\gamma z,\gamma z)\,(\sqrt{-1})^n\, \eta \wedge \overline{\eta}$ we deduce that $\sqrt{-1}\,\partial\overline{\partial} \log K(z,z) = \sqrt{-1}\,\partial\overline{\partial} \log K(\gamma z,\gamma z)$, i.e., the Bergman metric is invariant under $\mathrm{Aut}(\Omega)$. ▎

In general the Bergman metric on a bounded domain does not have very nice curvature properties. However, in case Ω is a homogeneous domain, i.e., $\mathrm{Aut}(\Omega)$ acts transitively on Ω, we have

PROPOSITION 3

Let Ω be a homogeneous bounded domain. Then, the Bergman metric ds_B^2 is Kähler–Einstein.

Proof:

By Prop.2 the volume form ω^n of the Bergman metric ds_B^2 on Ω is invariant under $\mathrm{Aut}(\Omega)$. Since $\mathrm{Aut}(\Omega)$ acts transitively on Ω, there is at most one invariant (n,n)–form on Ω up to a scaling constant. As both $\kappa(z,z)$ and ω^n are invariant under $\mathrm{Aut}(\Omega)$ it follows that $\omega^n(z) \equiv c\kappa(z,z)$ on Ω for some positive constant c. Write $\omega^n = (\sqrt{-1})^n\, V\, dz^1 \wedge \ldots \wedge dz^n \wedge d\bar{z}^1 \wedge \ldots \wedge d\bar{z}^n$. By [Ch.2, (2.4), Prop.1] we have $\mathrm{Ric}(\Omega,\omega) = -\sqrt{-1}\,\partial\bar{\partial} \log V$. By the definition of the Bergman metric it follows that $\mathrm{Ric}(\Omega,\omega) = -\omega$, so that (Ω,ω) is Kähler–Einstein. ▌

Let Ω be a bounded symmetric domain. Equipped with the Bergman metric (Ω,ω) becomes a Hermitian symmetric manifold. As Ω is a homogeneous bounded domain, (Ω,ω) is of constant negative Ricci curvature by Prop.3, so that (Ω,ω) is a Hermitian symmetric manifold of non–compact type. Let $\Gamma \subset \mathrm{Aut}(\Omega)$ be a discrete subgroup acting on Ω without fixed points such that $X = \Omega/\Gamma$ is compact. By BOREL [BO] such quotient manifolds X always exist. The $\mathrm{Aut}(\Omega)$–invariant Bergman metric descends to a quotient metric on X. We will denote the quotient metric and its associated Kähler form by the same symbols ds_B^2 and ω resp. (X,ω) is therefore a compact Kähler–Einstein manifold with holomorphic bisectional curvature ≤ 0, by [Ch.3, (1.2), Prop.2] and [Ch.2, (3.3), Prop.1]. Our first result will be to show that if (Ω,ω) is irreducible as a Hermitian symmetric manifold and Ω is not biholomorphic to the unit Euclidean ball B^n, then in fact the Bergman metric is the only Kähler metric of seminegative holomorphic bisectional curvature up to a scaling constant. This will be attained by some integral formula on a complex submanifold $\mathcal{S}$ on the projectivized tangent bundle $\mathbb{P}T_X$ of X.

(1.3) We will be studying in [Ch.6, (5.1)] holomorphic mappings between quotients of bounded symmetric domains of finite volume. For that purpose we will need an Ahlfors–Schwarz lemma for such holomorphic mappings. In general, the intrinsic Bergman metric ds_B^2 does not behave well under holomorphic transformations. We introduce here another intrinsic metric. First, by a (possibly degenerate) Finsler metric on a smooth manifold X we mean a continuous function $\|.\|$: $T_X^{\mathbf{R}} \to \mathbf{R}^+ \cup \{0\}$ on the (real) tangent bundle $T_X^{\mathbf{R}}$ satisfying (*) $\|\lambda v\| = |\lambda|\ \|v\|$ for a real number λ. Given $\|.\|$, we can measure lengths of smooth curves and introduce a distance function d on $X \times X$ by defining $d(x,y)$ to be the infimum of lengths of smooth curves joining x to y, as in Riemannian Geometry. On a complex manifold M under the identification $T_M^{1,0} \cong T_M^{\mathbf{R}}$ given by $v \mapsto 2Re(v)$ one can define Finsler metrics by using $T_M^{1,0}$. If $\|.\|$: $T_M^{1,0} \to \mathbf{R}^+ \cup \{0\}$ satisfies (*) for complex numbers λ we call $\|.\|$ a complex Finsler metric. We have

DEFINITION – PROPOSITION 1

Let M be a complex manifold. Denote by $\mathcal{F} = \mathcal{F}(M)$ the space of holomorphic mappings into the unit disc $\Delta \subset \mathbf{C}$. Let $\|.\|_\Delta$ denote the Poincaré metric of constant Gaussian curvature -1. For each $x \in M$ we define $\|.\|_c$ on $T_x(M)$ by the formula $\|v\|_c = \sup_{f \in \mathcal{F}} \|df(v)\|_\Delta$. Then, $\|.\|_c$ is a complex Finsler metric on M called the Carathéodory metric.

Proof:

The only thing that needs to be verified is that $\|.\|_c$ is finite and continuous on $T_M^{1,0}$. Let $x \in M$ and U be an open neigborhood of x biholomorphic to a bounded domain in some $\mathbf{C}^n$. By the homogeneity of $(\Delta, \|.\|_\Delta)$ it suffices to consider only those $f \in \mathcal{F}$ such that $f(x) = o$. By considering the restriction $f|_U$: $U \to \Delta$ and applying Cauchy estimates one obtains uniform bounds on first and second derivatives of f on compact subsets of U in terms of Euclidean coordinates. The finiteness and continuity of $\|.\|$ follow readily. ∎

Denote by d_c the distance function on M defined by the Carathéodory metric. As an almost immediate consequence of the definition, we have

PROPOSITION 2

The Carathéodory metric on a complex manifold M is invariant under biholomorphic self–mappings Aut(M). It satisfies the distance–decreasing property with respect to holomorphic mappings in the sense that for any two complex manifolds M and N and any holomorphic mapping $\Psi: M \longrightarrow N$,

$$d_c(\Psi(x),\Psi(y)) \le d_c(x,y).$$

Proof:

The distance–decreasing property follows immediately from the fact that for any $f \in \mathcal{F}(N)$, $f \circ \Psi \in \mathcal{F}(M)$. Applied to Ψ and Ψ^{-1} in Aut(M) it follows that the Carathéodory metric is invariant under Aut(M). ▌

For bounded homogeneous domains Ω any complex Finsler metric invariant under Aut(Ω) is uniquely determined by its value at one point. It follows therefore that any two non–degenerate complex Aut(Ω)–invariant Finsler metric are uniformly equivalent. Consequently, we have

PROPOSITION 3

Let $\Psi: \Omega \longrightarrow \Omega'$ be a holomorphic mapping between two bounded homogeneous domains Ω, Ω'. Denoting their Bergman metrics by ds^2_Ω resp. $ds^2_{\Omega'}$, we have

$$\Psi^*(ds^2_{\Omega'}) \le C\, ds^2_\Omega$$

for some constant C depending only on the domains Ω and Ω'.

§2 Classical Bounded Symmetric Domains

(2.1) The first example of a bounded symmetric domain is the unit disc $\Delta \subset \mathbf{C}$ equipped with the Poincaré metric $2Re\,\dfrac{dz \otimes d\bar{z}}{(1 - |z|^2)^2}$, of constant negative Gaussian curvature. The symmetry σ_0 at the origin is given by $\sigma_0(z) = -z$. Aut(Δ) acts transitively on Δ via Möbius transformations $\Psi(z) = e^{i\theta}\,\dfrac{z - \alpha}{1 - \bar{\alpha}z}$, for any $\alpha \in \Delta$ and any real number θ. In matrix form we write $\Psi(z) = \begin{bmatrix} a & b \\ c & d \end{bmatrix}(z)$, where the matrix $\begin{bmatrix} a & b \\ c & d \end{bmatrix}$ acts on Δ as fractional linear transformations $\dfrac{az + b}{cz + d}$. Thinking of $\mathbf{C}$ as part of $\mathbb{P}^1$ using the embedding $z \longrightarrow [1,z]$, with

$[w_0,w_1]$ denoting homogeneous coordinates on $\mathbb{P}^1$, Ψ acts as a projective linear transformation on $\mathbb{P}^1$. We have $\Delta = \{[w_0,w_1]\in\mathbb{P}^1 : |w_0|^2 - |w_1|^2 > 0\}$ in homogeneous coordinates. Let $H_{p,q}$ denote the indefinite Hermitian bilinear form on $\mathbf{C}^{p+q}$, $p, q \geq 1$, defined by $H_{p,q}(w,w) = \Sigma_{1\leq i\leq p}|w_i|^2 - \Sigma_{p+1\leq j\leq p+q}|w_j|^2$. Denote the group of linear isometries of $H_{p,q}$ by $U(p,q) \subset GL(p+q,\mathbf{C})$ and call it the unitary group of (the indefinite form) $H_{p,q}$. Those of determinant 1 constitute the special unitary group $SU(p,q)$ of $H_{p,q}$. It is clear that $SU(1,1)$, acting as projective linear transformations on $\mathbb{P}^1$, preserves Δ. Conversely, the Möbius transformation $\Psi(z) = \dfrac{z + \alpha}{1 + \bar{\alpha}z}$ can be written in homogeneous coordinates as

$$\Psi([w_0,w_1]) = [w_0 + \bar{\alpha}w_1, \alpha w_0 + w_1].$$

Putting $\beta = \sqrt{1 - |\alpha|^2}$, we can write

$$\Psi([w_0,w_1]) = [(w_0 + \bar{\alpha}w_1)/\beta, (\alpha w_0 + w_1)/\beta].$$

In other words Ψ is represented by the matrix $M = \begin{bmatrix} 1/\beta & \bar{\alpha}/\beta \\ \alpha/\beta & 1/\beta \end{bmatrix}$ of determinant $(1 - |\alpha|^2)/\beta^2 = 1$. $M \in SU(1,1)$ because

$$\overline{M}^t \begin{bmatrix} 1 & 0 \\ 0 & -1 \end{bmatrix} M = \begin{bmatrix} 1/\beta & \bar{\alpha}/\beta \\ \alpha/\beta & 1/\beta \end{bmatrix} \begin{bmatrix} 1 & 0 \\ 0 & -1 \end{bmatrix} \begin{bmatrix} 1/\beta & \bar{\alpha}/\beta \\ \alpha/\beta & 1/\beta \end{bmatrix}$$

$$= \begin{bmatrix} 1/\beta & -\bar{\alpha}/\beta \\ \alpha/\beta & -1/\beta \end{bmatrix} \begin{bmatrix} 1/\beta & \bar{\alpha}/\beta \\ \alpha/\beta & 1/\beta \end{bmatrix} = \frac{1}{\beta^2}\begin{bmatrix} 1-|\alpha|^2 & 0 \\ 0 & -1+|\alpha|^2 \end{bmatrix} = \begin{bmatrix} 1 & 0 \\ 0 & -1 \end{bmatrix}.$$

We have thus represented $\mathrm{Aut}(\Delta)$ as $SU(1,1)$, except that two such matrices A and B represent the same Möbius transformation if $A = \pm B$. This yields the realization $\mathrm{Aut}(\Delta) \cong SU(1,1)/\{\pm 1\}$. Inside $SU(1,1)$ there is the circle group $U(1)$ consisting of $\begin{bmatrix} \alpha & 0 \\ 0 & 1/\alpha \end{bmatrix}$, with $|\alpha| = 1$. The isotropy group at $o \in \Delta$ is given by $U(1)/\{\pm 1\}$ with this identification. We have therefore the presentation $\Delta \cong (SU(1,1)/\{\pm 1\})/(U(1)/\{\pm 1\}) \cong SU(1,1)/U(1)$ as a homogeneous space.

To proceed further we need the following theorem of Cartan on automorphism groups of bounded domains obtained by the method of iterations. We say that a domain Ω is circular if and only if it is invariant under the action of the circle group S^1 given by $(z_1,\dots,z_n) \to (e^{i\theta}z_1,\dots,e^{i\theta}z_n)$. A circular domain Ω is said to be complete if it contains the origin o. We have the following

THEOREM 1 (CARTAN, cf. NARASIMHAN [NA3])
Let $\Omega \subset\subset \mathbb{C}^n$ be a bounded complete circular domain. Then, the isotropy group K at the origin consists entirely of linear transformations.

Proof:
Consider first of all $\Omega \subset\subset \mathbb{C}^n$ an arbitrary bounded domain containing o, and $f \in \mathrm{Isot}_o(\Omega)$ admitting a power series expansion

$$f(z) = z + P_2(z) + \dots + P_r(z) + \dots, \tag{1}$$

where $P_r(z)$ is a homogeneous polynomial in $z = (z_1,\dots,z_n)$ of degree r. We are going to show that $f(z) \equiv z$. Suppose otherwise. Let P_N be the first non–zero term in the expansion (1) of f. By iteration we have

$$f^k(z) = z + kP_N(z) + \dots \tag{2}$$

As the iterates $\{f^k\}_{1\le k<\infty}$ are all automorphisms of the bounded domain Ω, by Cauchy estimates for any partial derivation $\partial_I \equiv \partial_1^{i_1}\dots\partial_n^{i_n}$ we have a uniform bound on $\{\partial_I f^k(o)\}$. Applying this to (2) for some I of total degree $i_1 + \dots i_n = N$ we obtain immediately a contradiction, proving our assertion that $f(z) \equiv z$. We now make use of the fact that Ω is circular. For θ real we have the automorphism $k_\theta(z) = e^{i\theta}z$. The automorphism $g = k_{-\theta} f k_\theta f^{-1}$ is an isotropy at o such that $dg(o) = \mathrm{id}$. It follows that g admits the power series expansion (1) so that $g(z) \equiv z$. In other words we have $k_{-\theta} f k_\theta = f$. Writing now

$$f(z) = P_1(z) + \dots + P_r(z) + \dots \tag{3}$$

for the power series expansion of f into homogeneous polynomials we obtain

$$k_{-\theta} f k_\theta(z) = e^{-i\theta}\sum e^{ri\theta}P_r(z) = \sum e^{(r-1)i\theta}P_r(z) \tag{4}$$

Equating the power series expansions (3) and (4) we conclude that in fact $P_r(z) \equiv 0$ for $r \ge 2$, proving that f is linear. ∎

There are two immediate generalizations of the unit disc Δ to higher dimensions. The first is given by polydiscs Δ^n. The group of product Möbius transformations $(\mathrm{Aut}(\Delta))^n$ acts on Δ^n factor by factor. By Thm.1 it is easy to deduce that $\mathrm{Aut}(\Delta^n)$ consists of compositions of $(\mathrm{Aut}(\Delta))^n$ and permutations $\Phi(z_1,\ldots,z_n) = (z_{\tau(1)},\ldots,z_{\tau(n)})$ for permutations τ of $\{1,\ldots,n\}$. We have therefore $\mathrm{Aut}_o(\Delta^n) \cong (\mathrm{Aut}(\Delta))^n$ for the identity component $\mathrm{Aut}_o(\Delta^n)$ of $\mathrm{Aut}(\Delta^n)$.

The second generalization is the unit ball $B^n = \{(z_1,\ldots,z_n)\in\mathbb{C}^n: \Sigma|z_i|^2 < 1\}$. Embedding $\mathbb{C}^n$ as part of $\mathbb{P}^n$ via $(z_1,\ldots,z_n) \to [1,z_1,\ldots,z_n]$ in terms of homogeneous coordinates, we have $B^n = \{[w_0,\ldots,w_n]\in\mathbb{P}^n: |w_0|^2 - \Sigma_{1\le i\le n}|w_i|^2 > 0\}$. The homogeneity of B^n can be seen by considering the automorphisms

$$\Psi(z_1,\ldots,z_n) = \left[\frac{z_1 + \alpha}{1 + \bar{\alpha}z_1}, \frac{\sqrt{1 - |\alpha|^2}}{1 + \bar{\alpha}z_1} z_2, \ldots, \frac{\sqrt{1 - |\alpha|^2}}{1 + \bar{\alpha}z_1} z_n\right],$$

where $|\alpha| < 1$. Given any $x \in B^n$ there exists an automorphism Φ_x such that $\Phi_x(o) = x$ of the form $\Phi_x(z) = \Psi(U(z))$, where U is a unitary transformation and Ψ is of the form above for some choice of α. At the origin we have the symmetry $\sigma_o(z) = -z$. From the homogeneity of B^n it follows that B^n is a bounded symmetric domain. In fact, $\Phi_x\sigma_o\Phi_x^{-1}$ is a holomorphic symmetry at x. Using homogeneous coordinates as in the case of Δ we see that $SU(1,n)$ acts transitively on B^n with a finite kernel $\{\mu_{n+1}I\}$ consisting of diagonal matrices ϵI, where ϵ is an $(n+1)$–th root of unity and $I = I_{n+1}$ is the identity matrix of rank $n+1$. The transformations Ψ above are represented in homogeneous coordinates by

$$\Psi([w_0,\ldots,w_n]) = \left[w_0 + \bar{\alpha}w_1, \alpha w_0 + w_1, \sqrt{1 - |\alpha|^2}\, w_2, \ldots, \sqrt{1 - |\alpha|^2}\, w_n\right].$$

Writing $\beta = \sqrt{1 - |\alpha|^2}$, Ψ is represented by the matrix

$$\Psi([w_0,\ldots,w_n]) = [(w_0 + \bar{\alpha}w_1)/\beta, (\alpha w_0 + w_1)/\beta, w_2, \ldots, w_n], \qquad (1)$$

so that Ψ is represented by the matrix $\begin{bmatrix} 1/\beta & \bar{\alpha}/\beta & 0 \\ \alpha/\beta & 1/\beta & 0 \\ 0 & 0 & I_{n-1} \end{bmatrix} \in SU(1,n)$. We have more precisely $\mathrm{Aut}(B^n) \cong SU(1,n)/\{\mu_{n+1}I\}$ as a consequence of Thm.1.

To find all automorphisms of B^n using Thm.1, it suffices to find the group K of all linear transformations preserving B^n, which is simply the unitary group $U(n)$. A unitary transformation U of C^n can be represented by a projective unitary transformation of the form $[w_0,\ldots,w_n] \to [\lambda w_0, \lambda U(w_1,\ldots,w_n)] \in SU(1,n)$, where $\lambda^{n+1} = 1/\det(U)$. In this representation we have $K \cong U(n+1)/\{\mu_{n+1}I\}$. Since K and any Ψ of the form (1) is represented by an element of $SU(1,n)$ it follows therefore that $SU(1,n\}/\{\mu_{n+1}I\}$ exhausts all possible automorphisms of B^n. We have the presentation $B^n \cong (SU(1,n)/\{\mu_{n+1}I\})/(U(n+1)/\{\mu_{n+1}I\}) \cong SU(1,n)/U(n+1)$ as a homogeneous space.

(2.2) Let $M(p,q;C)$ denote the space of matrices with complex entries of order $p \times q$. As a generalization of B^n we consider

$$D^I_{p,q} = \{Z \in M(p,q;C) \cong C^{pq}\colon\ I_q - \bar{Z}^t Z > 0\}.$$

For $p = 1$, $D^I_{p,1} = \{(z_1,\ldots,z_p)\colon 1 - \Sigma|z_i|^2 > 0\}$ is simply the unit ball B^p. As in the above for this and ensuing examples we will show that the bounded domains Ω are symmetric by exhibiting a homogeneous group of automorphisms G and showing that Ω is symmetric at the origin. We will not determine the automorphism group directly. It will however follow from the general theory of Riemannian symmetric manifolds (cf. HELGASON [HEL, Ch.V, Thm.(4.1), p.243]) that G is the identity component of $\mathrm{Aut}(\Omega)$. To describe $\Omega = D^I_{p,q}$ we embed it as an open subset of the Grassmannian $G(q,p)$ of complex q–planes in C^{p+q}, with Euclidean coordinates $(w_1,\ldots,w_{p+q})$. The standard coordinate covering of $G(q,p)$ consists of copies of C^{pq}. We describe one of them in terms of some choices of Euclidean coordinates on C^{p+q}. Let $Z \in M(p,q)$. Z corresponds to the $(p+q)$ by q matrix $\begin{bmatrix} Z \\ I_q \end{bmatrix}$, which represents the complex q–dimensional vector space V_Z generated by its q C–linearly independent column vectors. Since the correspondence $Z \leftrightarrow Z^t$ establishes a biholomorphism between $D^I_{p,q}$ and $D^I_{q,p}$, we will assume that $p \leq q$. The action of $GL(p+q,C)$ on C^{p+q} induces automorphisms on $G(q,p)$. We are going to descibe a subgroup which keeps $D^I_{p,q}$ invariant. The condition $Z \in \Omega = D^I_{p,q}$ can be written as $[\,\bar{Z}^t\ I_q\,]\begin{bmatrix} Z \\ -I_q \end{bmatrix} < 0$ as Hermitian matrices. Equivalently, we write

$$[\,\bar{Z}^t\ I_q\,]\begin{bmatrix} I_p & 0 \\ 0^p & -I_q \end{bmatrix}\begin{bmatrix} Z \\ I_q \end{bmatrix} < 0. \tag{1}$$

Recall that $SU(p,q)$ is the special unitary group for the indefinite form $H_{p,q} = \Sigma_{1\leq i\leq p}|w_i|^2 - \Sigma_{p+1\leq j\leq p+q}|w_j|^2$. Let $M \in SU(p,q)$ and write $M\begin{bmatrix} Z \\ I_q \end{bmatrix} = \begin{bmatrix} W \\ E \end{bmatrix}$. Since $M \in SU(p,q)$ we have $\bar{M}^t I_{p,q} M = I_{p,q}$ for $I_{p,q} = \begin{bmatrix} I_p & 0 \\ 0^p & -I_q \end{bmatrix}$. The condition (1) therefore implies

$$[\,\bar{W}^t\ \bar{E}^t\,]\begin{bmatrix} I_p & 0 \\ 0^p & -I_q \end{bmatrix}\begin{bmatrix} W \\ E \end{bmatrix} < 0, \text{ i.e.,}$$

$$\bar{E}^t E - \bar{W}^t W > 0 \tag{2}$$

as Hermitian matrices. In particular E is invertible. We say that two matrices F, F' of order $(p+q) \times q$ are equivalent, written $F \approx F'$, if and only if they are of rank q and the vector subspaces generated by the q column vectors are the same. We have $\begin{bmatrix} W \\ E \end{bmatrix} \approx \begin{bmatrix} WE^{-1} \\ I_q \end{bmatrix}$. In other words, $\Phi(Z) = WE^{-1}$ represents an automorphism $D^I_{p,q} \subset M(p,q;C) \cong C^{pq}$ which extends to an automorphism of $G(q,p)$. We will from now on write also Φ for the matrix representative. In coordinates if we write $\Phi = \begin{bmatrix} A & B \\ C & D \end{bmatrix}$, where A and D are square matrices of order p and q resp., etc., we have

$$\Phi(Z) = (AZ + B)(CZ + D)^{-1}. \tag{3}$$

To see that Φ is in fact an automorphism of $\Omega = D^I_{p,q}$ we check that by (2), $\bar{E}^t E - \bar{W}^t W > 0$, so that $I_q - (\overline{WE^{-1}})^t(WE^{-1}) > 0$, i.e.,

$$I_q - \overline{\Phi(Z)}^t\Phi(Z) > 0$$

so that $\Phi(Z) \in \Omega$ whenever $Z \in \Omega$. We have therefore shown that SU(p,q) gives rise to automorphisms on Ω. Furthermore, Φ acts trivially on Ω if and only if it acts as scalar multiplication on C^{p+q}, giving rise to a finite kernel $\{\mu_{p+q}I\}$. The reflection $Z \longrightarrow -Z$ gives a holomorphic symmetry at the zero matrix 0. To show that Ω is a bounded symmetric domain it suffices to show that SU(p,q) acts transitively on Ω. To this end we examine the isotropy group of SU(p,q) at the origin 0. The condition $\begin{bmatrix} A & B \\ C & D \end{bmatrix} \in SU(p,q)$ means $\begin{bmatrix} \overline{A} & \overline{B} \\ \overline{C} & \overline{D} \end{bmatrix}^t \begin{bmatrix} I_p & 0 \\ 0^p & -I_q \end{bmatrix} \begin{bmatrix} A & B \\ C & D \end{bmatrix}$ $= \begin{bmatrix} I_p & 0 \\ 0^p & -I_q \end{bmatrix}$, which translates readily to the set of conditions

$$\begin{cases} \overline{A}^tA - \overline{C}^tC = I_p; \\ \overline{B}^tB - \overline{D}^tD = -I_q; \\ \overline{A}^tB - \overline{C}^tD = 0. \end{cases} \tag{4}$$

By (3) $\Phi = \begin{bmatrix} A & B \\ C & D \end{bmatrix} \in SU(p,q)$ represents $\Phi(Z) = (AZ + B)(CZ + D)^{-1}$, so that it fixes the origin if and only if $B \equiv 0$. From (4) it follows that $\overline{D}^tD = I_q$ and hence that D is invertible, implying again by (4) that $C \equiv 0$. As a consequence, we have $M = \begin{bmatrix} U & 0 \\ 0 & V \end{bmatrix}$, with U and V unitary of order p and q resp. Given $Z \in \Omega$ to construct $\Phi \in Aut(\Omega)$ such that $\Phi(0) = Z$ we show first

LEMMA 1

Given $X \in M(p,q)$, $p \le q$ there exist unitary matrices U and V of order p and q resp. such that

$$UXV = \begin{bmatrix} \alpha_1 & & & \\ & \ddots & & 0 \\ & & \alpha_p & \end{bmatrix}.$$

<u>Proof:</u>

Consider X as a linear mapping $C^p \longrightarrow C^q$. Take $v_1 \in C^p$ of unit length such that $\|Xv_1\| = \sup_{\|v\|=1} \|Xv\|$. From the maximality of $\|Xv_1\|$ it follows readily that whenever $v \perp v_1$, we have $Xv \perp Xv_1$. Hence one can choose orthonormal bases $\{v_1,\dots,v_p\}$ and $\{u_1,\dots,u_q\}$ of C^p and C^q resp. so that $Xv_i = \alpha_i u_i$ for some complex number α_i. The latter statement is equivalent to the lemma. ▌

By abuse of terminology we will call a matrix W of the form given in the lemma a diagonal matrix and write $W = \mathrm{diag}_{p,q}(\alpha_1,\ldots,\alpha_p)$ $(p \le q)$. When there is no danger of confusion we will drop the suffices. As a consequence of the lemma, to show that $\Omega = D^I_{p,q}$ is a bounded symmetric domain it suffices to show that given any diagonal matrix $\alpha \in \Omega$, there exists $\Psi \in \mathrm{Aut}(\Omega)$ such that $\Psi(0) = \alpha$. The condition $\alpha \in \Omega$ is equivalent to having $|\alpha_i| < 1$ for $1 \le i \le p$. In other words, the set of diagonal matrices in $\Omega = D^I_{p,q}$ is precisely a unit polydisc Δ^p. To construct Ψ one simply imitates the construction of Ψ in the case of $B^n \cong D^I_{1,n}$. In other words one constructs a product Möbius transformation on Δ^p sending 0 to Z which extends to an automorphism of $D^I_{p,q}$. Write $\beta_i = \sqrt{1 - |\alpha_i|^2}$. Define $\Psi = \begin{bmatrix} A & B & 0 \\ \overline{B} & A & 0 \\ 0 & 0 & I_{q-p} \end{bmatrix}$, where $A = \mathrm{diag}_{p,p}(1/\beta_1,\ldots,1/\beta_p)$, $B = \mathrm{diag}_{p,p}(\alpha_1/\beta_1,\ldots,\alpha_p/\beta_p)$. The fact that $\Psi \in SU(p,q)$ follows readily from the fact that for $1 \le i \le p$, $\begin{bmatrix} 1/\beta_i & \overline{\alpha}_i/\beta_i \\ \alpha_i/\beta_i & 1/\beta_i \end{bmatrix}$ represents a Möbius transformation, thus belonging to $SU(1,1)$. Clearly $\Psi(0) = \mathrm{diag}(\alpha_1,\ldots,\alpha_p)$, as desired.

(2.3) In this section we describe two types of classical domains which are submanifolds of $D^I_{n,n}$ for some n. We define

$$\begin{aligned} D^{II}_n &= \{Z \in D^I_{n,n} : Z^t = -Z\}, \\ D^{III}_n &= \{Z \in D^I_{n,n} : Z^t = Z\}. \end{aligned}$$

We begin with the domains D^{III}_n (of type III). On $\mathbb{C}^{2n}$ consider a symplectic (i.e., non–degenerate alternating) 2–form J. One can choose a basis $\{e_i\}_{1\le i\le 2n}$ and corresponding Euclidean coordinates $\{w_i\}_{1\le i\le 2n}$ of $\mathbb{C}^{2n}$ such that in terms of the dual basis $\{e^i\}_{1\le i\le 2n}$, we have $J = e^1 \wedge e^{n+1} + \ldots + e^n \wedge e^{2n}$. (Any two non–degenerate alternating forms on $\mathbb{C}^{2n}$ are conjugate over $\mathbb{C}$). In terms of matrices, J is represented by the matrix $J_n = \begin{bmatrix} 0 & I_n \\ -I_n & 0 \end{bmatrix}$. Define $G^{III}(n,n) \subset$

$G(n,n)$ to be the set of complex n–planes V in $\mathbb{C}^{2n}$ such that $J|_V \equiv 0$. Representing V in the form $\begin{bmatrix} Z \\ I_n \end{bmatrix}$ as before the latter condition is equivalent to $[\,Z^t\, I_n\,]\begin{bmatrix} 0 & I_n \\ -I_n & 0 \end{bmatrix}\begin{bmatrix} Z \\ I_n \end{bmatrix} = 0$, i. e., $Z^t = Z$. Hence we may write $D_n^{III} \subset M_s(n,\mathbb{C}) \to G^{III}(n,n)$, where $M_s(n,\mathbb{C})$ denotes the complex vector space of symmetric matrices of order $n \times n$, so that $\dim_{\mathbb{C}} M_s(n,\mathbb{C}) = \frac{1}{2}(n^2+n)$. Define $Sp(n,\mathbb{C})$ to be the complex Lie group preserving J. In matrix form, $M \in Sp(n,\mathbb{C})$ if and only if $M^tJ_nM = J_n$. Since M must also preserve the 2n–form $J^n = n!\,(e_1 \wedge \ldots \wedge e_{2n})$ we have $\det(M) = 1$. $Sp(n,\mathbb{C})$ acts holomorphically on $G^{III}(n,n)$. As $D_n^{III} = D^I_{n,n} \cap G^{III}(n,n)$, it is invariant under the group $G_0 = Sp(n,\mathbb{C}) \cap U(n,n) \subset SU(n,n)$. Writing $M \in G_0$ in the form $\begin{bmatrix} A & B \\ C & D \end{bmatrix}$ for n by n square matrices A, B, C, D, we have from $M^tJ_nM = J_n$ the equation

$$M^{-1} = J_n^{-1}M^tJ_n = \begin{bmatrix} 0 & -I_n \\ I_n & 0 \end{bmatrix}\begin{bmatrix} A^t & C^t \\ B^t & D^t \end{bmatrix}\begin{bmatrix} 0 & I_n \\ -I_n & 0 \end{bmatrix} = \begin{bmatrix} D^t & -B^t \\ -C^t & A^t \end{bmatrix}. \quad (1)$$

On the other hand since $X \in U(n,n)$ we deduce from $\overline{M}^tI_{n,n}M = I_{n,n}$ the relation

$$M^{-1} = I_{n,n}^{-1}\overline{M}^tI_{n,n} = \begin{bmatrix} I_n & 0 \\ 0 & -I_n \end{bmatrix}\begin{bmatrix} \overline{A}^t & \overline{C}^t \\ \overline{B}^t & \overline{D}^t \end{bmatrix}\begin{bmatrix} I_n & 0 \\ 0 & -I_n \end{bmatrix} = \begin{bmatrix} \overline{A}^t & -\overline{C}^t \\ -\overline{B}^t & \overline{D}^t \end{bmatrix}. \quad (2)$$

From (1) and (2) it follows that

$$G_0 = \left\{ M = \begin{bmatrix} A & B \\ \overline{B} & \overline{A} \end{bmatrix} : M \in SU(n,n) \right\}. \quad (3)$$

It is obvious that $Z \to -Z$ is a symmetry of $\Omega = D_n^{III}$ at the origin. To show that G_0 acts transitively on Ω we examine first of all the isotropy subgroup K of G_0 at the origin 0. From (2.2) and the description (3) of G_0 it follows that we have

$$K = \left\{ X = \begin{bmatrix} U & 0 \\ 0 & \overline{U} \end{bmatrix} : U \in U(n) \right\}.$$

Thus, $U \in U(n) \cong K$ acts on Ω by $X \to UXU^t$ since $\overline{U}^{-1} = U^t$. We have

LEMMA 1

Let X be an arbitrary complex symmetric matrix of order $n \times n$. Then, there exist nonnegative real numbers $\{\alpha_1,\dots,\alpha_n\}$ and a unitary matrix U such that $UXU^t = \text{diag}(\alpha_1,\dots,\alpha_n)$.

Proof:

For any matrix X, $\overline{X}^tX$ is Hermitian. Thus, for $X \in M_s(n,C)$ there exists nonnegative real numbers $\{a_1,\dots,a_n\}$ and a unitary matrix $V \in U(n)$ such that $\overline{V}^t(\overline{X}^tX)V = \text{diag}(a_1,\dots,a_n)$. Write $Y = V^tXV$. Then, Y is symmetric and $\overline{Y}^tY = \overline{V}^t\overline{X}^t(\overline{V}V^t)XV = \overline{V}^t(\overline{X}^tX)V = \text{diag}(a_1,\dots,a_n)$. Write $Y = A + \sqrt{-1}\,B$ with A and B real. We have $\overline{Y}Y = (A^2 + B^2) + \sqrt{-1}(AB - BA)$. Since $\overline{Y}Y = \overline{Y}^tY$ is real A and B must commute. There exists an orthogonal matrix $T \in O(n)$ such that $TAT^t = \text{diag}(b_1,\dots,b_n)$ with $\{b_i\}$ real satisfying $b_1 \leq \dots \leq b_n$. Replacing X by TYT^t we may therefore assume without loss of generality that $X = C + \sqrt{-1}\,D$ with C, D real such that $C = \text{diag}(b_1,\dots,b_n)$ satisfying $b_1 \leq \dots \leq b_n$ and such that $CD = DC$. Since C and D commute $D = \begin{bmatrix} D_1 & & \\ & \ddots & \\ & & D_k \end{bmatrix}$, where each D_j, $1 \leq j \leq k$, is a square $n(i) \times n(i)$ matrix corresponding to a chain of identical b_i's in $(b_1,\dots,b_n)$. In other words, we have (writing $i = \sqrt{-1}$)

$$X = \begin{bmatrix} c_1I_{n(1)} + i\,D_1 & & \\ & \ddots & \\ & & c_kI_{n(k)} + i\,D_k \end{bmatrix}$$

for some real numbers $(c_1,\dots,c_k)$ and for some real symmetric matrices $D_1, \dots D_k$. We may now choose an orthogonal matrix $S \in O(n)$ such that SXS^t is diagonal. Finally, there exists a diagonal matrix $R = \text{diag}(\gamma_1,\dots,\gamma_n)$ with γ_i of unit length such that $R(SXS^t)R^t$ is a real diagonal matrix, proving the lemma. ∎

From Lemma 1 it follows that given any $Z \in \Omega = D_n^{III}$ there exists $U \in U(n) \cong K$ such that $UZU^t = \text{diag}(\alpha_1,\dots,\alpha_n)$ with $|\alpha_i| \leq 1$. To show that G_0 acts transitively on Ω we are going to construct $\Phi \in G_0$ such that $\Psi(0) =$

$\text{diag}(\alpha_1,\ldots,\alpha_n)$. In the construction we do not need the assumption that α_i's are real. It suffices in fact to take the Ψ that one constructs in (2.2) and observe that Ψ keeps $\Omega = D_n^{III} \subset D_{n,n}^{I}$ invariant. Write $\beta_i = \sqrt{1 - |\alpha_i|^2}$. In terms of the coordinates $\{w_i\}_{1\leq i\leq 2n}$ of C^{2n}, Ψ is represented by $\begin{bmatrix} A & B \\ \bar{B} & \bar{A} \end{bmatrix}$, where $A = \text{diag}(1/\beta_1,\ldots,1/\beta_n)$ and $B = \text{diag}(\alpha_1/\beta_1,\ldots,\alpha_n/\beta_n)$, showing that $\Psi \in Sp(n,C) \cap SU(n,n) = G_0$.

We have shown that $D_n^{III} = G_0/K$ with $G_0 = Sp(n,C) \cap SU(n,n)$ and $K \cong U(n) \hookrightarrow G_0$. In more familiar form we have an isomorphism $G_0 \cong Sp(n,\mathbf{R}) = Sp(n,C) \cap GL(2n,\mathbf{R})$. This isomorphism can be seen by making a unitary change of variables on C^{2n} given by the matrix $\frac{1}{\sqrt{2}}\begin{bmatrix} I_n & iI_n \\ iI_n & I_n \end{bmatrix} = U$ in such a way that the alternating form J is now represented by the matrix $\begin{bmatrix} iI_n & 0 \\ 0 & -iI_n \end{bmatrix} = iI_{n,n}$. (This corresponds to a decomposition of the complexification of a real vector space with an "almost complex structure" into vectors of types (1,0) and (0,1).) At the same time $U \in Sp(n,C)$. Then, we have the embedding $Sp(n,\mathbf{R}) \hookrightarrow Sp(n,C)$ given by $M \longrightarrow U^{-1}MU = \bar{U}^tMU := N$ such that $N^t(iI_{n,n})N = iI_{n,n}$. In other words, $N \in Sp(n,C) \cap SU(n,n)$ so that $Sp(n,\mathbf{R}) \hookrightarrow G_0$. One sees that this is in fact an isomorphism by checking that the inverse mapping $N \longrightarrow UNU^{-1}$ maps $N \in G_0$ into $Sp(n,\mathbf{R})$. Since $U \in Sp(n,C)$ the only thing to check is that $U^{-1}NU$ is real. This follows from

$$\frac{1}{2}\begin{bmatrix} I_n & iI_n \\ iI_n & I_n \end{bmatrix}\begin{bmatrix} A & B \\ \bar{B} & \bar{A} \end{bmatrix}\begin{bmatrix} I_n & -iI_n \\ -iI_n & I_n \end{bmatrix} = \begin{bmatrix} Re A + Im B & Re B + Im A \\ Re B - Im A & Re A - Im B \end{bmatrix}.$$

Identifying G_0 with $Sp(n,\mathbf{R})$ this way we have the presentation $D_n^{III} \cong Sp(n,\mathbf{R})/U(n)$ as a homogeneous space.

(2.4) To discuss $D_n^{II} \subset D_{n,n}^{I}$ we consider a non–degenerate complex symmetric bilinear form Σ on C^{2n}. We choose Euclidean coordinates on C^{2n} such that Σ is represented by the complex symmetric matrix $\begin{bmatrix} 0 & I_n \\ I_n & 0 \end{bmatrix}$. (Any two non–degenerate complex bilinear forms in $\mathbf{C}^m$ are conjugate to each other.) Define $G^{II}(n,n) \subset G(n,n)$ to be the set of complex n–planes V in $\mathbf{C}^{2n}$ such that $\Sigma|_V \equiv 0$. On the affine piece of $G(n,n)$ consisting of $Z \in M(n,n;\mathbf{C})$ the condition $\Sigma|_V \equiv 0$ is equivalent to $[\,Z^t\ I_n\,]\begin{bmatrix} 0 & I_n \\ I_n & 0 \end{bmatrix}\begin{bmatrix} Z \\ I_n \end{bmatrix} = 0$, i.e., $Z^t = -Z$, so that $D_n^{II} \subset M_a(n,\mathbf{C}) \subset G^{II}(n,n)$, where $M_a(n,\mathbf{C})$ stands for the complex vector space of skew-symmetric matrices of order $n \times n$ $(\dim_{\mathbf{C}}(M_a(n,n)) = \frac{1}{2}(n^2 - n))$. Let $O(n,\mathbf{C})$ be the complex orthogonal group with respect to S_n. $M \in O(n,\mathbf{C})$ if and only if $M^tS_nM = S_n$ for the matrix $S_n = \begin{bmatrix} 0 & I_n \\ I_n & 0 \end{bmatrix}$. $O(n,\mathbf{C})$ operates holomorphically on $G(n,n)$ and $G^{II}(n,n)$ is invariant under $O(n,\mathbf{C})$. Define $G_0 = O(n,\mathbf{C}) \cap SU(n,n)$. Writing $M \in G_0$ in the form $\begin{bmatrix} A & B \\ C & D \end{bmatrix}$ for square matrices A, B, C, D of order $n \times n$, we have from $M^tS_nM = S_n$ the equation

$$M^{-1} = S_n^{-1}M^tS_n = \begin{bmatrix} 0 & I_n \\ I_n & 0 \end{bmatrix}\begin{bmatrix} A^t & B^t \\ C^t & D^t \end{bmatrix}\begin{bmatrix} 0 & I_n \\ I_n & 0 \end{bmatrix} = \begin{bmatrix} D^t & B^t \\ C^t & A^t \end{bmatrix}. \tag{1}$$

On the other hand since $X \in U(n,n)$ we have from $\overline{M}^tI_{n,n}M = I_{n,n}$ the relation

$$M^{-1} = I_{n,n}^{-1}\overline{M}^tI_{n,n} = \begin{bmatrix} I_n & 0 \\ 0 & -I_n \end{bmatrix}\begin{bmatrix} \overline{A}^t & \overline{C}^t \\ \overline{B}^t & \overline{D}^t \end{bmatrix}\begin{bmatrix} I_n & 0 \\ 0 & -I_n \end{bmatrix} = \begin{bmatrix} \overline{A}^t & -\overline{C}^t \\ -\overline{B}^t & \overline{D}^t \end{bmatrix}. \tag{2}$$

From (1) and (2) it follows that

$$G_0 = \left\{ M = \begin{bmatrix} A & B \\ -\overline{B} & \overline{A} \end{bmatrix} : M \in SU(n,n) \right\}. \tag{3}$$

Clearly the mapping $Z \longrightarrow -Z$ is a holomorphic symmetry of $\Omega = D_n^{II}$. To show that Ω is a bounded symmetric domain we are going to show that G_0 acts transitively on Ω. First, the isotropy subgroup $K \subset G_0$ at the origin 0 is given by

$$K = \left\{ M = \begin{bmatrix} U & 0 \\ 0 & \overline{U} \end{bmatrix} : M \in SU(n,n) \right\}.$$

Thus, $U \in U(n) \cong K$ acts on Ω by $Z \longrightarrow UZU^t$ since $\overline{U}^{-1} = U$. Recall that J_1

$= \begin{bmatrix} 0 & 1 \\ -1 & 0 \end{bmatrix}$. For $n = 2p$ even we write $\mu(\alpha_1,\ldots,\alpha_p) := \begin{bmatrix} \alpha_1 J_1 & & \\ & \ddots & \\ & & \alpha_p J_1 \end{bmatrix}$. For $n = 2p + 1$ odd we write $\mu(\alpha_1,\ldots,\alpha_p) := \begin{bmatrix} \alpha_1 J_1 & & & \\ & \ddots & & \\ & & \alpha_p J_1 & \\ & & & 0 \end{bmatrix}$. We have

LEMMA 1

Let X be an arbitrary complex skew–symmetric matrix of order $n \times n$. Then, there exist nonnegative real numbers $\{\alpha_1,\ldots,\alpha_p\}$ and a unitary matrix U such that $UXU^t = \mu(\alpha_1,\ldots,\alpha_p)$, where $n = 2p$ or $2p + 1$.

Proof:

We also use X to denote the corresponding skew–symmetric bilinear form. Consider first of all the case when X is real. In this case $\sqrt{-1}X$ is Hermitian and there exists a unitary matrix V such that $V(\sqrt{-1}X)V^t = \mathrm{diag}(a_1,\ldots,a_n)$ for some real numbers a_i, $1 \leq i \leq n$. As X is real v is an eigenvector with $\sqrt{-1}Xv = \lambda v$ (λ real) if and only if $\sqrt{-1}X\bar{v} = -\lambda\bar{v}$. We group non–zero eigenvalues (counting multiplicities) in pairs $\pm b_i$, $1 \leq i \leq r$, with unit eigenvectors v_i, $\bar{v}_i$. Since $v_i \perp \bar{v}_i$ we have $\|Rev_i\| = \|Imv_i\|$. For $1 \leq i \leq m$ write $e'_{2i-1} := \sqrt{2}Rev_i$, $e'_{2i} := \sqrt{2}Imv_i$. We have, $X(e'_{2i-1}) = b_i e'_{2i}$ and $X(e'_{2i}) = -b_i e'_{2i-1}$. Define $\{e'_j\}_{2r+1\leq j\leq n}$ to be a basis of zero–eigenvectors of X. Then, in terms of the basis $\{e'_k\}_{1\leq k\leq n}$ X is represented by $\mu(b_1,\ldots,b_r,0,\ldots,0)$. This proves the lemma in case X is real. In the general case as in the proof of [(2.3), Lemma 1] we may assume without loss of generality that $X = A + \sqrt{-1}B$ such that A and B commute. From $AB = BA$ one can find a simultaneous orthonormal basis of eigenvectors of the Hermitian bilinear forms $\sqrt{-1}A$ and $\sqrt{-1}B$. It follows from the proof of the real case that there exists an orthogonal matrix T such that $TXT^t = \mu(\beta_1,\ldots,\beta_p)$ for some complex numbers β_i. It is now clear that there exists a diagonal matrix $R = \mathrm{diag}(\gamma_1,\ldots,\gamma_n)$ with γ_i of unit length such that $R(TXT^t)R^t = \mu(\alpha_1,\ldots,\alpha_p)$ for some nonnegative real numbers α_i, proving the lemma. ▎

To show that G_0 acts transitively on $\Omega = D_n^{II}$ given $\{\alpha_i\}_{1\le i\le n}$ with $|\alpha_i| < 1$ (α_i not necessarily real) we are going to construct $\Psi \in G_0$ such that $\Psi(0) = \mu(\alpha_1,\ldots,\alpha_n)$. Unlike the case of D_n^{III} the point $\mu(\alpha_1,\ldots,\alpha_n)$ does not lie on the distinguished polydisc $\Delta^n \subset D_{n,n}^I$ of diagonal matrices. We write down Ψ in case of $n = 1$. Clearly $D_2^{III} \cong \Delta$, the unit disc. Given $\begin{bmatrix} 0 & \alpha \\ -\alpha & 0 \end{bmatrix}$ we look for an automorphism of $D_{2,2}^I$ extending a Möbius transformation on $D_2^{III} \cong \Delta$. This is given by $\Psi = \frac{1}{\beta}\begin{bmatrix} 1 & 0 & 0 & \alpha \\ 0 & 1 & -\alpha & 0 \\ 0 & -\overline{\alpha} & 1 & 0 \\ \overline{\alpha} & 0 & 0 & 1 \end{bmatrix}$, where $\beta = \sqrt{1 - |\alpha|^2}$. $\Psi \in SU(2,2)$ since both $\frac{1}{\beta}\begin{bmatrix} 1 & -\alpha \\ -\overline{\alpha} & 1 \end{bmatrix}$ (the inner block) and $\frac{1}{\beta}\begin{bmatrix} 1 & \alpha \\ \overline{\alpha} & 1 \end{bmatrix}$ (the outer block) belong to $SU(1,1)$. Moreover, restricted to D_2^{III}, Ψ is given by

$$\Psi\left(\begin{bmatrix} 0 & z \\ -z & 0 \end{bmatrix}\right) = \left(\begin{bmatrix} 0 & z \\ -z & 0 \end{bmatrix} + \begin{bmatrix} 0 & \alpha \\ -\alpha & 0 \end{bmatrix}\right)\left(\begin{bmatrix} \overline{\alpha}z & 0 \\ 0 & \overline{\alpha}z \end{bmatrix} + \begin{bmatrix} 1 & 0 \\ 0 & 1 \end{bmatrix}\right)^{-1}$$
$$= \begin{bmatrix} 0 & \dfrac{z+\alpha}{1+\overline{\alpha}z} \\ -\dfrac{z+\alpha}{1+\overline{\alpha}z} & 0 \end{bmatrix}$$

It is clear how one can define Ψ similarly for a general n. Thus, we have shown that G_0 acts transitively on $\Omega = D_n^{II}$. This proves that Ω is a bounded symmetric domain. Recall that

$$G_0 = \{M \in SL(2n,\mathbb{C}) : \overline{M}^t I_{n,n} M = I_{n,n};\ M^t S_n M = S_n\}.$$

In more conventional form $G_0 \cong SO^*(2n)$, which is defined by

$$SO^*(2n) = \{M \in SL(2n,\mathbb{C}) : M^t M = I_{2n};\ \overline{M}^t J_n M = J_n\}$$

(cf. HELGASON [HEL, Ch.X, p.445]). The isomorphism $G_0 \cong SO^*(2n)$ is given by $M \to UMU^{-1}$, where $U = \frac{1}{\sqrt{2}}\begin{bmatrix} I & iI \\ iI & I \end{bmatrix}$ is unitary and complex symmetric, as can be easily checked using $U^2 = UU^t = \sqrt{-1}\, S_n$ and $UI_{n,n}U^{-1} = -2\sqrt{-1}\, J_n$.

(2.5) In order to describe bounded symmetric domains of type IV it is convenient to start with its compact dual. Consider on $V \cong \mathbf{R}^{n+2}$ with basis $\{e_i\}_{1\le i\le n+2}$ and corresponding Euclidean coordinates $(u_i)_{1\le i\le n+2}$ the indefinite quadratic form Q of signature (n,2) defined by $Q(u,u) = u_1^2 + \dots + u_n^2 - u_{n+1}^2 - u_{n+2}^2$. (Any two real quadratic forms of the same signature are conjugate over the reals.) Write $V_+ = \Sigma_{1\le i\le n} \mathbf{R}e_i$ and $V_- = \mathbf{R}e_{n+1} + \mathbf{R}e_{n+2}$. Extend by complexification to $V^{\mathbf{C}} \cong \mathbf{C}^{n+2}$ with Euclidean coordinates $(w_i)_{1\le i\le n+2}$, $w_i = u_i + \sqrt{-1}v_i$. Consider on $\mathbf{C}^{n+2}$ the space of complex lines L such that $Q|_L \equiv 0$. (The space of such L is in one–to–one correspondence with the space of 2–planes W in $\mathbf{R}^{n+2}$ such that $Q|_W = 0$, cf. SATAKE [SA2, App. §6, p.286ff.].) This space can be identified with the hyperquadric $Q^n \subset \mathbf{P}^{n+1}$ defined by the homogeneous equation $w_1^2 + \dots + w_n^2 - w_{n+1}^2 - w_{n+2}^2 = 0$. (Any 2 non–degenerate complex symmetric forms are conjugate over $\mathbf{C}$.) Associated to Q is a Hermitian symmetric form H on $\mathbf{C}^{n+2}$ given by $H(w,w') = Q(w,\overline{w'})$. Define $\Omega_0 := \{L \in Q^n: H|_L < 0\}$. The condition $L \in \Omega_0$ means

$$\sum_{1\le i\le n} w_i^2 - w_{n+1}^2 - w_{n+2}^2 = 0 \qquad (1)$$

$$\sum_{1\le i\le n} |w_i|^2 < |w_{n+1}|^2 + |w_{n+2}|^2. \qquad (2)$$

We make the unitary change of coordinates on $\mathbf{C}^{n+2}$ by defining $z_i = w_i$ for $1 \le i \le n$, $z_{n+1} = \frac{1}{\sqrt{2}}(w_{n+1} + \sqrt{-1}\, w_{n+2})$ and $z_{n+2} = \frac{1}{\sqrt{2}}(w_{n+1} - \sqrt{-1}\, w_{n+2})$. We write $\{e'_{n+1}, e'_{n+2}\}$ for the corresponding basis of $V_-^{\mathbf{C}} = V_- \otimes_{\mathbf{R}} \mathbf{C}$. From now on we will use the z coordinates for $V^{\mathbf{C}}$ unless otherwise specified. The two conditions (1) and (2) translate into

$$\sum_{1\le i\le n} z_i^2 - 2z_{n+1}z_{n+2} = 0. \qquad (3)$$

$$\sum_{1\le i\le n} |z_i|^2 < |z_{n+1}|^2 + |z_{n+2}|^2. \qquad (4)$$

On the subset Q^n defined by $z_{n+1} \neq 0$ we can identify $(z_1,...,z_n) \in \mathbf{C}^n$ with the point $\left[z_1, ..., z_n, 1, \frac{1}{2}\Sigma_{1\leq i\leq n} z_i^2\right]$. The conditions (3) and (4) imply that $|z_{n+2}| \neq 1$. Otherwise we would have $|\Sigma_{1\leq i\leq n} z_i^2| = 2$ while $\Sigma_{1\leq i\leq n} |z_i|^2 < 2$. Define $\Omega \subset \Omega_o$ to be the connected component containing the point $[0,...,0,1,0] \in Q^n$, which corresponds to $o \in \mathbf{C}^n$. Thus,

$$\Omega = \{z = (z_1,...,z_n) \in \mathbf{C}^n : \|z\|^2 < 2 \text{ and } \|z\|^2 < 1 + |\tfrac{1}{2}\Sigma z_i^2|^2\}.$$

Let $G_0 = SO_o(n,2)$ be the identity component of the real Lie group $SO(n,2)$ acting on $\mathbf{R}^{n+2}$ equipped with the quadratic form Q. The action extends to $\mathbf{C}^{n+2}$ and hence to Q^n. We are going to show that G_0 acts transitively on Ω. To see this consider first the special case of $n = 2$. We assert

LEMMA 1

The hyperquadric Q^2 is biholomorphic to $\mathbb{P}^1 \times \mathbb{P}^1$. Moreover, the domain $\Omega \subset Q^2$ is biholomorphic to Δ^2.

<u>Proof</u>:

Consider the Segre embedding $\sigma: \mathbb{P}^1 \times \mathbb{P}^1 \hookrightarrow \mathbb{P}^3$ defined by $\sigma([\xi_o,\xi_1];[\zeta_o,\zeta_1]) = [\xi_o\zeta_o,\xi_1\zeta_1,\xi_o\zeta_1,\xi_1\zeta_o] = [\eta_1,\eta_2,\eta_3,\eta_4]$. Clearly we have $\eta_1\eta_2 - \eta_3\eta_4 = 0$. Since the symmetric bilinear form on $\mathbf{C}^4$ defined by the quadratic polynomial $\eta_1\eta_2 - \eta_3\eta_4$ is non–degenerate the image of σ is biholomorphic to Q^2. Explicitly consider the linear change of variables $z_1 = \lambda(\eta_1 - \sqrt{-1}\,\eta_2)$, $z_2 = \lambda(\sqrt{-1}\eta_1 - \eta_2)$ with $\lambda^2 = \frac{\sqrt{-1}}{2}$; $z_3 = \eta_3$, $z_4 = \eta_4$; the equation $z_1^2 + z_2^2 - 2z_1z_2 = 0$ is transformed into $2\eta_1\eta_2 - 2\eta_3\eta_4 = 0$. Under this change of variables the condition (4) translates into

$$|\eta_1|^2 + |\eta_2|^2 < 1 + |\eta_1\eta_2|^2,$$
$$\text{i.e., } (1 - |\eta_1|^2)(1 - |\eta_2|^2) > 0, \qquad (5)$$

which is equivalent to the conditions $|\eta_1|, |\eta_2| < 1$ or $|\eta_1|, |\eta_2| > 1$. The connected component of o is given by the unit polydisc $\Delta^2 = \{(\eta_1,\eta_2): |\eta_1|, |\eta_2| < 1\}$, proving the lemma. ∎

The change of variables in fact induces an isomorphism $SO_0(2,2)/\{\pm 1\} \cong (SU(1,1)/\{\pm 1\})^2$. To see this it is clear from the definition of Ω that $SO_0(2,2)$ acts as automorphisms of Ω_0. The kernel of the action of $SO_0(2,2)$ on Ω_0 is $\{\pm 1\}$. The Lie algebra of $SO(2,2)$ is defined by $\mathfrak{so}(2,2) = \{M \in M(4,4;\mathbf{R}) : M^tQ + QM = 0\}$ for $Q = \begin{bmatrix} I_2 & 0 \\ 0 & -I_2 \end{bmatrix}$. A dimension count shows that $SO(2,2)$ is of 6 real dimensions, as $(SU(1,1))^2$, implying that $SO_0(2,2)/\{\pm 1\} \cong Aut_0(\Omega) \cong (SU(1,1)/\{\pm 1\})^2$. It is now clear from this isomorphism that given any $x \in D_2^{IV} \subset D_n^{IV}$ with $n \geq 3$, there exists $\Psi \in SO(n,2)$ such that $\Psi(o) = x$. It suffices to use the obvious embedding $SO_0(2,2) \hookrightarrow SO_0(n,2)$. To prove that $G_0 = SO_0(n,2)$ acts transitively on Ω we examine the isotropy subgroup $K \subset SO_0(n,2)$. In terms of the coordinates $(u_i)_{1 \leq i \leq n+2}$ on $V = \mathbf{R}^{n+2} = V_+ \oplus V_-$ and $w_i = u_i + \sqrt{-1}\, v_i$ on $V^{\mathbf{C}} \cong \mathbf{C}^{n+2}$, $SO(n) \times SO(2)$ acts on Ω in the obvious way fixing the point $[0,\ldots,0,1,i]$ (in w–coordinates) which corresponds to the origin in the coordinates $(z_i)_{1 \leq i \leq n}$. On the other hand since $SO_0(n,2) \subset GL(V,\mathbf{R}) \hookrightarrow GL(V^{\mathbf{C}},\mathbf{C})$ by complexification any $\Phi \in SO_0(n,2)$ fixing $[0,\ldots,0,1,i]$ must also fix the conjugate point $[0,\ldots,0,1,-i]$. In other words Φ must stabilize V_- and hence V_+ so that Φ lies in $SO(V_+) \times SO(V_-) \cong SO(n) \times SO(2)$. We have thus proved that $K = SO(n) \times SO(2)$. Denote by $\rho_\theta \in SO(2)$ the rotation map given by $\rho_\theta(e_{n+1}) = (\cos\theta)e_{n+1} + (\sin\theta)e_{n+2}$, $\rho_\theta(e_{n+2}) = -(\sin\theta)e_{n+1} + (\cos\theta)e_{n+2}$. We have $\rho_\theta(e'_{n+1}) = e^{-i\theta}e'_{n+1}$ and $\rho_\theta(e'_{n+2}) = e^{i\theta}e'_{n+2}$. In terms of the z–coordinates on $V^{\mathbf{C}}$, ρ_θ acts on $\Omega \subset \mathbf{P}(V^{\mathbf{C}})$ by $\rho_\theta([z_1,\ldots,z_n,1,z_{n+2}]) = \left[z_1,\ldots,z_n,e^{-i\theta},e^{i\theta}z_{n+2}\right] = \left[e^{i\theta}z_1,\ldots,e^{i\theta}z_n,1,e^{2i\theta}z_{n+2}\right]$. Hence, in terms of the embedding $\Omega \hookrightarrow \mathbf{C}^n$, $(T,\rho_\theta) \in SO(n) \times SO(2)$ acts on Ω as $(T,\rho_\theta)(z) = e^{i\theta}(Tz)$. Write $e'_1 = \lambda(e_1 + \sqrt{-1}e_2)$ and $e'_2 = \lambda(e_1 - \sqrt{-1}e_2)$ with $\lambda^2 = \frac{\sqrt{-1}}{2}$. We prove

LEMMA 2

Given any $z \in \Omega = D_n^{IV}$ there exists $\Psi \in SO(n) \times SO(2)$ such that $\Psi(z) = \alpha_1 e'_1 + \alpha_2 e'_2 \in D_2^{IV}$ with α_1, α_2 real such that $0 \leq \alpha_1, \alpha_2 < 1$.

Proof:
Write $z = x + \sqrt{-1}y$ with x, y real. There exists an orthogonal matrix $T \in O(n)$ such that $T(x) = (a,0,\ldots,0)$. Thus, $Tz = (a,0,\ldots,0) + i(b_1,\ldots,b_n)$, with $a; b_1, \ldots, b_n \in \mathbf{R}$. There exists an orthogonal transformation S fixing $(1,0,\ldots,0)$ such that $S(b_1,\ldots,b_n) = (b_1,b,0,\ldots,0)$. Consequently, $STz = (\mu,\nu,0,\ldots,0)$ with μ, ν complex. The action of $R_{\theta,\varphi} = (\rho_\theta,\rho_\varphi) \in SO(2)\times SO(2) \hookrightarrow SO(n)\times SO(2)$ (with the first SO(2) corresponding to $\mathbf{R}e_1 + \mathbf{R}e_2$) is given by $R_{\theta,\varphi}(\eta_1 e_1' + \eta_2 e_2') = e^{i\varphi}(e^{-i\theta}e_1' + e^{i\theta}e_2')$. It is clear that one can now choose θ and φ so that $R_{\theta,\varphi}STz = \alpha_1 e_1' + \alpha_2 e_2'$ with α_1, α_2 real such that $0 \le \alpha_1, \alpha_2 < 1$, as desired. ▌

We have presented $D_n^{IV} \cong SO_0(n,2)/SO(n)\times SO(2)$ as a homogeneous space. Since $z \longrightarrow -z$ is a holomorphic symmetry at o it follows that D_n^{IV} is a bounded symmetric domain.

(2.6) We have presented the four types of classical symmetric domains $\Omega = D_{p,q}^{I}, D_n^{II}, D_n^{III}, D_n^{IV}$. In all cases we have inclusions $\Omega \subset\subset \mathbf{C}^N \subset X$ for some compact complex manifold X. The inclusion $\Omega \subset\subset X$ is in fact the Borel embedding. We are going to discuss this in the case of $D_{p,q}^{I}$. For more details cf. WOLF [WOL2, Part I]. The inclusion $\Omega \subset\subset \mathbf{C}^N$ is the Harish–Chandra embedding, which will be discussed in Chapter 5.

In case of $\Omega = D_{p,q}^{I}$, $p, q \ge 1$, we have $D_{p,q}^{I} \subset\subset \mathbf{C}^{pq} \subset G(p,q)$. Recall that $\Omega \cong SU(p,q)/S(U(p)\times U(q))$. The complex Lie group $SL(p+q,\mathbf{C})$ acts on the Grassmannian $G(q,p)$ holomorphically. On the affine part $\mathbf{C}^{pq}$ the action is given by $\Phi(Z) = (AZ + B)(CZ + D)^{-1}$ for $\Phi = \begin{bmatrix} A & B \\ C & D \end{bmatrix} \in SL(p+q,\mathbf{C})$. The subgroup $SU(p+q)$ acts transitively on $G(q,p)$. Thus $G(q,p) \cong SU(p+q)/K$, where K is the isotropy subgroup of $G(q,p)$ at the origin 0 in $\mathbf{C}^{pq}$. We have

$$K = \left\{\begin{bmatrix} A & 0 \\ C & D \end{bmatrix} : \begin{bmatrix} A & 0 \\ C & D \end{bmatrix} \in SU(p+q)\right\} = \left\{\begin{bmatrix} U & 0 \\ 0 & V \end{bmatrix} \in SU(p+q) : U \in U(p),\ V \in U(q)\right\}.$$

We have the dual pair of symmetric manifolds $D^{I}_{p,q} = SU(p,q)/S(U(p)\times U(q))$, $G(q,p) = SU(p+q)/S(U(p)\times U(q))$. The simply-connected connected Lie groups $G_0 = SU(p,q)$ and $G_c = SU(p+q)$ are non-compact resp. compact real forms of $SL(p+q,\mathbb{C})$. To see this recall that the Lie algebra $\mathfrak{su}(p+q)$ consists of skew-Hermitian matrices with zero traces. Any $X \in \mathfrak{sl}(p+q,\mathbb{C})$ can be decomposed as $X = \frac{1}{2}(X - \overline{X}^t) + \frac{1}{2}(X + \overline{X}^t) \in \mathfrak{su}(p+q) + \sqrt{-1}\,\mathfrak{su}(p+q)$, while on the other hand $X \in \mathfrak{su}(p+q) \cap \sqrt{-1}\,\mathfrak{su}(p+q)$ implies that $X + \overline{X}^t = \sqrt{-1}\,X - \sqrt{-1}\,\overline{X}^t = 0$, i.e., $X = 0$, showing that $\mathfrak{su}(p+q)$ is a compact real form of $\mathfrak{sl}(p+q,\mathbb{C})$. Let θ_c be the inner automorphism $X \longrightarrow I^{-1}_{p,q}XI_{p,q}$ on $\mathfrak{su}(p+q)$. Explicitly $\theta_c\left(\begin{bmatrix} A & B \\ C & D\end{bmatrix}\right) = \begin{bmatrix} A & -B \\ -C & D\end{bmatrix}$. θ_c is obviously a Lie algebra involution whose fixed point set is precisely $\mathfrak{k} = \mathfrak{s}(\mathfrak{u}(p)+\mathfrak{u}(q))$. The (-1)-eigenspace of θ_c, which we denote by $\sqrt{-1}\mathfrak{m}$, consists of skew-Hermitian matrices of the form $\begin{bmatrix} 0 & B \\ C & 0\end{bmatrix}$, $C = -\overline{B}^t$.

On the other hand the Lie algebra $\mathfrak{su}(p,q)$ of $SU(p,q)$ consists of matrices $X = \begin{bmatrix} A & B \\ C & D\end{bmatrix}$ satisfying $\overline{X}^t I_{p,q} + I_{p,q}X = 0$, i.e., $\begin{bmatrix} \overline{A}^t & -\overline{C}^t \\ \overline{B}^t & -\overline{D}^t\end{bmatrix} + \begin{bmatrix} A & B \\ -C & -D\end{bmatrix} = 0$. Obviously $\begin{bmatrix} A & B \\ C & D\end{bmatrix} \in \mathfrak{su}(p+q)$ if and only if $\begin{bmatrix} A & iB \\ iC & D\end{bmatrix} \in \mathfrak{su}(p+q)$. Thus $\mathfrak{su}(p,q) = \mathfrak{k} + \mathfrak{m}$, so that $\mathfrak{su}(p,q)$ is a non-compact real form of $\mathfrak{sl}(p+q,\mathbb{C})$. $\mathfrak{k} = \mathfrak{s}(\mathfrak{u}(p)+\mathfrak{u}(q))$ is a maximal compact subalgebra of $\mathfrak{su}(p,q)$ and the inner automorphism θ_0: $\mathfrak{su}(p,q) \longrightarrow \mathfrak{su}(p,q)$ defined by $\theta_0(X) = I^{-1}_{p,q}XI_{p,q}$ is a Cartan involution on the simple non-compact Lie algebra $\mathfrak{su}(p,q)$.

We have thus for $\mathfrak{g}_0 = \mathfrak{su}(p,q)$, $\mathfrak{g}_c = \mathfrak{su}(p+q)$ the dual orthogonal symmetric Lie algebras $(\mathfrak{g}_0,\mathfrak{k};\theta_0)$, $(\mathfrak{g}_c,\mathfrak{k};\theta_c)$ and their corresponding canonical decompositions $\mathfrak{g}_0 = \mathfrak{k} + \mathfrak{m}$ and $\mathfrak{g}_c = \mathfrak{k} + \sqrt{-1}\mathfrak{m}$. On $\mathfrak{su}(p,q)$ we have the J-operator given by $J\left(\begin{bmatrix} A & B \\ C & D\end{bmatrix}\right) = \begin{bmatrix} A & iB \\ -iC & D\end{bmatrix}$. The decomposition $\mathfrak{m}^{\mathbb{C}} = \mathfrak{m}^+ + \mathfrak{m}^-$ is accordingly given by $\mathfrak{m}^+ = \left\{\begin{bmatrix} 0 & B \\ 0 & 0\end{bmatrix} : B \in M(p,q,\mathbb{C})\right\}$, $\mathfrak{m}^- = \left\{\begin{bmatrix} 0 & 0 \\ C & 0\end{bmatrix} : C \in M(q,p,\mathbb{C})\right\}$. The parabolic subgroup $P \subset G^{\mathbb{C}} = SL(p+q)$ corresponding to the complex Lie subalgebra $\mathfrak{p} = \mathfrak{k}^{\mathbf{C}} + \mathfrak{m}^{-}$ is given by $P = \left\{ \begin{bmatrix} A & 0 \\ C & D \end{bmatrix} : A \in \mathfrak{gl}(p,\mathbf{C}),\ D \in \mathfrak{gl}(q,\mathbf{C}),\ \mathrm{Tr}(A) + \mathrm{Tr}(D) = 0 \right\}$. $G^{\mathbf{C}} = SL(p+q,\mathbf{C})$ acts holomorphically on $\mathbf{C}^{p+q}$ and hence on the Grassmannian $G(q,p)$ such that $P \subset SL(p+q,\mathbf{C})$ is the isotropy subgroup at $0 \in M(p,q;\mathbf{C}) \subset G(q,p)$. $\Omega = D^{I}_{p,q}$ is the orbit of 0 under $SU(p,q) = G_0 \subset G^{\mathbf{C}}$. The embedding $D^{I}_{p,q} \hookrightarrow SL(p+q,\mathbf{C})/P \cong G(q,p)$ is the Borel embedding.

The description of the Borel embeddings $D^{II}_{n} \hookrightarrow G^{II}(n,n)$ and $D^{II}_{n} \hookrightarrow G^{II}(n,n)$ are very similar. We defer the discussion on D^{IV}_{n} to (3.2), where we will use duality to compute curvature on the domain.

§3 Curvatures of Classical Bounded Symmetric Domains

(3.1) Let Ω be a classical bounded symmetric domain, ω be the Kähler form of the Bergman metric on Ω. We are going to compute the curvature tensor of (Ω,ω) explicitly. Our main concern is to examine the set of zeros of the bisectional curvatures. Let $\kappa(Z,W)$ be the Bergman kernel form on Ω. In case of types I – III we are going to determine $\kappa(Z,Z)$ up to scaling constants. This will allow us to compute the Bergman metric explicitly. In case of domains of type IV it is more convenient to compute the curvature tensor of their compact duals, the hyperquadrics, and use duality. For the precise determination of the Bergman kernel form $\kappa(Z,W)$ on all classical bounded symmetric domains we refer the reader to HUA [HUA, Ch.IV, p.77ff.].

We use the embeddings $\Omega \hookrightarrow \mathbf{C}^N$ as given in §2. Let $d\lambda$ denote the Euclidean volume form. Up to a scaling constant $\kappa(Z,Z)$ is the unique volume form on Ω invariant under $\mathrm{Aut}(\Omega)$. Consider $D^{I}_{p,q} \cong G_0/K = SU(p,q)/S(U(p)\times U(q)) \subset M(p,q;\mathbf{C}) \cong \mathbf{C}^{pq}$. We assume as usual $p \leq q$. For τ a real number consider the volume form $\mu_\tau = (\det(I - \bar{Z}^t Z))^{-\tau}\, d\lambda$. We claim that for any τ, μ_τ is K–invariant. On the one hand clearly $K = S(U(p)\times U(q))$ acts as unitary transformations on $M(p,q;\mathbf{C})$ so that $d\lambda$ is invariant under K. On the other hand for $U \in U(p)$ and $V \in U(q)$ we have $I - (\bar{V}^t \bar{Z}^t \bar{U}^t)(UZV) = \bar{V}^t(I - \bar{Z}^t Z)V$ so that the function $\det(I - \bar{Z}^t Z)$ is K–invariant. Consequently, μ_τ is K–invariant for any α, as claimed. We assert furthermore that for some τ to be determined μ_τ is invariant under G_0. By [(2.2), Lemma 1] it suffices to consider the special automorphisms Ψ_α for $\alpha = \mathrm{diag}(\alpha_1,\dots,\alpha_p) \in \Delta^p \subset D^I_{p,q}$ for Δ^p the polydisc of diagonal matrices in $D^I_{p,q}$. In general for $\Phi(Z) = (AZ + B)(CZ + D)^{-1}$ we have $d\Phi(0)(X) = AXD^{-1} - BD^{-1}CXD^{-1} = (A - BD^{-1}C)XD^{-1}$. Write $\sqrt{1 - |\alpha_i|^2} = \beta_i$. For $\Phi = \Psi_\alpha$ we have from (2.2) $A = \mathrm{diag}_{p,p}(1/\beta_1,\dots,1/\beta_p)$; $B = \mathrm{diag}_{p,q}(\alpha_1/\beta_1,\dots,\alpha_p/\beta_p)$, $C = \bar{B}^t$ and $D = \begin{bmatrix} A & 0 \\ 0 & I_{q-p} \end{bmatrix}$. Hence $A - BD^{-1}C = \mathrm{diag}_{p,p}[(1 - \alpha_1\bar{\alpha}_1)/\beta_1,\dots,(1 - \alpha_p\bar{\alpha}_p)/\beta_p] = \mathrm{diag}_{p,p}(\beta_1,\dots,\beta_p)$, yielding

$$d\Psi_\alpha(0)(X) = \begin{bmatrix} \beta_1 & & \\ & \ddots & \\ & & \beta_p \end{bmatrix} X \begin{bmatrix} \beta_1 & & & \\ & \ddots & & \\ & & \beta_p & \\ & & & I_{q-p} \end{bmatrix}.$$

We have

$$(\Psi_*(d\lambda))(\alpha) = \Pi_{1\le i\le p}\, \beta_i^{2p+2q}\, d\lambda(\alpha)$$
$$= (1 - |\alpha_1|^2)^{p+q} \dots (1 - |\alpha_p|^2)^{p+q}\, d\lambda(\alpha); \text{ and}$$
$$\det(I - \alpha\bar{\alpha}^t) = (1 - |\alpha_1|^2)\dots(1 - |\alpha_p|^2).$$

It follows readily that the volume form $\mu_{p+q} = (\det(I - \bar{Z}^t Z))^{-(p+q)}\, d\lambda$ is invariant under K and $\{\Psi_\alpha : \alpha \in \Delta^p\}$, hence under $G_0 = SU(p,q)$. Since Ω is homogeneous it follows that for the Bergman kernel form κ^I on $D^I_{p,q}$ we have $\kappa^I(Z,Z) = c\mu_{p+q}$ for some positive constant c.

For the cases of $\Omega = D^{II}_n$, D^{III}_n one obtains by the same method the following formulas for the Bergman kernel forms κ^{II} resp. κ^{III}.

$$\kappa^{II}(Z,Z) = c(\det(I + \bar{Z}Z))^{-(n-1)}\, d\lambda$$
$$\kappa^{III}(Z,Z) = c(\det(I - \bar{Z}Z))^{-(n+1)}\, d\lambda$$

for some positive constants c depending on Ω. Here we use the embedding $\Omega \hookrightarrow \mathbf{C}^N$ as given in (2.3) and (2.4) resp.

In the case of $\Omega = D_n^{IV}$ we determine instead the canonical metric on its compact dual Q^n. Recall that we have an (n+2)–dimensional real vector space $V = V_+ + V_-$, with basis $\{e_i\}_{1 \le i \le n}$ and $\{e_{n+1}, e_{n+2}\}$ resp., equipped with the indefinite quadratic form E given by $u_1^2 + \dots + u_n^2 - u_{n+1}^2 - u_{n+2}^2$, where (u_i) are Euclidean coordinates of V in terms of $\{e_i\}$. Define $V' = V_+ + \sqrt{-1}V_-$ and write $V^{\mathbf{C}} = V' \otimes_{\mathbf{R}} \mathbf{C}$. The compact dual $SO(n+2)$ of $SO(n,2)$ acts on V' in the standard way and hence on $V^{\mathbf{C}}$ and Q^n. Recall that $w_i = u_i + \sqrt{-1}v_i$ are complex Euclidean coordinates on $V^{\mathbf{C}}$. Writing $w_i = \xi_i$ for $1 \le i \le n$ and $\xi_{n+j} = \sqrt{-1}w_{n+j}$ for $j = 1, 2$, Q^n is defined by $\Sigma\xi_i^2 = 0$. In the ξ–coordinates $SO(n+2)$ acts as complexifications of real orthogonal transformations. The action of $SO(n+2)$ on Q^n is transitive as a consequence of the following lemma:

LEMMA 1

Let $\xi = (\xi_1,\dots,\xi_N) \in \mathbf{C}^N$ be such that $\Sigma\xi_i^2 = 0$. Then, there exists an orthogonal transformation S of $\mathbf{R}^N$ such that $S\xi = \alpha(1,i,0,\dots,0)$ for a complex number α.

Proof:
As proved in [(2.5), Lemma 1] there always exists S such that $S\xi = (\mu,\nu,0,\dots,0)$. Regarding ξ as a column vector the equation $\Sigma\xi_i^2 = \xi^t\xi = 0$ is preserved by S. It follows that $\mu = \pm i\nu$. If $\mu = i\nu$ the lemma is proved. If $\mu = -i\nu$ one can simply switch the two variables ξ_1, ξ_2. ▌

By the same consideration as in (2.5) one sees that the isotropy subgroup of $SO(n+2)$ at $[0,\dots,0,1,i]$ (in w–coordinates) is $SO(n)\times SO(2)$. Thus, the pair $D_n^{IV} \cong SO_0(n,2)/SO(n)\times SO(2)$ and $Q^n \cong SO(n+2)/SO(n)\times SO(2)$ are dual Hermitian symmetric manifolds arising from real forms of the simple complex Lie group $O(n+2,C) \cong \{M \in GL(n+2,C)\colon M^tM = I_{n+2}\}$. In terms of the homogeneous coordinates $[z_1,\ldots,z_n]$ on $Q^n \subset \mathbb{P}(V^{\mathbb{C}}) \cong \mathbb{P}^{n+1}$ and the embedding $D_n^{IV} \hookrightarrow \mathbb{C}^n \hookrightarrow Q^n$ given in (2.5), we have

$$Q^n = \{[z_1,\ldots,z_{n+2}]: \sum_{1\le i\le n} z_i^2 - 2z_{n+1}z_{n+2} = 0\}.$$

$$D_n^{IV} = \{(z_1,\ldots,z_n) \in \mathbb{C}^n: \|z\|^2 < 2 \text{ and } \|z\|^2 < 1 + |\tfrac{1}{2}\Sigma z_i^2|^2\},$$

where the embedding $\mathbb{C}^n \hookrightarrow Q^n$ is given by $(z_1,\ldots,z_n) \longrightarrow [z_1, \ldots, z_n, 1, \frac{1}{2}\Sigma z_i^2]$. The composite embedding $D_n^{IV} \hookrightarrow Q^n$ is the Borel embedding. The nonnegative (1,1)–form $\sqrt{-1}\,\partial\bar\partial \log \|z\|^2$ gives rise to a Fubini–Study metric g on $\mathbb{P}(V^{\mathbb{C}}) \cong \mathbb{P}^{n+1}$. We assert that $(Q^n, g|_{Q^n})$ is a Hermitian symmetric manifold. Observe first of all that $SO(n+2)$ acts as a group of unitary transformations on $V^{\mathbb{C}}$, hence as holomorphic isometries of $(Q^n, g|_{Q^n})$. To prove the assertion it suffices to write down a holomorphic isometric symmetry σ_o at o ([0,...,0,1,0] in homogeneous z–coordinates.) This is given by

$$\sigma_o([z_1, \ldots, z_{n+2}]) = [-z_1,\ldots,-z_n,z_{n+1},z_{n+2}],$$

which when restricted to the affine part $\mathbb{C}^n \supset D_n^{IV}$ is simply given by $(z_1,\ldots,z_n) \longrightarrow (-z_1,\ldots,-z_n)$. To write down the Kähler form ω_c of $(Q^n, g|_{Q^n})$ on the affine part $\mathbb{C}^n$ it suffices to write

$$\begin{aligned}\omega_c &= \sqrt{-1}\,\partial\bar\partial \log \Sigma_{1\le i\le n+2} |z_i|^2 \\ &= \sqrt{-1}\,\partial\bar\partial \log \left[1 + \Sigma_{1\le i\le n} |z_i|^2 + |\tfrac{1}{2}\Sigma_{1\le i\le n} z_i^2|^2\right].\end{aligned}$$

(3.2) For any classical bounded symmetric domain $\Omega \hookrightarrow \mathbb{C}^N$ we write $g_o = 2Re(\Sigma g_{i\bar j} dz_i \otimes d\bar z^j)$ for the Bergman metric, normalized so that $g_{i\bar j}(o) = \delta_{ij}$. Write ω for the Kähler form of g_o. In all cases the symmetry σ_o at the origin o is given by $\sigma_o(z) = -z$. Since σ_o is an isometry the functions $g_{i\bar j}$ are even. In particular, $dg_{i\bar{j}}(o) = 0$ for $1 \leq i, j \leq N$, so that the Euclidean coordinates (z_i) are complex geodesic coordinates at the origin for the (Ω,ω). By [Ch.2, Prop.1] the curvature tensor is given by $R_{i\bar{j}k\bar{l}}(o) = -\partial_i\partial_{\bar{j}}g_{k\bar{l}}(o)$. We write down case by case the curvature tensor as given in SIU [SIU2].

Domains of type I

In (3.1) we determined $\kappa(Z,Z) = K(Z,Z)\, d\lambda$ up to a normalizing constant. We have $\omega = \sqrt{-1}\,\partial\bar{\partial}\,\Phi$ with $\Phi(Z) = c \log K(Z,Z) = c' \log \det (I - Z\bar{Z}^t)$. We put $c' = 1$ so as to get the normalization $g_{ij,\overline{kl}} = \delta_{ij,kl}$. We have

$$\Phi = \sum |z_{ij}|^2 + \frac{1}{2}\sum z_{ik}\bar{z}_{il}z_{jl}\bar{z}_{jk} + \text{higher order terms},$$

so that

$$R_{ij,\overline{kl},pq,\overline{rs}} = -\frac{\partial^4\Phi}{\partial z_{ij}\partial\bar{z}_{kl}\,\partial z_{pq}\partial\bar{z}_{rs}}$$

$$= -\,\delta_{ik}\,\delta_{pr}\,\delta_{js}\,\delta_{lq} - \delta_{ir}\,\delta_{pk}\,\delta_{jl}\,\delta_{qs}$$

For the tangent vectors of type (1,0) $X = \sum X^{ij}\frac{\partial}{\partial Z_{ij}}$, $Y = \sum Y^{ij}\frac{\partial}{\partial Z_{ij}}$, identified with matrices $X = (X^{ij})$ and $Y = (Y^{ij})$ the bisectional curvatures are given by

$$B(X,Y) = R_{ij,\overline{kl},pq,\overline{rs}}X^{ij}\,\overline{X^{kl}}\,Y^{pq}\,\overline{Y^{rs}} = -\sum X^{ij}\,\overline{Y^{iq}}\,Y^{pq}\,\overline{X^{pj}} - \sum X^{ij}\,\overline{Y^{pj}}\,Y^{pq}\,\overline{X^{iq}}$$

$$= -\sum_{j,q}\Big|\sum_i X^{ij}\overline{Y^{iq}}\Big|^2 - \sum_{i,q}\Big|\sum_j X^{ij}\overline{Y^{pj}}\Big|^2 = -\,\|X\bar{Y}^t\|^2 - \|X^t\bar{Y}\|^2. \qquad (1)$$

Writing $X = (X^{ij})$, etc. formula (1) implies that $B(X,Y) = 0$ if and only if any row of X is orthogonal to any row of Y and that any column of X is orthogonal to any column of Y. For any X write $\mathcal{N}_X$ for the null–space of the Hermitian bilinear form $H_X(V,W) = R_{X\bar{X}V\bar{W}}$ at 0. We are going to determine $\dim_{\mathbb{C}} \mathcal{N}_X$ for any X. We have

PROPOSITION 1

For $X \in T_0^{1,0}(D^I_{p,q}) \cong M(p,q;C)$ denote by $r(X)$ the rank of X as a matrix. Then,

$$\dim_{\mathbf{C}} \mathcal{N}_X = (p - r(X))\,(q - r(X)).$$

In particular for $X \neq 0$, $\dim_{\mathbf{C}} \mathcal{N}_X$ is maximal if and only if $r(X) = 1$.

Proof:

Recall that $K = S(U(p)\times U(q))$ acts as the isotropy subgroup at $0 \in D^I_{p,q}$ and that given any tangent vector X there exists $\gamma \in K$ such that $\gamma X = \mathrm{diag}(\alpha_1,...,\alpha_p)$ $(q \geq p)$. We may assume without loss of generality that all α_i's are real and nonnegative such that $\alpha_1 \geq ... \geq \alpha_p$. Since K leaves the Bergman metric invariant and $r(X)$ is clearly invariant under K to prove the proposition it suffices to consider the special case when X is in the normal form $\mathrm{diag}(\alpha_1,...,\alpha_p)$. In this case as $B(X,Y) = -\|XY^t\|^2 - \|X^tY\|^2$, we have $B(X,Y) = 0$ if and only if $Y = \begin{bmatrix} 0 & 0 \\ 0 & V \end{bmatrix}$, where $V \in M(p-r(X),q-r(X);C)$, proving the proposition. ▌

Domains of types II and III

To prove the analogue of Prop. 1 for D^{II}_n and D^{III}_n we start with

LEMMA 1

The inclusions $D^{II}_n \hookrightarrow D^I_{n,n}$ and $D^{III}_n \hookrightarrow D^I_{n,n}$ realize D^{II}_n and D^{III}_n, equipped with their canonical metrics (up to normalizing constants), as totally geodesic submanifolds of $(D^I_{n,n},\omega)$.

Proof:

Since D^{II}_n is homogeneous under the action of $SO^*(2n)$, which is the subgroup of $SU(n,n)$ keeping D^{II}_n invariant, to prove the lemma it suffices to consider the origin 0. Clearly, by the description of the Bergman kernel form in (3.1) the restriction of ω to D^{II}_n gives the canonical metric on the submanifold. As the potential function Φ for ω is even the Euclidean coordinates on $D^I_{n,n}$ and D^{II}_n serve as complex geodesic coordinates at 0 for ω and $\omega|D^{II}_n$. Consequently the second fundamental form of the embedding $D^{II}_n \hookrightarrow D^I_{n,n}$ vanishes at 0 (cf. [Ch.2, (3.2)]), as desired. The proof of the lemma for the case of D^{III}_n is the same. ▌

It follows from Lemma 1 that the curvature tensor on D^{II}_n and D^{III}_n are simply the restriction of the curvature tensor of $(D^I_{n,n},\omega)$. From Prop.1 and the proofs of [(2.3), Lemma 1] and [(2.4), Lemma 1] we have

PROPOSITION 2

For any $X \in T^{1,0}_0(D^{II}_n) \cong M_a(n;C)$ we have

$$\dim_C \mathcal{N}_X = \frac{1}{2}(n - r(X))\,(n - r(X) - 1).$$

In particular, for $X \neq 0$, $\dim_C \mathcal{N}_X$ is maximal if and only if $r(X) = 2$.

PROPOSITION 3

For any $X \in T^{1,0}_0(D^{III}_n) \cong M_s(n;C)$ we have

$$\dim_C \mathcal{N}_X = \frac{1}{2}(n - r(X))\,(n - r(X) + 1).$$

In particular, for $X \neq 0$, $\dim_C \mathcal{N}_X$ is maximal if and only if $r(X) = 1$.

<u>Domains of type IV</u>

For the domains D^{IV}_n we compute first of all the curvature of the compact dual (Q^n,ω_c). Recall that $\omega_c = \sqrt{-1}\,\partial\bar\partial\,\Phi$ with

$$\begin{aligned}\Phi &= \log\left(1 + \Sigma|z_i|^2 + \left|\frac{1}{2}\Sigma z_i^2\right|^2\right)\\ &= \Sigma|z_i|^2 + \left|\frac{1}{2}\Sigma z_i^2\right|^2 - (\Sigma|z_i|^2)^2 + \text{higher order terms.}\end{aligned}$$

in terms of the Euclidean coordinates $(z_i)_{1\le i\le n}$ on $D^{IV}_n \subset C^n \subset Q^n$. Φ is an even function and (z_i) serves as complex geodesic coordinates for (Q^n,ω_c) at the origin. Denote by R^c the curvature tensor of (Q^n, ω_c). We have

$$R^c_{i\bar{j}k\bar{l}}(o) = -\frac{\partial^4 \Phi}{\partial z_i \partial \bar{z}_j \partial z_k \partial \bar{z}_l} = (\delta_{ij}\,\delta_{kl} + \delta_{il}\,\delta_{jk} - \delta_{ik}\,\delta_{jl}).$$

We denote by R the curvature tensor of (D^{IV}_n, ω). The Borel embedding $D^{IV}_n \hookrightarrow Q^n$ induces an isometry at the origin $o = eK$ when both symmetric manifolds are equipped with the canonical metrics arising from the Killing form. Furthermore, the curvature tensors at o are opposite to each other, by [Ch.3, (1.3), proof of Prop.2]. We have therefore

$$R_{i\bar{j}k\bar{l}}(o) = -(\delta_{ij}\,\delta_{kl} + \delta_{il}\,\delta_{jk} - \delta_{ik}\,\delta_{jl}).$$

Let $\xi = \xi^i \frac{\partial}{\partial z_i}$ and $\eta = \eta^i \frac{\partial}{\partial z_i}$ be tangent vectors of type (1,0) at $o \in D^{IV}_n$. Then, we have

$$\begin{aligned} R_{\xi\bar{\xi}\eta\bar{\eta}} &= -(\Sigma|\xi^i|^2|\eta^i|^2 + |\Sigma\xi^i\bar{\eta}^i|^2 - |\Sigma\xi^i\eta^i|^2) \\ &= -(\|\xi\|^2\|\eta\|^2 - |(\xi,\bar{\eta})|^2) - 2|(\xi,\eta)|^2 \\ &\leq -|(\xi,\eta)|^2 \end{aligned}$$

by the Cauchy–Schwarz inequality. Here $(.,.)$ and $\|.\|$ denote the standard Euclidean Hermitian inner product and norm on $\mathbf{C}^n$ resp., and $\bar{\eta}$ means conjugation on $\mathbf{C}^n$ (not on $T^{\mathbf{C}}_0$). Hence, $B(\xi,\eta) = 0$ if and only if ξ and $\bar{\eta}$ are parallel while ξ and η are orthogonal. In this case, there exists some complex number γ, such that $\bar{\eta}_i = \gamma\xi_i$ for $1 \leq i \leq n$ and $\Sigma\xi^i\eta^i = \bar{\gamma}\,\Sigma(\xi^i)^2 = 0$. We have thus proved

PROPOSITION 4

For any non–zero vector $\xi = \xi^i \frac{\partial}{\partial z_i} \in T^{1,0}_0(D^{IV}_n)$, we have $\dim_{\mathbf{C}} \mathscr{N}_\xi \leq 1$. Moreover, $\dim_{\mathbf{C}} \mathscr{N}_\xi = 1$ if and only if $\Sigma(\xi^i)^2 = 0$.

CHAPTER 5 BOUNDED SYMMETRIC DOMAINS — GENERAL THEORY

§1 The Polydisc Theorem (and the Polysphere Theorem)

(1.1) In this chapter we will discuss some general theory of bounded symmetric domains Ω, culminating in the Harish–Chandra Embedding Theorem. We start with the Polydisc and the Polysphere Theorems (cf. WOLF [WOL2, Part I]). To set up the terminology let $G^{\mathbb{C}}$ be a connected centerless simple Lie group over $\mathbb{C}$, $G_0 \subset G^{\mathbb{C}}$ be a non–compact real form of Hermitian type, $K \subset G_0$ be a maximal compact subgroup and (X_o,g_o), $X_o = G_0/K$ be the corresponding irreducible Hermitian symmetric manifold of non–compact type. Let $G_c \supset K$ denote a compact real form of $G^{\mathbb{C}}$, (X_c,g_c) be the compact dual of (X_o,g_o); and $G_0/K \cong X_o \hookrightarrow X_c \cong G_c/K$ be the Borel embedding. Then, we have

THEOREM 1 (THE POLYDISC AND POLYSPHERE THEOREMS)
Suppose (X_o,g_o) is of rank r as a Riemannian symmetric manifold. Then there exists a totally geodesic complex submanifold D such that $(D,g_o|_D)$ is isometric to a Poincaré polydisc (Δ^r,ρ_o) and $X_o = \bigcup_{\gamma\in K}\gamma D$. Moreover, there exists a totally geodesic complex submanifold S of (X_c,g_c) containing D as an open subset such that $(S,g_c|_S)$ is isometric to a polysphere $((\mathbb{P}^1)^r,\rho_c)$ equipped with a product Fubini–Study metric ρ_c.

For the definition of the rank of a Riemannian symmetric manifold cf. [App.II.2, Prop.4]. Before giving a proof of Thm.1 we note that the description of the classical bounded symmetric domains Ω ($\cong X_o \cong G_0/K$) $\hookrightarrow \mathbb{C}^N \hookrightarrow X_c$ and their Borel embeddings in [Ch.4, (2.2)–(2.5)] furnish explicit examples for Thm.1. In all cases we have exhibited a polydisc D which is the intersection of Ω with a Euclidean subspace of $\mathbb{C}^N$ such that the K–orbits of D cover Ω. By construction D is homogeneous under the subgroup of G_0 keeping D invariant. The fact that D is totally geodesic in Ω (and that S is totally geodesic in X_c) follows from using Euclidean coordinates as complex geodesic coordinates, as in [Ch.4, (3.2)], observing that the symmetry at the origin is given by the reflection $z \longrightarrow -z$.

We start with some preliminary discussions on root space decompositions. For the terminology and general facts regarding root systems, cf. [App I]. We use the notations on Lie algebras as in [Ch.3, (1.2)]. The dual pair (X_o,X_c) corresponds to the pair $(\mathfrak{g}_o,\mathfrak{k};\theta_o)$ and $(\mathfrak{g}_c,\mathfrak{k};\theta_c)$ of orthogonal symmetric Lie algebras with canonical decompositions $\mathfrak{g}_o = \mathfrak{k} + \mathfrak{m}$, $\mathfrak{g}_c = \mathfrak{k} + \sqrt{-1}\mathfrak{m}$. Write θ for the corresponding involution on $\mathfrak{g}^C$. Recall that we have the decomposition $\mathfrak{m}^C = \mathfrak{m}^+ + \mathfrak{m}^-$ into $(\pm\sqrt{-1})$–eigenspaces of the J–operator on $\mathfrak{m}^C$ and the identifications $T_o^{1,0}(X_o)$, $T_o^{1,0}(X_c) \cong \mathfrak{m}^+$ at the origin $o = eP$ in the Borel embedding $X_o \hookrightarrow X_c \cong G^C/P$. Write τ_o and τ_c for the conjugations on $\mathfrak{g}^C$ with respect to $\mathfrak{g}_o = \mathfrak{k} + \mathfrak{m}$ and $\mathfrak{g}_c = \mathfrak{k} + \sqrt{-1}\mathfrak{m}$ resp. We have $\tau_o(\mathfrak{m}^+) = \tau_c(\mathfrak{m}^+) = \mathfrak{m}^-$.

By [Ch.3, (3.1), Prop.3] one can define the J–operator by some central element z of $\mathfrak{k}$ (whose center $\mathfrak{z}$ is one–dimensional), i.e., $Jv = [z,v]$ on $\mathfrak{m}^C$. We have $\mathfrak{k} = \mathfrak{k}_s + \mathfrak{z}$, where $\mathfrak{k}_s := [\mathfrak{k},\mathfrak{k}]$ is the semisimple part of $\mathfrak{k}$. Fix a Cartan subalgebra $\mathfrak{h}_s \subset \mathfrak{k}_s$ and define $\mathfrak{h} = \mathfrak{h}_s + \mathfrak{z}$. $\mathfrak{h}$ is a Cartan subalgebra of $\mathfrak{g}_o$. We have therefore $\operatorname{rank}(\mathfrak{g}_o) = \operatorname{rank}(\mathfrak{k}) + 1$. Write $\mathfrak{h}^* = \operatorname{Hom}(\mathfrak{h},\mathbf{R})$ and $\mathfrak{h}^*_{\mathbf{R}} = \sqrt{-1}\mathfrak{h}^*$. $\mathfrak{h}^*_{\mathbf{R}}$ can be identified with $\mathfrak{h}_{\mathbf{R}} := \sqrt{-1}\mathfrak{h}$ by duality using the restriction of the Killing form B of $\mathfrak{g}^C$ to $\mathfrak{h}_{\mathbf{R}}$ ($B|\mathfrak{h}_{\mathbf{R}}$ is positive definite). Let $\Delta \subset \mathfrak{h}^*_{\mathbf{R}}$ denote the space of $\mathfrak{h}^C$–roots ρ of $\mathfrak{g}^C$. The root space belonging to ρ is one–dimensional over C, generated by a root vector e_ρ. Thus, $[h,e_\rho] = \rho(h)e_\rho$ for any $h \in \mathfrak{h}$ with $\rho \in \Delta$. Since the involution θ is a Lie–algebra automorphism fixing $\mathfrak{k}^C$ we have $[h,\theta(e_\rho)] = \rho(h)\theta(e_\rho)$, i.e., $\theta(e_\rho)$ is also a root vector belonging to the root ρ, so that e_ρ must be an eigenvector of θ. It follows that there is a decomposition of the roots Δ into $\Delta_K \cup \Delta_M$ of compact roots and non–compact roots, with root spaces $Ce_\rho \subset \mathfrak{k}^C$ and $\mathfrak{m}^C$ resp. Clearly, $\rho \in \Delta_K$ if and only if $\rho|\mathfrak{z} \equiv 0$. Pick $\sqrt{-1}y \in \mathfrak{h}_{\mathbf{R}} \cong \mathfrak{h}^*_{\mathbf{R}}$ such that $\sqrt{-1}y$ is in the interior of some Weyl chamber C containing $\sqrt{-1}z$. Choose C (thought of as a subset of $\mathfrak{h}^*_{\mathbf{R}}$) as the positive Weyl chamber of $\mathfrak{g}^C$–roots. Thus, the space of roots Δ decompose into the subsets Δ^+ and Δ^- of positive resp. negative roots where $\Delta^+ = \{\rho \in \Delta: \rho(-\sqrt{-1}y) > 0 \text{ for } y \in \sqrt{-1}\mathfrak{h}\}$. We have $\rho(-\sqrt{-1}z) \geq 0$ for $\rho \in \Delta^+$.

Write $\Delta_M^+ = \Delta_M \cap \Delta^+$ and call it the set of non–compact positive roots, etc. For each compact $\mathfrak{k}_{\mathfrak{s}}$–root γ with respect to $\mathfrak{h}_{\mathfrak{s}}$, we extend γ to an $\mathfrak{h}$–root of $\mathfrak{g}^C$ by defining $\gamma|\mathfrak{z} \equiv 0$. We will make no distinction between γ and its extension. The choice of $\sqrt{-1}y \in \mathfrak{h}_R$ defines a fundamental system $\{\rho_2,\ldots,\rho_s\}$, $s = \mathrm{rank}(\mathfrak{g}_o)$ of positive $\mathfrak{k}_{\mathfrak{s}}$–roots. We observe that $\gamma \in \Delta_K^+$ is indecomposable in Δ_K^+ if and only if it is indecomposable in Δ^+. In fact, if $\gamma = \alpha + \beta$ with $\alpha, \beta \in \Delta^+$, then $0 = \gamma(-\sqrt{-1}z) = \alpha(-\sqrt{-1}z) + \beta(-\sqrt{-1}z) \geq 0$ implies that $\alpha(-\sqrt{-1}z) = \beta(-\sqrt{-1}z) = 0$, i.e., that α and β are compact roots. Thus the fundamental system Σ of simple $\mathfrak{g}^C$–roots is given by $\Sigma = \{\rho_1,\ldots,\rho_s\}$ for some $\rho_1 \in \Delta_M^+$ such that $-\sqrt{-1}\rho_1(z) > 0$. We use the lexicographic ordering $\geq$ of roots with respect to the ordered fundamental system of roots $\Phi = (\rho_1,\ldots,\rho_s)$. Thus, any root $\varphi \in \Delta_M^+$ dominates any $\gamma \in \Delta_K$.

Let $\varphi \in \Delta_M$ be a non–compact root. It follows from $[h,Je_\varphi] = [h,[z,e_\varphi]] = [z,[h,e_\varphi]]$ (from Jacobi identity) $= J(\varphi(h)e_\varphi) = \varphi(h)Je_\varphi$ that e_φ is an eigenvector of J. Thus, $e_\varphi \in \mathfrak{m}^+$ or $\mathfrak{m}^-$. We have $\tau_o(\mathfrak{g}^\varphi) = \tau_c(\mathfrak{g}^\varphi) = \mathfrak{g}^{-\varphi}$. Since $[z,v] = \sqrt{-1}v$ (resp. $-\sqrt{-1}v$) on $\mathfrak{m}^+$ (resp. $\mathfrak{m}^-$) we have $\sqrt{-1}\varphi(z) > 0$ if and only if $\varphi \in \Delta_M^+$. Given any two roots ρ, σ we know that $[e_\rho,e_\sigma] = N_{\rho,\sigma}\, e_{\rho+\sigma}$, where $e_{\rho+\sigma}$ is set to be 0 if $\rho+\sigma \notin \Delta$. Let R^o and R^c denote the curvature tensor of (X_o,g_o) and (X_c,g_c) resp. From the formulas in [Ch.3, (1.3)] on computing curvatures it follows that for $\varphi, \psi \in \Delta_M^+$ and for $e_{-\varphi} = \tau_o(e_{-\varphi})$, $R^o(e_\varphi,e_{-\varphi};e_\psi,e_{-\psi}) = \text{const.}\, B_o([e_\varphi,e_{-\psi}],[e_{-\varphi},e_\psi]) = 0$ if $\varphi \neq \psi$ and $\varphi - \psi \notin \Delta$ (similarly for R^c). We define

DEFINITION 1

Two roots $\rho, \sigma \in \Delta$ are said to be strongly orthogonal, written $\rho \perp\!\!\!\perp \sigma$, if and only if $\rho \pm \sigma \notin \Delta$.

By [App.(I.3), Prop.3] $\rho \perp\!\!\!\perp \sigma$ implies that they are orthogonal in the usual sense. In case ρ, σ are positive non–compact roots from $[e_\rho,e_\sigma] = N_{\rho,\sigma}e_{\rho+\sigma}$

(with $N_{\rho,\sigma} \neq 0$ if $\rho + \sigma$ is a root) and $[\mathfrak{m}^+,\mathfrak{m}^+] = 0$ we conclude that $\rho + \sigma$ cannot be a root, so that $\rho \amalg \sigma$ if and only if $\rho - \sigma$ (and hence $\sigma - \rho$) is not a root. The key point of the proof of Thm.1 is the following

PROPOSITION 1 (HARISH–CHANDRA [HA])
There exist $r = \operatorname{rank}(X_0,g_0)$ linearly independent positive non–compact roots ψ_1, ...,ψ_r (with respect to to the Cartan subalgebra $\mathfrak{h}$) such that $\psi_i \amalg \psi_j$ for $1 \leq i < j \leq r$.

Proof:

Let $\psi_1 \in \Delta_M^+$ be the highest root in Δ. If the set $S(\psi_1) := \{\varphi \in \Delta_M^+ : \varphi \amalg \psi_1\}$ is non–empty choose a highest vector ψ_2 in $S(\psi_1)$, etc. until we arrive at a set $\Psi = \{\psi_1,...,\psi_s\}$ of non–compact positive roots such that $S(\Psi) := \{\varphi \in \Delta_M^+ : \varphi \amalg \psi_i$ for all i, $1 \leq i \leq s\}$ is empty. We claim that $s = r$ $(= \operatorname{rank}(X_0,g_0))$. For $1 \leq i \leq s$ we write e_i for e_{ψ_i}, x_i for x_{ψ_i}, etc. Consider the R–vector subspace $\mathfrak{a} \subset \mathfrak{m}$ generated by $\{x_i: 1 \leq i \leq s\}$. From the definition of Ψ it is clear that $[\mathfrak{a},\mathfrak{a}] = 0$, i.e., $\mathfrak{a} \subset \mathfrak{m}$ is an abelian subalgebra of $\mathfrak{g}_0$. We claim that $\mathfrak{a}$ is in fact a maximal abelian subalgebra, i.e., $\dim_{\mathbf{R}}\mathfrak{a} = \operatorname{rank}(X_0,g_0) = r$. Suppose otherwise. Then, there exists some $v \in \mathfrak{m}$ orthogonal to $\mathfrak{a}$ such that $\mathfrak{a} + \mathbf{R}v$ is abelian. We have a direct sum deompositions $\mathfrak{m}^+ = \Sigma_{\varphi\in\Delta_M^+} \mathbf{C}e_\varphi$, $\mathfrak{m} = \Sigma_{\varphi\in\Delta_M^+} (\mathbf{R}x_\varphi + \mathbf{R}y_\varphi)$. We can thus write $v = \Sigma_{\varphi\in S} (\alpha_\varphi x_\varphi + \beta_\varphi y_\varphi)$, $(\alpha_\varphi,\beta_\varphi)$ real and $\neq 0$ for some subset $S \subset \Delta_M^+ - \Psi$. We can equivalently write $v = \Sigma_{\varphi\in S} (\gamma_\varphi e_\varphi + \overline{\gamma_\varphi} e_{-\varphi})$, γ_φ complex and non–zero. We assert that for any $\varphi \in S$, we have $\varphi \amalg \Psi$, contradicting the maximality of Ψ. Since $\mathfrak{a} + \mathbf{R}v$ is abelian, for the highest root $\psi_1 \in \Psi$ we have in particular

$$0 = [x_1,v] = [e_1+e_{-1}, \Sigma_{\varphi\in S} (\gamma_\varphi e_\varphi + \overline{\gamma_\varphi} e_{-\varphi})] \tag{1}$$

To make use of (1) we observe first of all

(*) $\psi_1 - \varphi$ and $\varphi' - \psi_1$ cannot be identical.

In fact, if $\psi_1 - \varphi$ and $\varphi' - \psi_1$ were identical we would have $2\psi_1 = \varphi + \varphi'$ contradicting the fact that ψ_1 is the highest root. It now follows from (1) and (*) that in fact $\Sigma_{\varphi\in S} \overline{\gamma_\varphi}[e_1,e_{-\varphi}] = 0$ and hence $[e_1,e_{-\varphi}] = 0$ for all $\varphi \in S$ since γ_φ

$\neq 0$, proving that $\varphi \,\|\, \psi_1$ for all $\varphi \in S$. From the choice of ψ_2 we have $\psi_2 \geq \varphi$ for all $\varphi \in S$. The proof of Prop.1 can now be completed by induction. ▌

Regarding totally geodesic submanifolds on a symmetric manifold we have

LEMMA 1

Let (M,h) be a Riemannian symmetric manifold and $N \subset M$ be a closed submanifold. Then, N is totally geodesic in M if and only if for any $x \in N$ and for the symmetry σ_x of (M,h) at x, we have $\sigma_x(N) = N$. In particular, if $N = Ex$ is the orbit of a connected subgroup E of $\mathrm{Aut}_0(M,h) = F$ and $\mathfrak{e} = \mathrm{Lie}(E)$ is invariant under the involution of θ of $\mathfrak{f} = \mathrm{Lie}(F)$ defining σ_x, $N = Ex$ is totally geodesic in (M,h).

Proof:

Suppose $N \subset M$ is totally geodesic. As (M,h) is complete, for any $x \in N$ and any $v \in T_x(N)$, $\exp_x(v) \in N$. Since any $y \in N$ can be written $y = \exp_x(v)$ for some $v \in T_x(N)$, and $\sigma_x(\exp(v)) = \exp_x(-v) \in N$, we see that N is invariant under the symmetries σ_x for $x \in N$. Conversely suppose N is invariant under such symmetries. Write S for the second fundamental form of N in X. We have, for $u, v \in T_x(N)$, $S(\sigma u, \sigma v) = \sigma(S(u,v))$ giving $S \equiv -S$, i.e., $S \equiv 0$, meaning that N is totally geodesic in (M,h). Consider now the special case $N = Ex$. Write $\nu\colon F \longrightarrow F$ for the involution on F such that $\theta = d\nu$. $\theta(\mathfrak{e}) = \mathfrak{e}$ implies that E is invariant under ν. On the other hand writing L for the isotropy subgroup at x, the involution σ_x is given by $\sigma_x(fL) = \nu(f)L$, so that necessarily N is invariant under σ_x. As E acts as a group of isometries of (M,h) the same applies to any point x' on N, implying therefore that N is totally geodesic in (M,h), as desired. ▌

We proceed now to give a proof of Thm.1. For $\varphi \in \Delta_M^+$ choose a non-zero root vector $e_\varphi \in \mathfrak{g}^\varphi$ and define $e_{-\varphi}$ to be $\tau_0(e_\varphi)$. We define $x_\varphi = e_\varphi + e_{-\varphi}$, $y_\varphi = \sqrt{-1}(e_\varphi - e_{-\varphi}) \in \mathfrak{m}$. We have $J(x_\varphi) = y_\varphi$, $J(y_\varphi) = -x_\varphi$ and $T_0^{\mathbb{R}}(X_0) \cong \mathfrak{m} = \Sigma \mathbb{R} x_\varphi + \Sigma \mathbb{R} y_\varphi$. For the compact dual $X_c = G_c/K$ in terms of the conjugation τ_c defined by $\mathfrak{g}_c$ we have $\tau_c(e_\varphi) = -e_{-\varphi}$. We define $x_{\varphi,c} = \sqrt{-1}(e_\varphi + e_{-\varphi})$, $y_{\varphi,c} = -(e_\varphi - e_{-\varphi}) \in \sqrt{-1}\mathfrak{m}$. We have $J(x_{\varphi,c}) = y_{\varphi,c}$ and $J(y_{\varphi,c}) = -x_{\varphi,c}$ and

$T_o^{\mathbf{R}}(X_c) \cong \sqrt{-1}\mathfrak{m} = \Sigma \mathbf{R}x_{\varphi,c} + \Sigma \mathbf{R}y_{\varphi,c}$. For any positive non-compact root φ define $h_\varphi = \sqrt{-1}[e_\varphi, e_{-\varphi}] \neq 0$. h_φ is real and lies in $\mathfrak{h} \subset \mathfrak{k}$.

Proof of Theorem 1:

Consider the vector subspaces $\mathfrak{g}^{\mathbf{C}}(\varphi) \subset \mathfrak{g}^{\mathbf{C}}$, $\mathfrak{g}_o(\varphi) \subset \mathfrak{g}_o$, $\mathfrak{g}_c(\varphi) \subset \mathfrak{g}_c$ defined by

(a) $\mathfrak{g}^{\mathbf{C}}(\varphi) = \mathbf{C}e_\varphi + \mathbf{C}e_{-\varphi} + \mathbf{C}h_\varphi \subset \mathfrak{g}^{\mathbf{C}}$,

(b) $\mathfrak{g}_o(\varphi) = \mathbf{R}x_\varphi + \mathbf{R}y_\varphi + \mathbf{R}h_\varphi = \mathfrak{g}^{\mathbf{C}}(\varphi) \cap \mathfrak{g}_o$,

(c) $\mathfrak{g}_c(\varphi) = \mathbf{R}x_{\varphi,c} + \mathbf{R}y_{\varphi,c} + \mathbf{R}h_\varphi = \mathfrak{g}^{\mathbf{C}}(\varphi) \cap \mathfrak{g}_c$.

We have $[h_\varphi, e_\varphi] = \varphi(h_\varphi)e_\varphi$, $[h_\varphi, e_{-\varphi}] = -\varphi(h_\varphi)e_{-\varphi} \neq 0$. In addition, $[h_\varphi, h_\varphi] = [e_\varphi, e_\varphi] = [e_{-\varphi}, e_{-\varphi}] = 0$. Clearly $\mathfrak{g}^{\mathbf{C}}(\varphi) \subset \mathfrak{g}^{\mathbf{C}}$ is a complex Lie subalgebra isomorphic to $\mathfrak{sl}(2,\mathbf{C})$. Moreover $\mathfrak{g}_o(\varphi) = \mathfrak{g}^{\mathbf{C}}(\varphi) \cap \mathfrak{g}_o$ and $\mathfrak{g}_c(\varphi) = \mathfrak{g}^{\mathbf{C}}(\varphi) \cap \mathfrak{g}_c$ are the non-compact resp. compact real forms of $\mathfrak{g}^{\mathbf{C}}(\varphi)$. The same construction works for a set Φ of strongly orthogonal positive non-compact roots (of cardinality $|\Phi|$). One defines $\mathfrak{g}^{\mathbf{C}}(\Phi) = \Sigma_{\varphi\in\Phi} (\mathbf{C}e_\varphi + \mathbf{C}e_{-\varphi} + \mathbf{C}h_\varphi) \subset \mathfrak{g}^{\mathbf{C}}$. We have $\mathfrak{g}^{\mathbf{C}}(\Phi) \cong (\mathfrak{sl}(2,\mathbf{C}))^{|\Phi|}$. It suffices to observe that for $\varphi, \psi \in \Phi$ with $\varphi \neq \psi$, $[e_{\pm\varphi}, e_{\pm\psi}] = 0$ since $\varphi \amalg \psi$; that $[h_\varphi, h_\psi] = 0$ since $\mathfrak{h}$ is abelian; and that moreover $[h_\varphi, e_{\pm\psi}] = \sqrt{-1}[[e_\varphi, e_{-\varphi}], e_{\pm\psi}] = -\sqrt{-1}[e_{-\varphi}, [e_\varphi, e_{\pm\psi}]] = 0$. We define $\mathfrak{g}_o(\Phi) = \mathfrak{g}_o \cap \mathfrak{g}^{\mathbf{C}}(\Phi)$ resp. $\mathfrak{g}_c(\Phi) = \mathfrak{g}_c \cap \mathfrak{g}^{\mathbf{C}}(\Phi)$, which are the non-compact and compact real forms of $\mathfrak{g}^{\mathbf{C}}(\Phi)$. Define also $\mathfrak{k}(\Phi) = \Sigma_{\varphi\in\Phi} \mathbf{R}h_\varphi$ and $\mathfrak{p}(\Phi) = \Sigma_{\varphi\in\Phi} (\mathbf{C}e_{-\varphi} + \mathbf{C}h_\varphi) \subset \mathfrak{g}^{\mathbf{C}}(\Phi)$. Write $G^{\mathbf{C}}(\Phi) \subset G^{\mathbf{C}}$ for the connected complex subgroup corresponding to $\mathfrak{g}^{\mathbf{C}}(\Phi)$, etc. We have the Borel embedding ([Ch.3, (3.1)])

$$\mu\colon \Delta^{|\Phi|} \cong G_0(\Phi)/K(\Phi) \hookrightarrow G^{\mathbf{C}}(\Phi)/P(\Phi) \cong G_c(\Phi)/K(\Phi) \cong (\mathbf{P}^1)^{|\Phi|}.$$

As $\mathfrak{p}(\Phi) = \mathfrak{p} \cap \mathfrak{g}^{\mathbf{C}}(\Phi)$, $P(\Phi)$ is the identity component of $P \cap G^{\mathbf{C}}(\Phi)$, so that we have a holomorphic immersion $\nu\colon (\mathbf{P}^1)^{|\Phi|} \cong G^{\mathbf{C}}(\Phi)/P(\Phi) \longrightarrow G^{\mathbf{C}}/P \cong X_c$ with image $S(\Phi) \subset X_c$, which is a compact complex submanifold. Since $S(\Phi) = G_c(\Phi)(o)$ and $\mathfrak{g}_c(\Phi)$ is invariant under θ_c, $S(\Phi)$ is totally geodesic in (X_c, g_c) by Lemma 1. The same applies to $D(\Phi) = G_0(\Phi)(o)$. Write $\mathfrak{a}_o = \Sigma \mathbf{R}x_\varphi$ and $\mathfrak{a}_c = \Sigma \mathbf{R}x_{\varphi,c}$.

$\mathfrak{a}_o$ and $\mathfrak{a}_c$ are maximal abelian spaces in $\mathfrak{m} = \mathfrak{m}_o$ and $\sqrt{-1}\mathfrak{m} = \mathfrak{m}_c$ resp. Let $A_o \subset G_0$ and $A_c \subset G_c$ be the corresponding abelian subgroups. Write $S = S(\Psi)$ and $D = D(\Psi)$. (Recall that $|\Psi| = \text{rank}(X_o, \mathfrak{g}_o)$. Since $KA_oK = G_0$ and $KA_cK = G_c$ (cf. [App.II, (1.2), Prop.4]) we have clearly $\cup_{\gamma \in K} \gamma D = X_o$ and $\cup_{\gamma \in K} \gamma S = X_c$ resp. The rest of the statements in Thm.1 are obvious. ▌

§2 The Harish–Chandra Embedding Theorem

(2.1) We are now ready to formulate the Harish–Chandra Embedding Theorem proved in HARISH–CHANDRA [HA]. The presentation here follows mostly WOLF [WOL2, Part I] (cf. also SATAKE [SA2, Ch.II, §4, p.56ff.]). We have

THEOREM 1 (HARISH–CHANDRA EMBEDDING THEOREM [HEL])
The holomorphic map $F\colon M^+ \times K^{\mathbb{C}} \times M^- \to G^{\mathbb{C}}$ defined by $F(m^+,k,m^-) = m^+km^-$ is a biholomorphism of $M^+ \times K^{\mathbb{C}} \times M^-$ onto a dense open subset of $G^{\mathbb{C}}$ containing G_0. In particular, the map

$$\eta\colon \mathfrak{m}^+ \to G^{\mathbb{C}}/P \cong X_c \quad \text{given by} \quad \eta(\mathrm{m}^+) = \exp(\mathrm{m}^+)P$$

is a biholomorphism onto a dense open subset of X_c containing $G_0/K \cong X_o$. Furthermore, $\eta^{-1}(X_o)$ is a bounded domain on $\mathfrak{m}^+ \cong \mathbb{C}^N$.

Recall that $\mathfrak{p} = \mathfrak{k}^{\mathbb{C}} + \mathfrak{m}^- \subset \mathfrak{g}^{\mathbb{C}}$ and that the corresponding parabolic subgroup $P = K^{\mathbb{C}}M^-$. To start with we prove

PROPOSITION 1 (cf. HELGASON [HEL Ch.VIII, Lemma (7.9), p.388])
We have $M^+ \cap P = \{1\}$. Moreover the holomorphic map F is a biholomorphism of $M^+ \times K^{\mathbb{C}} \times M^-$ onto an open subset of $G^{\mathbb{C}}$.

Proof:
To prove $M^+ \cap P = \{1\}$ recall that $[\mathfrak{m}^+,\mathfrak{m}^+] = [\mathfrak{m}^-,\mathfrak{m}^-] = 0$, $[\mathfrak{m}^+,\mathfrak{k}^{\mathbb{C}}] \subset \mathfrak{m}^+$, $[\mathfrak{m}^+,\mathfrak{m}^-] \subset \mathfrak{k}^{\mathbb{C}}$. We have hence $[\mathrm{m}^+,[\mathrm{m}^+,\mathfrak{k}^{\mathbb{C}}]] = 0$, $[\mathrm{m}^+,[\mathrm{m}^+,[\mathrm{m}^+,\mathrm{m}^-]]] = 0$. In particular, $ad(\mathrm{m}^+)$ acts on $\mathfrak{g}^{\mathbb{C}}$ as nilpotent transformations such that $ad^3(\mathrm{m}^+) \equiv 0$. Suppose $\mathrm{m}^+ \in \mathfrak{m}^+$ is such that $m^+ = \exp(\mathrm{m}^+) \in M^+ \cap P$. Given $\mathrm{m}^- \in \mathfrak{m}^-$ we have $ad(m^+)(\mathrm{m}^-) \in \mathfrak{m}^-$ since $m^+ \in P = K^{\mathbb{C}}M^-$ and $[\mathfrak{k}^{\mathbb{C}},\mathfrak{m}^-] = \mathfrak{m}^-$. On the

other hand

$$ad(m^+)(m^-) = \exp(ad(m^+))(m^-) = \sum_{\nu\geq 0} \frac{1}{\nu!}(ad(m^+))^\nu(m^-)$$
$$= m^- + [m^+,m^-] + \frac{1}{2}[m^+,[m^+,m^-]] \in \mathfrak{m}^- + (\mathfrak{k}^C + \mathfrak{m}^+) \qquad (1)$$

showing that $ad(m^+)(m^-) = m^-$. As this is true for any $m^- \in \mathfrak{m}^-$ it follows from (1) that $[m^+,m^-] = 0$, a contradiction unless $m^+ = 0$ (otherwise $\mathfrak{a}_0 + R(Re\, m^+) \subset \mathfrak{m}_0$ would be an abelian subspace of dimension bigger than the rank of X_0), i.e., $m^+ = 1$ and $M^+ \cap P = \{1\}$. We proceed to show that $F: M^+ \times K^C \times M^- \rightarrow G^C$ is injective. If $m_1^+k_1m_1^- = m_2^+k_2m_2^-$ we have $(m_2^+)^{-1}m_1^+ = (k_2m_2^-)(k_1m_1^-)^{-1} \in P$. From $M^+ \cap P = \{1\}$ we conclude that $m_1^+ = m_2^+$. Similarly $m_1^- = m_2^-$, and so $k_1 = k_2$, proving the injectivity of F. Finally, as F is equi–dimensional and injective it is a biholomorphism onto its image (cf. e.g. FISCHER [FI]). Alternatively one can show that F is a local biholomorphism by computing the differential of F. Identifying tangent vectors with left–invariant vector fields, denoted by [...], we have $dF(m^+,k,m^-)\,[u,v,w] = [ad(k)ad(m^-)u + ad(m^-)v + w]$, showing that $dF(m^+,k,m^-)\,[\mathfrak{g}^C] = [ad(k)\mathfrak{m}^+ + \mathfrak{k}^C + \mathfrak{m}^-] = [\mathfrak{g}^C]$, so that F is a local biholomorphism.

To prove the Harish–Chandra Embedding Theorem (Thm.1) we are going to write down explicitly the embedding in the special case of $\mathfrak{g}^C = \mathfrak{sl}(2,C)$ and apply the Polydisc and Polysphere Theorems [(1.1), Thm.1] to the general case.

<u>Proof of Thm.1</u>:
We use the notations in the proof of [(1.1), Thm.1]. Let $\psi \in \Psi$ be arbitrary. We consider the case $G^C = G^C[\psi]$. We have $\mathfrak{g}^C = \mathfrak{sl}(2,C)$. In this case we can take $\mathfrak{k} = \mathfrak{h} = \mathfrak{s}(\mathfrak{u}(1)+\mathfrak{u}(1))$, $\mathfrak{g}_0 = \mathfrak{sl}(2,R)$, $\mathfrak{g}_c = \mathfrak{su}(2)$. We have thus $K = \{k_\theta\colon \theta \in R\}$ for $k_\theta = \begin{bmatrix} e^{i\theta} & 0 \\ 0 & e^{-i\theta}\end{bmatrix}$. We define

$$e_\psi = \begin{bmatrix} 0 & 1 \\ 0 & 0\end{bmatrix},\ e_{-\psi} = \begin{bmatrix} 0 & 0 \\ 1 & 0\end{bmatrix},\ H_\psi = \begin{bmatrix} 1 & 0 \\ 0 & -1\end{bmatrix} \in \sqrt{-1}\mathfrak{h}, \text{ so that}$$

$$x_\psi = \begin{bmatrix} 0 & 1 \\ 1 & 0\end{bmatrix},\ x_{\psi,c} = \begin{bmatrix} 0 & i \\ i & 0\end{bmatrix}.$$

We have $A_o = \exp(\mathbf{R}x_\psi)$, $A_c = \exp(\mathbf{R}x_{\psi,c})$ and

$$\exp(tx_\psi) = \begin{bmatrix}\cosh t & \sinh t\\ \sinh t & \cosh t\end{bmatrix} = \begin{bmatrix}1 & \tanh t\\ 0 & 1\end{bmatrix}\begin{bmatrix}(\cosh t)^{-1} & 0\\ 0 & \cosh t\end{bmatrix}\begin{bmatrix}1 & 0\\ \tanh t & 1\end{bmatrix}$$

$$= \exp([\tanh t]e_\psi)\ \exp([-\log\cosh t]H_\psi)\ \exp([\tanh t]e_{-\psi}) \in M^+K^{\mathbf{C}}M^-. \quad (1)$$

For $\cos t \neq 0$ we have

$$\exp(tx_{\psi,c}) = \begin{bmatrix}\cosh it & \sinh it\\ \sinh it & \cosh it\end{bmatrix} = \begin{bmatrix}\cos t & i\sin t\\ i\sin t & \cos t\end{bmatrix}$$

$$= \begin{bmatrix}1 & i\tan t\\ 0 & 1\end{bmatrix}\begin{bmatrix}(\cos t)^{-1} & 0\\ 0 & \cos t\end{bmatrix}\begin{bmatrix}1 & 0\\ i\tan t & 1\end{bmatrix}$$

$$= \exp([i\tan t]e_\psi)\ \exp([-\log\cos t]H_\psi)\ \exp([i\tan t]e_{-\psi}) \in M^+K^{\mathbf{C}}M^-. \quad (2)$$

As $KA_cK = G_c$ and $ad(k_\theta)e_\psi = e^{2i\theta}e_\psi$; we have $k_\theta^{-1}\exp(ze_\psi)k_\theta = \exp(e^{2i\theta}ze_\psi)$ for $z \in \mathbf{C}$. By (2) it follows that

$$X_c = \{k_\theta \exp(\mathbf{R}x_{\psi,c})\ mod\ P:\ \theta \in \mathbf{R}\},$$

$$\eta(m^+) = \exp(\mathbf{C}e_\psi)\ mod\ P = \left\{\begin{bmatrix}1 & z\\ 0 & 1\end{bmatrix} mod\ P : z \in \mathbf{C}\right\}$$

$$= \left\{\begin{bmatrix}1 & ie^{i\theta}\tan t\\ 0 & 1\end{bmatrix} mod\ P:\ \theta, t \in \mathbf{R};\ \cos t \neq 0\right\}$$

$$= \{k_\theta \exp(tx_{\psi,c})\ mod\ P : \theta, t \in \mathbf{R};\ \cos t \neq 0\} \subset X_c$$

as a dense open subset. On the other hand from $G_o = KAK$ (cf. [App.II.2, Prop.3]) and by (1) we have

$$X_o = G_o\ mod\ P = \{k_\theta \exp(tx_\psi)\ mod\ P : t \in \mathbf{R}\}$$

$$= \{k_\theta \exp([\tanh t]e_\psi)\ mod\ P : t \in R\}$$

$$= \left\{\begin{bmatrix}1 & z\\ 0 & 1\end{bmatrix} mod\ P : z \in \mathbf{C},\ |z| < 1\right\},$$

exhibiting

$$\eta^{-1}(X_o) \cong \Delta = \{z \in \mathbf{C},\ |z| < 1\} \subset \mathbf{C} \cong m^+,$$

which is the standard realization of $X_o = SL(2,\mathbb{R})/S^1$ as a bounded symmetric domain. This proves Thm.1 for the case of $\mathfrak{g}^{\mathbb{C}} = \mathfrak{sl}(2,\mathbb{C})$. To complete the proof of Thm.1 it suffices to use the obvious generalization to the case of $\mathfrak{g}^{\mathbb{C}} = (\mathfrak{sl}(2,\mathbb{C}))^r$, $r = \mathrm{rank}(X_o,g)$ and apply the Polydisc and Polysphere Theorems [(1.1), Thm.1]. Then K acts as unitary transformations on $\mathfrak{m}^+ \cong \mathbb{C}^N$ and $\eta^{-1}(X_o) = \cup_{\gamma\in K}\gamma(\Delta^r) \subset\subset \mathfrak{m}^+ \cong \mathbb{C}^N$. Moreover, by (1) in the case of $G^{\mathbb{C}}[\Psi]$ we have $G_0 = KA_oK = KG_o[\Psi]K \subset KM^+K^{\mathbb{C}}M^-K = M^+K^{\mathbb{C}}M^- \subset G^{\mathbb{C}}$ since $K^{\mathbb{C}}$ normalizes M^+ and M^-. The rest of the statements in Thm.1 are either obvious or consequences of Prop.1. ▌

REMARKS

The map $\mathfrak{m} \longrightarrow \Omega$ defined by $m \in \mathfrak{m} \longrightarrow \exp(m)(o)$ is the exponential map $\exp_o$ of (X_o,g_o) at o. In particular, in case of the Poincaré disc (Δ,d_P) the formula $|z| = \tanh t$ gives the formula $d_P(0,z) = \log \dfrac{1 + |z|}{1 - |z|}$ for geodesic distances.

We call $\eta^{-1}(X_o) \subset\subset \mathbb{C}^N$ the Harish–Chandra realization of X_o. The Euclidean coordinates on $\mathbb{C}^N$ will be called Harish–Chandra coordinates. In case of classical bounded symmetric domains $\Omega \cong X_o$, the standard realizations as given in Chap.4 are in fact up to scaling and a choice of orthogonal basis of $\mathfrak{m}^+$ the Harish–Chandra realizations of X_o, as can be readily verified from the description of Ω given there (where we exhibited a distinguished polydisc through the origin explicitly). It can be readily checked that these domains Ω are convex. It turns out that the same is true for the exceptional case. Denote by B the Killing form on $\mathfrak{g}^{\mathbb{C}}$ and consider the Hermitian inner product on $\mathfrak{g}^{\mathbb{C}}$ given by $(u,v) = B(u,\tau_c(v))$. As $B|\mathfrak{g}_c$ is (real and) negative definite, $(.,.)$ is positive definite on $\mathfrak{g}^{\mathbb{C}}$. Define a norm $\|.\|$ on $ad(\mathfrak{g})$ by setting

$$\|ad(u)\| = \sup\{|ad(u)(v)| : v \in \mathfrak{g}^{\mathbb{C}} \text{ and } (v,v) = 1\}$$

Then, we have,

THEOREM 2 (HERMANN CONVEXITY THEOREM, cf. WOLF [WOL2, (I.4), p.286])
Let $\Omega \subset\subset \mathfrak{m}^+ \cong \mathbb{C}^N$ be the Harish–Chandra realization of $X_0 \cong G_0/K$. For any $\xi \in \mathfrak{m}^+$ denote by $Re(\xi) \in \mathfrak{m} = \mathfrak{m}_0$ the real part of ξ with respect to the non–compact real form $\mathfrak{g}_0$ of $\mathfrak{g}^{\mathbb{C}}$. Then,

$$\Omega = \{\xi \in \mathfrak{m}^+ : \|ad(Re(\xi))\| < 1\}.$$

In particular, Ω is convex in $\mathfrak{m}^+$.

The proof of the Hermann Convexity Theorem relies on some very explicit information on the root systems of $\mathfrak{g}^{\mathbb{C}}$, as given by the Restricted Root Theorem of HARISH–CHANDRA [HA] and MOORE [MOO], which we state in the Appendix [App.(III.1) Thm.1]. The same theorem implies that for a distinguished polydisc $D = G_0[\Psi](o)$ and for $G_0'[\Psi]$ denoting the restriction to D of the automorphisms of X_0 fixing D, $G_0'[\Psi]$ acts as the full group of automorphisms of $(D, g_0|_D)$, i.e., $G_0'[\Psi]$ permutes the direct factors arbitrarily (cf. [App.III.1, Cor.1]), a fact useful in the study of higher characteristic bundles of X_0, to be defined and studied in [App.III.4] and used in [App.IV.2] in the formulation of the Dual Generalized Frankel Conjecture.

The fact that the explicit bounded convex realizations of classical domains agree with the Harish–Chandra realizations is not a coincidence. Very recently, we proved

THEOREM 3 (MOK–TSAI [MT])
Let X_0 be a complex N–dimensional irreducible Hermitian symmetric manifold of non–compact type and of rank ≥ 2. Suppose $F: X_0 \to \Omega \subset\subset \mathbb{C}^N$ is a biholomorphism of X_0 onto a bounded convex domain Ω. Then, up to a complex affine transformation on $\mathbb{C}^N$, F is the Harish–Chandra realization of X_0.

The results of [MT] also cover the case of unbounded convex realizations. We proved that up to complex affine transformations such realizations are always obtained from the Harish–Chandra realizations by (partial) Cayley transforms (cf. KORANYI & WOLF [KW]).

CHAPTER 6 THE HERMITIAN METRIC RIGIDITY THEOREM FOR COMPACT QUOTIENTS

§1 The Characteristic Bundle $\mathcal{S}$

(1.1) Let Ω be an irreducible bounded symmetric domain of rank ≥ 2, $X = \Omega/\Gamma$ a compact quotient by a torsion–free discrete group of automorphisms Γ, and g be the canonical metric on X. In this chapter we study Hermitian metrics of seminegative curvature on X. By [Ch.3, (1.3), Prop.2], (X,g) is of seminegative bisectional curvature. We recall the first metric rigidity theorem as given in [Ch.1, (2.1), Thm.1].

THEOREM 1 (MOK [MOK3, 1987])

Let X be a compact quotient of an irreducible bounded symmetric domain of rank ≥ 2. Let h be a Hermitian metric on X of seminegative curvature in the sense of Griffiths. Then, h is necessarily a constant multiple of the canonical metric g.

Let $(L,\hat{h})$ be the tautological line bundle over the projectivized holomorphic tangent bundle $\mathbb{P}T(X)$ associated to (X,h). By [Ch.2, (4.2), Prop.1], the condition that h is of seminegative curvature (in the sense of Griffiths) is equivalent to the condition that the line bundle $(L,\hat{h})$ is of seminegative curvature. From the local irreducibility of X it suffices for the proof of Thm.1 to show that h is parallel as a tensor on (X,g). By [Ch.2, (3.2), Prop.2], given h one can deform the canonical metric g to get $g(t) = g + th$, $t \geq 0$, which is also of seminegative curvature. Heuristically, the idea of the proof of Thm.1 is to show that zeros of holomorphic bisectional curvature of (X,g) are stable under deformation. If this can be done by considering the isometric diagonal embedding δ: $(X,g(1)) \hookrightarrow (X,g) \times (X,h)$ one would obtain partial vanishing of the second fundamental form σ of δ by [Ch.2, (3.2)]. One could hope that the partial vanishing is enough to imply the vanishing of σ and hence that h is parallel on (X,g). In practice it will only be necessary to consider the triple (g,h,g+h) of Hermitian metrics on X and a subset of the zeros of holomorphic bisectional curvatures of (X,g). By the explicit formula relating $c_1(L,\hat{g})$ to the curvature R of (X,g) [Ch.2, (4.2), Prop.1] given $\mu \in T_x(X)$, the nonpositive (1,1)–form $c_1(L,\hat{g})$ has a zero eigenvalue at $[\mu] \in \mathbb{P}T(X)$ if and only if the $R_{\mu\bar{\mu}\zeta\bar{\zeta}} = 0$ for some non–zero vector $\zeta \in T^{1,0}(X)$

(which we identify with $T(X)$), i.e., the null space $\mathcal{N}_\mu$ of the seminegative Hermitian bilinear form $H_\mu(\xi,\eta) := R_{\mu\bar{\mu}\xi\bar{\eta}}$ is positive–dimensional. We will call a vector $\zeta \in \mathcal{N}_\mu$ a null–vector associated to μ. For a generic choice of μ, H_μ is actually negative definite. Since the first Chern class $[c_1(L,\theta)] \in H^2(\mathbf{P}T(X),\mathbf{R})$ is independent of the choice of the Hermitian metric θ one can obtain integral formulas of the form $I(g) = I(h)$ where I is some integral involving powers of $c_1(L)$. Such an integral formula can only be useful if the integrand is a semidefinite top form and $I(g) = 0$ for the canonical metric g. However, as $c_1(L,\hat{g})$ is negative definite for a generic choice of μ one has to work with some submanifold S instead. In order to use the properties $c_1(L,\hat{g})$, $c_1(L,\hat{h}) \leq 0$, $\mathcal{S}$ has to be complex–analytic. We construct such a submanifold S by exploiting the structure of some of the zeros of bisectional curvature of Ω.

The Polydisc Theorem [Ch.5, (1.1), Thm.1] provides some information about the structure of the zeros of holomorphic bisectional curvatures of (X,g). Write $X = \Omega/\Gamma$, where Ω is an irreducible bounded symmetric domain of rank ≥ 2 and Γ is a torsion–free discrete subgroup of $\mathrm{Aut}(\Omega)$. Equip Ω with the Bergman metric g_0 and write $\Omega = G_0/K$ for the standard presentation of Ω as a homogeneous manifold. By the Polydisc Theorem there exists a distinguished totally geodesic Poincaré polydisc D of dimension $\mathrm{rank}(\Omega) \geq 2$ such that $\Omega = \cup_{\gamma\in K}\gamma D$. Since D is totally geodesic for $x \in D$ and for $\xi, \eta \in T_x(X)$, we have $R_{\xi\bar{\xi}\eta\bar{\eta}} = R^D_{\xi\bar{\xi}\eta\bar{\eta}}$, where R^D denote the curvature tensor of the Kähler submanifold $(D,g_0|_D)$. In particular, since $\mathrm{rank}(\Omega) \geq 2$ it follows that (Ω,g_0) is not of strictly negative holomorphic bisectional curvature (at any point).

(1.2) We are going to construct on Ω a complex submanifold $\mathcal{S}(\Omega)$ of $\mathbf{P}T(\Omega)$ homogeneous under the natural action of G_0. For $X = \Omega/\Gamma$ one can then take the quotient $\mathcal{S} = \mathcal{S}(X) = \mathcal{S}(\Omega)/\Gamma$. Recall that in [Ch.5, (1.1)] we constructed the distinguished polydisc D by constructing a maximal strongly orthogonal set Ψ of positive non–compact roots starting with a highest (non–compact) root of $\psi = \psi_1$. Let $\alpha = e_\psi$ be a corresponding non–zero root vector. It follows from the Polydisc

Theorem that $\mathcal{N}_\alpha \neq \{0\}$. Roughly speaking, we are going to define $\mathcal{S}$ to consist of such $[\alpha]$.

Let $K_s \subset K$ be the connected subgroup cooresponding to $\mathfrak{k}_s = [\mathfrak{k},\mathfrak{k}]$. Then, K_s is semisimple. Consider the restriction of the (irreducible) isotropy K–representation ρ_0 on $T_0^{1,0}(\Omega) \cong T_0(\Omega)$ to the semisimple part K_s. Any Cartan subalgebra $\mathfrak{h}$ of $\mathfrak{k}$ is of the form $\mathfrak{z} + \mathfrak{h}_s$, where $\mathfrak{h}_s$ is a Cartan subalgebra of $\mathfrak{k}_s$. In the way that we defined a lexicographic ordering in [Ch.5, (1.1)], it is easy to see that $\psi \in \Delta_M^+$ is the highest (non–compact) root if and only if $\psi|\mathfrak{h}_s$ is the dominant root of $\rho_0|K_s$ (since $\varphi(c) = \sqrt{-1}$ for any $\varphi \in \Delta_M^+$ in the notations there). Clearly, e_ψ is also a root vector of $\rho_0|K_s$ belonging to $\psi|\mathfrak{h}_s$. We will call e_ψ a dominant root vector of the representation ρ_0. The same convention will apply to finite–dimensional irreducible complex representations of K in general. We define

DEFINITION 1

At $x \in \Omega$ let $K_x \cong K$ denote the isotropy subgroup at x of G_0. Denote by ρ_x: $K_x \to GL(T_x;\mathbb{C})$ the isotropy complex representation at x on $T_x(\Omega)$. We call a non–zero vector $\alpha \in T_x(\Omega)$ a characteristic vector at x if and only if α is a dominant weight vector with respect to some choice of Cartan subalgebra $\mathfrak{h}_x$ of the Lie algebra $\mathfrak{k}_x$. The submanifold $\mathcal{S}_x = \{[\alpha]$: α is a characteristic vector at x$\} \subset \mathbb{P}T_x(\Omega)$ is called the characteristic variety at x. $\mathcal{S} := \bigcup_{x\in\Omega} \mathcal{S}_x \subset \mathbb{P}T(\Omega)$ is called the characteristic bundle over Ω.

We prove

PROPOSITION 1

For any $x \in \Omega$, the characteristic variety $\mathcal{S}_x$ at x is a connected complex submanifold of $\mathbb{P}T_x(\Omega)$.

EXAMPLES

Consider classical bounded symmetric domains. We use the information given in [Ch.4, (2.2) & App.III.3]. In case of $D^I_{p,q}$, $\mathcal{S}_0$ is given by $\{[X]: X \in M(p,q;\mathbb{C})$ is a matrix of rank 1$\}$ which is the $S(U(p)\times U(q))$–orbit of the $[E_{11}]$, where E_{rs}

denotes the p by q matrix defined by $(E_{rs})_{ij} = \delta_{ri}\delta_{sj}$. As K acts transitively on $\mathcal{S}_0$, the latter is a compact manifold. On the other hand $\mathcal{S}_0$ is invariant under the action on $(P,Q) \in GL(p,\mathbb{C})\times GL(q,\mathbb{C})$ given by $Z \to PZQ^{-1}$, hence $\mathcal{S}_0$ must be a complex manifold. The description in the cases of D_n^{II} and D_n^{III} are very similar. In case of D_n^{IV}, $\mathcal{S}_0$ is $SO(n)\times SO(2)$–orbit of the point $[1,i,0,\dots,0]$, which is defined by the homogeneous equation $\Sigma z_i^2 = 0$, i.e., $\mathcal{S}_0 \cong Q^{n-2} \subset \mathbb{P}^{n-1}$ is a complex submanifold. For more precise information cf. App.III.3.

<u>Proof of Prop.1</u>:
Since G_0 acts as a group of holomorphic isometries on (Ω,g_0) it suffices to consider $x = o = eK$. In this case identifying $T_o(\Omega) \cong T_o^{1,0}(\Omega) \cong \mathfrak{m}^+$ the isotropy action ρ_o: $K \to GL(\mathfrak{m}^+,\mathbb{C})$ is given by adjoint action. Write $\mathfrak{k} = \mathfrak{z} + \mathfrak{k}_s$ for the decomposition of $\mathfrak{k}$ into the one–dimensional center $\mathfrak{z}$ and the semisimple part $\mathfrak{k}_s = [\mathfrak{k},\mathfrak{k}]$. Write $\mathfrak{h} = \mathfrak{z} + \mathfrak{h}_s$ for a Cartan subalgebra of $\mathfrak{k}$. Since ρ_o is a complex representation there is an induced action of K on $\mathbb{P}(\mathfrak{m}^+)$. Define first of all $\mathcal{S}_0'$ to be K–orbit of some $[\alpha]$ in $\mathbb{P}(\mathfrak{m}^+)$ for some characteristic vector α at o. In particular, $\mathcal{S}_0' \subset \mathbb{P}(\mathfrak{m}^+)$ is a smooth submanifold. Since there are at most a finite number of conjugacy classes of Cartan subalgebras of K_s under inner automorphisms (by the adjoint group of $\mathfrak{k}_s$) it is clear that $\mathcal{S}_0'$ is a connected component of $\mathcal{S}_0$. To prove Prop.1 we first show that $\mathcal{S}_0'$ is a complex manifold. To this end it suffices to show that $T_{[\alpha]}^{\mathbb{R}}(\mathcal{S}_0')$ of $\mathcal{S}_0' = \cup_{k\in K}[ad(k)\alpha]$ at $[\alpha]$ is J–invariant. Define f: $K \to T_o(\Omega) \cong \mathfrak{m}^+$ by $f(k) = ad(k)\alpha$. The differential at the identity $e \in K$ is given by $df(e)(k) = [k,\alpha]$ for $k \in K$. Identify $T_{[\alpha]}^{\mathbb{R}}(\mathbb{P}(\mathfrak{m}^+))$ as usual with the complex vector space $\mathfrak{m}^+/\mathbb{C}\alpha$. Under this identification the J–operator is given by multiplication by $\sqrt{-1}$. We have

$$T_{[\alpha]}^{\mathbb{R}}(\mathcal{S}_0') = (df(e)(\mathfrak{k}) + \mathbb{C}\alpha) / \mathbb{C}\alpha = ([\mathfrak{k},\alpha] + \mathbb{C}\alpha) / \mathbb{C}\alpha \qquad (1)$$

Write $\alpha = e_\psi$ for a dominant root ψ corresponding to some choice of Cartan subalgebra $\mathfrak{h} \subset \mathfrak{k}$. We use the same notation for roots as in [Ch.5, (1.1)]. Then, for any $\varphi \in \Delta_M^+$ we have $[e_\psi,e_{\pm\varphi}] = 0$. Define $\Delta_K^+ = \Delta_K \cap \Delta^+$ and write $\mathfrak{k} =$

$\Sigma_{\gamma\in\Delta_K^+} \mathbf{R}(e_\gamma + e_{-\gamma}) + \Sigma_{\gamma\in\Delta_K^+} \sqrt{-1}\mathbf{R}(e_\gamma - e_{-\gamma})$. We have $[\alpha, e_{\pm\gamma}] = [e_\psi, e_{\pm\gamma}] = N_{\pm\gamma,\psi}\, e_{\psi\pm\gamma}$ if $e_{\psi\pm\gamma}$ is a root and $[\alpha, e_{\pm\gamma}] = 0$ otherwise. Since ψ is the highest root and γ is a positive root $\psi + \gamma$ cannot be a root. Define $\Phi = \{\varphi \in \Delta_M^+ : \psi - \varphi = \gamma \in \Delta_K^+\}$. Then, in the notations of [App.I.3, Prop.3], for $\varphi \in \Phi$

$$[e_\gamma + e_{-\gamma}, e_\psi] = N_{-\gamma,\psi}\, e_\varphi,$$
$$[\sqrt{-1}\,(e_\gamma - e_{-\gamma}), e_\psi] = -\sqrt{-1}\, N_{-\gamma,\psi}\, e_\varphi,$$

so that by (1)

$$T^{\mathbf{R}}_{[\alpha]}(\mathcal{S}'_0) = (df(e)(\mathfrak{k}) + \mathbf{C}\alpha) / \mathbf{C}\alpha = \mathbf{C}\alpha + \Sigma_{\varphi\in\Phi} (\mathbf{R}e_\varphi + \sqrt{-1}\,\mathbf{R}e_\varphi) / \mathbf{C}\alpha$$

is invariant under multiplication by $\sqrt{-1}$, i.e., $T^{\mathbf{R}}_{[\alpha]}(\mathcal{S}'_0)$ is J–invariant, so that S_0 is complex–analytic. To prove Prop.1 it remains to show that $\mathcal{S}_0$ is connected. The complex Lie group $K^{\mathbf{C}} \subset G^{\mathbf{C}}$ corresponding to $\mathfrak{k}^{\mathbf{C}}$ acts on $\mathbf{P}(\mathfrak{m}^+)$ and hence on the complex submanifold $\mathcal{S}'_0$. All complex Cartan subalgebras of $\mathfrak{k}^{\mathbf{C}}_{\mathfrak{s}} = [\mathfrak{k}^{\mathbf{C}}, \mathfrak{k}^{\mathbf{C}}]$ are conjugate under inner automorphisms of $\mathfrak{k}^{\mathbf{C}}_{\mathfrak{s}}$. On the other hand, fixing a Cartan subalgebra $\mathfrak{h}_{\mathfrak{s}} \subset \mathfrak{k}_{\mathfrak{s}}$ and a choice of fundamental system Φ of positive roots, there is one and only one dominant weight ψ of ρ_0. Furthermore, all such choices of Φ are permuted by the Weyl group, which arise from restrictions of inner automorphisms of $\mathfrak{k}^{\mathbf{C}}_{\mathfrak{s}}$ fixing $\mathfrak{h}_{\mathfrak{s}}$ (cf. [AppI.3, Prop.6]). Thus, $\mathcal{S}_0$ is homogeneous under the action of $K^{\mathbf{C}}$, so that $\mathcal{S}_0 = \mathcal{S}'_0$, proving Prop.1. ▌

REMARKS

Prop.1 is a special case of (part of) the Borel–Weil Theorem (cf. [App.(I.6), Thm.1]).

As a consequence of Prop.1 we prove

PROPOSITION 2

The characteristic bundle $\mathcal{S}(\Omega) \to \Omega$ is holomorphic. Moreover, in terms of the Harish–Chandra embedding $\Omega \hookrightarrow \mathbf{C}^N$, $\mathcal{S}(\Omega)$ is parallel on Ω in the Euclidean sense, i.e., identifying $\mathbf{PT}(\Omega)$ with $\Omega \times \mathbf{P}^{N-1}$ using the Euclidean coordinates we have the identification $\mathcal{S}(\Omega) \cong \Omega \times \mathcal{S}_0$.

EXAMPLES

Consider the case of $\Omega = D^I_{p,q}$. Let $\Phi \in G_0$ be given by $\Phi(Z) = (AZ + B)(CZ + D)^{-1}$. At the origin, we have $d\Phi(0)(X) = (A - BD^{-1}C)XD^{-1}$. As Φ is an automorphism we have $d\Phi(0)(X) = PXQ$ for some $P, Q \in GL(p,C), GL(q,C)$ resp. Recall that $\mathcal{S}_0 = \{[X] : X \in M(p,q;C) \text{ is of rank } 1\}$. Obviously $d\Phi(0)$ preserves the rank of the matrix. Since $\Phi \in G_0$ is arbitrary, this implies readily $\mathcal{S}(\Omega) \cong \Omega \times \mathcal{S}_0$. The same argument applies to domains of types II and III.

Proof of Prop.2:

We use the Harish–Chandra and Borel embeddings $\Omega \hookrightarrow C^N \hookrightarrow G^{\mathbf{C}}/P \cong X_c$. Fix some $[\alpha] \in \mathcal{S}_0$. To prove that $\mathcal{S}(\Omega)$ is complex–analytic it suffices to show that $\mathcal{S}(\Omega)$ is the part of the $G^{\mathbf{C}}$–orbit of $[\alpha]$ sitting over Ω. Let $\mathcal{S}^c$ be the G_c–orbit of $[\alpha]$ in $\mathbf{P}T(X_c)$. Write $\mathcal{S}^c_x$ for the part of $\mathcal{S}^c$ sitting over x. Clearly at the origin $\mathcal{S}_0 = \mathcal{S}^c_0 = K[\alpha]$. We claim that $\mathcal{S}^c$ is invariant under $G^{\mathbf{C}}$. Since $G^{\mathbf{C}}/P \cong X_c$ and G_c acts transitively on X_c we have $G^{\mathbf{C}} = G_cP$. To establish the claim it suffices to show that $\mathcal{S}^c_0$ is invariant under the action of P. Recall that $P = K^{\mathbf{C}}M^-$. We already know from Prop.1 that $\mathcal{S}^c_0 = \mathcal{S}_0$ is $K^{\mathbf{C}}$–invariant. Let $f: P \to \mathbf{P}(\mathfrak{m}^+)$ defined by $f(\tau) = [\tau(\alpha)]$. We have the identification $T^{\mathbf{R}}_{[\alpha]}(\mathbf{P}(\mathfrak{m}^+)) \cong \mathfrak{m}^+/C\alpha$. The differential at the identity $e \in P$ is given by

$$df(e)(p) = [p,\alpha] \bmod (\mathfrak{p} + C\alpha) \in \mathfrak{g}^{\mathbf{C}}/(\mathfrak{p} + C\alpha) \cong \mathfrak{m}^+/C\alpha$$

for $p \in \mathfrak{p}$. To show that $\mathcal{S}^c_0$ is P–invariant it suffices to show that $df(e)(\mathfrak{p}) = df(e)(\mathfrak{k})$. As $df(e)(\mathfrak{k}^{\mathbf{C}}) = df(e)(\mathfrak{k})$ by the proof of Prop.1 it remains to show that $df(e)(\mathfrak{m}^-) = 0$. But this follows obviously from $[\mathfrak{m}^-,\mathfrak{m}^+] \subset \mathfrak{k}^{\mathbf{C}} \subset \mathfrak{p}$. We have proved that $\mathcal{S}^c$ is invariant under $G^{\mathbf{C}}$ so that $\mathcal{S}^c = G^{\mathbf{C}}[\alpha]$. We have $\mathcal{S} = G_0[\alpha] \subset G^{\mathbf{C}}[\alpha] = \mathcal{S}^c$. On the other hand we have $\mathcal{S}_0 = \mathcal{S}^c_0$, so that given $x \in \Omega$, there exists $\gamma \in G_0$ such that $\mathcal{S}_x = \gamma\mathcal{S}_0 = \gamma\mathcal{S}^c_0 = \mathcal{S}^c_x$ (as $\gamma \in G^{\mathbf{C}}$). Denoting by $\pi: \mathbf{P}T(X_c) \to X_c$ the base projection. We have shown that $\mathcal{S} = \mathcal{S}^c \cap \pi^{-1}(\Omega)$. To complete the proof of Prop.2 it suffices to observe that $\mathcal{S}^c$ is invariant under the subgroup $M^+ =$

$\exp(\mathfrak{m}^+)$, which acts as translations on $\mathbb{C}^N$ from the construction of the Harish–Chandra embedding, so that $\mathcal{S}^c \cap \pi^{-1}(\mathbb{C}^N) \cong \mathcal{S}^c_0 \times \mathbb{C}^N = \mathcal{S}_0 \times \mathbb{C}^N$ in terms of the isomorphism $\mathbb{P}T(\mathbb{C}^N) \cong \mathbb{C}^N \times \mathbb{P}^{N-1}$ using the Euclidean coordinates. The proof of Prop.2 is completed. ∎

We illustrate the key argument again by the example of $\Omega = D^I_{p,q}$. The group P is represented by matrices of the form $\begin{bmatrix} A & 0 \\ C & D \end{bmatrix}$, the group M^- by matrices $\begin{bmatrix} I & 0 \\ C & I \end{bmatrix}$. Then for $\Phi \in P$ we have $d\Phi(0)(X) = AXD^{-1}$ (preserving the rank of X), while for $\Phi \in M^-$, $d\Phi(0)(X) = X$, fixing X, so that for $X \neq 0$ and $f(p) = ad(p)\,[X]$, $f|_{M^-}$ is constant. In particular $df(e)|_{\mathfrak{m}^-} \equiv 0$.

There are other equivalent ways of characterizing $\mathcal{S}$ in algebraic terms. We will prove

PROPOSITION 3

The following conditions on a unit vector $\alpha \in T_x(\Omega)$ are equivalent.

(a) α is a characteristic vector ,

(b) α realizes the algebraic minimum of holomorphic sectional curvatures

(c) $\dim_{\mathbb{C}} \mathcal{N}_\alpha$ is maximum among non–zero vectors $\mu \in T_x(\Omega)$.

Here as in [Ch.5, (1.1)] $\mathcal{N}_\mu$ denotes the zero eigenspace of the Hermitian bilinear form $H_\mu(\xi,\eta) = R_{\mu\bar{\mu}\xi\bar{\eta}}$. The number $n(\Omega) := \dim_{\mathbb{C}} \mathcal{N}_\alpha$ in (c) will be called the null dimension of Ω. If $X = \Omega/\Gamma$ we write $n(X) = n(\Omega)$ and call it the null dimension of X. The proof of Prop.3, which will not be used in this chapter, will be given in [App.III.1, Prop.1]. For the purpose of a precise integral formula we will need

PROPOSITION 4

Write $n(X)$ for the null dimension of X and $\mathcal{S}$ for the characteristic bundle on X. Then, for $n = \dim_{\mathbb{C}}(X)$ we have

$$\dim_{\mathbb{C}}(\mathcal{S}) = 2n - 1 - n(X)$$

Proof:
We use the notations in the proof of Prop.1. Since $\pi: \mathcal{S} \to X$ realizes $\mathcal{S}$ as a bundle over X with fibers biholomorphic to $\mathcal{S}_0$ it suffices to show that $\dim_{\mathbf{C}}(\mathcal{S}_0) = n - 1 - n(X)$. From the proof of Prop.1 it follows that $\dim_{\mathbf{C}}(\mathcal{S}) = \mathrm{Card}(\Phi)$, where $\Phi = \{\varphi \in \Delta_M^+ : \psi - \varphi = \gamma \in \Delta_K^+\}$. (Recall that ψ is a dominant root.) We have $\Delta_M^+ = \{\psi\} \cup \Phi \cup N$, where N consists of non–compact positive roots ν such that $\psi - \nu$ is not a root. Clearly, $\dim_{\mathbf{C}}(\mathcal{S}_0) = n - 1 - \mathrm{Card}(N)$. On the other hand, for any a non–compact positive root ρ with unit root vector e_ρ we have $R(e_\psi, \overline{e_\psi}; e_\rho, \overline{e_\rho}) = N^2_{\psi,-\rho} \neq 0$ (resp. $= 0$) if $\rho = \varphi \in \Phi$ (resp. $\rho = \nu \in N$), while $R(e_\psi, \overline{e_\psi}; e_\rho, \overline{e_{\rho'}}) = 0$ for $\rho \neq \rho'$, by the curvature formula given in [Ch.3, (1.3), Prop.1] and by [App.I.3, Prop.3]. It follows that $\mathrm{Card}(N) = \dim_{\mathbf{C}} \mathcal{N}_\alpha = n(X)$ for $\alpha = e_\psi$, yielding $\dim_{\mathbf{C}}(\mathcal{S}_0) = n - 1 - n(X)$, as desired. ∎

§2 An Integral Formula on $\mathcal{S}$ and an Algebraic Deduction of the Hermitian Metric Rigidity Theorem for Compact Quotients

(2.1) We recall here the Hermitian metric rigidity theorem on compact quotients of bounded symmetric domains which we already stated in [Ch.1, Thm.1]

THEOREM 1
Let X be a compact quotient of an irreducible bounded symmetric domain of rank ≥ 2. Let h be a Hermitian metric of seminegative curvature in the sense of GRIFFITHS [GRI1]. Then, h is necessarily a constant multiple of the canonical metric g.

The basic technique of the proof is an integral formula on the characteristic bundle $\mathcal{S} = \mathcal{S}(X)$. Recall that (X,g) is of seminegative bisectional curvature, so that if $(L,\hat{g})$ denotes the Hermitian tautological line bundle on $\mathbb{P}T_X$ associated to (T_X,g), we have $c_1(L,\hat{g}) \leq 0$ as a (1,1)–form on $\mathbb{P}T_X$. The first idea is to show that the zeros of bisectional curvature are rigid in the sense that they are stable under the deformation $\{g + th: 0 \leq t \leq 1\}$ within the space of Hermitian metrics of seminegative curvature on X. Let $\pi: \mathbb{P}T_X \to X$ be the canonical base projection and ω be the Kähler form of (X,g). Then, we have on $\mathbb{P}T_X$ two semipositive closed (1,1)–forms $-c_1(L,\hat{g})$ and $\pi^*\omega$. Define $\nu = -c_1(L,\hat{g}) + \pi^*\omega$ on $\mathbb{P}T_X$. We have

LEMMA 1

$(\mathbb{P}T_X,\nu)$ is a Kähler manifold.

Proof:

It suffices to show that the closed (1,1)–form ν is positive. Suppose $\eta \in T_{[\xi]}(\mathbb{P}T_X)$ is in the kernel of ν. Then, $0 = \nu(\sqrt{-1}\,\eta \wedge \overline{\eta}) \geq \omega\,(\sqrt{-1}\,\pi_*\eta \wedge \overline{\pi_*\eta}) \geq 0$ so that $\pi_*\eta = 0$. But then η is tangent to some vertical fiber $\mathbb{P}T_x(X)$, so that $-c_1(L,\hat{g})\,(\sqrt{-1}\,\eta \wedge \overline{\eta}) > 0$ unless $\eta = 0$, proving the lemma. ▌

Let $S \subset \mathbb{P}T_X$ be a complex–analytic submanifold of dimension s. Then, for any two nonegative integers a and b with $a + b = s$, $\mu_{a,b}(g) := (-c_1(L,\hat{g}))^a \wedge \nu^b$ is a semipositive top–degree form on S. Defining $\mu_{a,b}(h)$ by replacing $\hat{g}$ by $\hat{h}$, we obtain cohomologous semipositive forms $\mu_{a,b}(g)$ and $\mu_{a,b}(h)$ whose integrals on S are therefore equal. Such an integral formula would be of use if in fact $\mu_{a,b}(g) \equiv 0$ on S, forcing therefore by $\int_S \mu_{a,b}(h) = \int_S \mu_{a,b}(g) = 0$ the identity $\mu_{a,b}(h) \equiv 0$. If we choose $S = \mathbb{P}T_X$ the difficulty is that $-c_1(L,\hat{g})[\xi] > 0$ at a generic point $[\xi]$ of $\mathbb{P}T_X$. To find an S with a useful integral formula we would like to have the properties: (i) that S is a complex submanifold, as said; (ii) that $-c_1(L,\hat{g})$ is not strictly positive at any point of S, i.e, that to any $[\xi] \in S$ there are associated zeros of bisectional curvature; and (iii) that at least some of the zero–eigenvectors of $-c_1(L,\hat{g})$ are tangential to the submanifold S. Using the description of the characteristic bundle $\mathcal{S} = \mathcal{S}(X)$ as given in [Ch.5, (1.2), Prop.2] we are going to show that $\mathcal{S}$ satisfies the requirements. We start with

PROPOSITION 1

Let (X,g) be a compact quotient of an irreducible bounded symmetric domain Ω

of rank ≥ 2. Let ω be the Kähler form of (X,g), $\pi\colon \mathbb{P}T_X \to X$ the canonical base projection, and h be any Hermitian metric on T_X. Write $\nu = -c_1(L,\hat{g}) + \pi^*\omega$. ($\nu$ is a Kähler form by Lemma 1). Let $\hat{h}$ denote the induced Hermitian metric on the tautological line bundle $L \to \mathbb{P}T_X$ and $c_1(L,\hat{h})$ be the first Chern form of L defined by $\hat{h}$. Write $q = n(X)$ for the null dimension of X. Then, on the characteristic bundle $\mathcal{S} := \mathcal{S}(X)$ over X we have

$$\int_{\mathcal{S}} (-c_1(L,\hat{h}))^{2n-2q} \wedge \nu^{q-1}$$
$$= \int_{\mathcal{S}} (-c_1(L,\hat{g}))^{2n-2q} \wedge \nu^{q-1} = 0$$

In particular, if (T_X,h) is of seminegative curvature, we have

$$c_1(L,\hat{h})^{2n-2q} \equiv 0 \quad \text{on } \mathcal{S}.$$

Proof:

Write $\mu(h) := (-c_1(L,\hat{h}))^{2n-2q} \wedge \nu^{q-1}$. To prove the integral formula it suffices to show that $\mu(g) = 0$. Since ν is a Kähler form it is equivalent to show

$$(*) \quad c_1(L,\hat{g})^{2n-2q} \equiv 0 \quad \text{on } \mathcal{S}.$$

Moreover, given the integral formula, it follows from $\mu(h) \geq 0$ that in fact $\mu(h) \equiv 0$, and hence $c_1(L,\hat{h})^{2n-2q} \equiv 0$ on $\mathcal{S}$ again because ν is a Kähler form. To show (*) we lift the differential forms to Ω and retain the same symbols for the lifted entities. We use the Harish–Chandra embedding $\Omega \subset\subset \mathbb{C}^n$ as usual. By the homogeneity of μ under the action of $G_o = \mathrm{Aut}_o(\Omega)$, it suffices to verify the identity (*) at the origin. Recall that the Euclidean coordinates on $\mathbb{C}^n$ serve as complex geodesic coordinates at the origin (for a suitable normalization of the canonical metric g). Let $[\alpha] \in \mathcal{S}_o$, α a unit vector with respect to g, be arbitrary. By a unitary change of Euclidean coordinates we may assume that $\alpha = \partial/\partial z_1$. Write a tangent vector of type (1,0) as $v = \Sigma\, v_i(\partial/\partial z_i)$ on Ω. Since the Euclidean coordinates are complex geodesic at o, the holomorphic fiber coordinates $(v_i)_{1\leq i\leq n}$ are special fiber coordinates for T_X at o. Denoting by $w_k = v_{k+1}/v_1$, $1 \leq i \leq n-1$, the fiber coordinates on $\mathbb{P}T_X$ at points $[v]$ with $v_1 \neq 0$, we have by [Ch.2, (4.2), Prop.1] the following formula for the first Chern form $c_1(L,\hat{g})$

$$2\pi\, c_1(L,\hat{g})([\alpha])$$

$$= -\sqrt{-1}\sum_{1\le k\le n-1} dw^k \wedge d\overline{w}^k \;+\; \sqrt{-1}\sum_{1\le i,j\le n} R_{1\bar{1}i\bar{j}}(g)\, dz^i \wedge d\overline{z}^j, \qquad (1)$$

where $R(g)$ denotes the curvature tensor of (X,g). From (1) it follows that a tangent vector $\tau \in T_{[\alpha]}\mathcal{S}(\Omega)$ is a zero eigenvector of $c_1(L,g)([\alpha])$ if and only if (i) τ is "horizontal" in the sense that for the trivialization $\mathbb{P}T_X \cong \Omega \times \mathbb{P}(\mathfrak{m}^+)$ given by the Euclidean coordinates we have $\tau = (\zeta,0)$ for a vector $\zeta \in T_0(\Omega) \cong \mathfrak{m}^+$ and (ii) $\zeta \in \mathcal{N}_\alpha$. Consequently the zero–eigenspace of $c_1(L,\hat{g})([\alpha])$ is tangential to the characteristic bundle $\mathcal{S}(\Omega)$. It follows that $c_1(L,\hat{g})([\alpha])|_{\mathcal{S}}$ is negative semidefinite of rank $\dim_{\mathbb{C}}\mathcal{S} - q = (2n-q-1) - q = 2n - 2q - 1$. As $[\alpha] \in \mathcal{S}$ is arbitrary we conclude that $(c_1(L,\hat{g}))^{2n-2q}|_{\mathcal{S}} \equiv 0$, as asserted in (*). In particular, we have $\mu(g) \equiv 0$ on $\mathcal{S}$, proving Prop.1. ▌

Since $(T_X,g+h)$ is also of seminegative curvature Prop.1 also applies to $g + h$. From this we will deduce that for any $[\alpha] \in \mathcal{S}$ and any $\zeta \in \mathcal{N}_\alpha$, (α,ζ) remains a zero of curvature with respect to any Hermitian metric h of seminegative curvature. We are going to exploit this further by considering the triple of Hermitian metrics $(g,h,g+h)$ and deduce as a consequence of Prop.1 the following partial vanishing theorem for ∇h.

PROPOSITION 2

In the notations of Prop.1 let h be a Hermitian metric on X such that (T_X,h) is of seminegative curvature. Regard h also as a convariant (1,1)–tensor on X. Denote by ∇ the connection on (X,g). Then, for any $[\alpha] \in \mathcal{S}_x$, $\zeta \in \mathcal{N}_\alpha$ and $\eta \in T_x(X)$ we have

$$\nabla_\zeta h_{\alpha\bar{\eta}} = 0.$$

Proof:

By Prop.1 we have the identity $\mu(h) := (-c_1(L,\hat{h}))^{2n-2q} \wedge \nu^{q-1} \equiv 0$. Since $(T_X,g+h)$ is also of seminegative curvature, we also have $\mu(g+h) \equiv 0$. For $[\alpha] \in \mathcal{S}$ denote by $\mathcal{N}(g)([\alpha])$ the zero–eigenspace of $c_1(L,\hat{g})[\alpha]$, etc. We have by [Ch.2,

(3.2), Prop.2] $\mathcal{N}(g+h)([\alpha]) \subset \mathcal{N}(g)([\alpha])$. We claim that in fact

(*) $\mathcal{N}(g+h)([\alpha]) = \mathcal{N}(g)([\alpha])$.

Suppose otherwise. Then, $\dim_{\mathbb{C}} \mathcal{N}(g+h)([\alpha]) < q = n(X)$ at some point $[\alpha] \in \mathcal{S}$, so that the seminegative (1,1)–form $c_1(L,\hat{g}+\hat{h})|_{\mathcal{S}}$ is of rank $\geq 2n - q - 1 - (q-1) = 2n - 2q$, contradicting with Prop.1. The statement (*) says considerably more, as the lifting of ζ as an elements of $\mathcal{N}_\alpha(g)$ and $\mathcal{N}_\alpha(g+h)$ are the same. This says in fact something about the second fundamental form of the diagonal embedding δ: $(X,g+h) \hookrightarrow (X,g) \times (X,h)$, which we are going to exploit. From the curvature–decreasing property of Hermitian subbundles we have

$$0 = R_{\alpha\bar{\alpha}\zeta\bar{\zeta}}(g+h) \leq R_{\alpha\bar{\alpha}\zeta\bar{\zeta}}(g) + R_{\alpha\bar{\alpha}\zeta\bar{\zeta}}(h) \leq 0, \qquad (1)$$

so that in particular

$$R_{\alpha\bar{\alpha}\zeta\bar{\zeta}}(h) = 0. \qquad (2)$$

We now make the second fundamental form σ of δ more precise. Let $x \in X$ be arbitrary and (z_i) be local holomorphic coordinates at x which are complex geodesic with respect to the canonical metric. Let $[\alpha] \in \mathcal{S}_x$, $\zeta \in \mathcal{N}_\alpha$ be arbitrary. For simplicity we write $R_{\alpha\bar{\alpha}\zeta\bar{\zeta}}(h)$ for $R_{\alpha\bar{\alpha}\zeta\bar{\zeta}}(h)(x)$, etc. We have

$$R_{\alpha\bar{\alpha}\zeta\bar{\zeta}}(g) = -\frac{\partial^2 g_{\alpha\bar{\alpha}}}{\partial z_\zeta\,\partial \bar{z}_\zeta},$$

$$R_{\alpha\bar{\alpha}\zeta\bar{\zeta}}(h) = -\frac{\partial^2 h_{\alpha\bar{\alpha}}}{\partial z_\zeta\,\partial \bar{z}_\zeta} + \sum_{\mu,\nu} h^{\mu\bar{\nu}} \frac{\partial h_{\alpha\bar{\nu}}}{\partial z_\zeta}\frac{\partial h_{\mu\bar{\alpha}}}{\partial \bar{z}_\zeta},$$

$$R_{\alpha\bar{\alpha}\zeta\bar{\zeta}}(g+h) = -\left[\frac{\partial^2 g_{\alpha\bar{\alpha}}}{\partial z_\zeta\,\partial \bar{z}_\zeta} + \frac{\partial^2 h_{\alpha\bar{\alpha}}}{\partial z_\zeta\,\partial \bar{z}_\zeta}\right] + \sum_{\mu,\nu} (g+h)^{\mu\bar{\nu}} \frac{\partial h_{\alpha\bar{\nu}}}{\partial z_\zeta}\frac{\partial h_{\mu\bar{\alpha}}}{\partial \bar{z}_\zeta}. \qquad (3)$$

Comparing the three expressions we have

$$R_{\alpha\bar{\alpha}\zeta\bar{\zeta}}(g+h) = R_{\alpha\bar{\alpha}\zeta\bar{\zeta}}(g) + R_{\alpha\bar{\alpha}\zeta\bar{\zeta}}(h) - S_{\alpha\bar{\alpha}\zeta\bar{\zeta}}, \quad \text{where}$$

$$S_{\alpha\bar{\alpha}\zeta\zeta} = \sum_{\mu,\nu} p^{\mu\bar{\nu}} \frac{\partial h_{\alpha\bar{\nu}}}{\partial z_\zeta} \frac{\partial h_{\mu\bar{\alpha}}}{\partial \bar{z}_\zeta}, \quad \text{with}$$

$$p^{\mu\bar{\nu}} = h^{\mu\bar{\nu}} - (g+h)^{\mu\bar{\nu}}. \tag{4}$$

From (2) and (4) we conclude that

$$S_{\alpha\bar{\alpha}\zeta\zeta} = \sum_{\mu,\nu} p^{\mu\bar{\nu}} \frac{\partial h_{\alpha\bar{\nu}}}{\partial z_\zeta} \frac{\partial h_{\mu\bar{\alpha}}}{\partial \bar{z}_\zeta} = 0. \tag{5}$$

By diagonalizing $(h_{\mu\bar{\nu}})$ with respect to $(g_{\mu\bar{\nu}})$ it is clear that $(p^{\mu\bar{\nu}})$ is a positive definite Hermitian matrix. Since (z_i) are complex geodesic coordinates at x the Riemann–Christoffel symbols for g vanish at the point x. It follows therefore that for any $\eta \in T_x(X)$

$$\nabla_\zeta h_{\alpha\bar{\eta}} = \frac{\partial h_{\alpha\bar{\eta}}}{\partial z_\zeta} = 0, \tag{6}$$

proving the proposition. ∎

We are now ready to finish the proof of the Hermitian metric rigidity theorem [(6.1), Thm.1]. Since (X,g) is locally irreducible, to prove that $h = cg$ it suffices to show that h is parallel as a covariant (1,1)–tensor on g. The problem is to show that by varying $[\alpha]$ and $\zeta \in \mathcal{N}_\alpha$ it will follow from Prop.1 that in fact $\nabla h \equiv 0$. In other words we need to establish

PROPOSITION 3

Let $W \subset m^+ \otimes m^+$ be the complex vector subspace given by

$$W = \sum_{[\alpha]\in\mathcal{S}_0,\, \zeta\in\mathcal{N}_\alpha} \mathbb{C}\,(\alpha \otimes \zeta).$$

Then, $W = m^+ \otimes m^+$.

<u>Proof</u>:

For a complex vector space V let $Gr(V,q)$ denote the Grassmannian of complex q–dimensional vector subspaces in V. Consider the mapping $\nu\colon \mathcal{S}_0 \to Gr(m^+, n(X))$ defined by $\nu([\alpha]) = \mathcal{N}_\alpha$. We claim that

(*) ν is anti–holomorphic.

To prove (*) we make use of the fact that (X,g) is of seminegative curvature in the dual Nakano sense. In other words, the Hermitian bilinear form Q on $\mathfrak{m}^+ \otimes \mathfrak{m}^-$ defined by $Q(\beta\otimes\bar{\gamma},\sigma\otimes\bar{\tau}) = R_{\beta\bar{\gamma}\tau\bar{\sigma}}$ and extended by Hermitian bilinearity is negative semidefinite. For any $[\alpha] \in \mathcal{S}_o$ and $\zeta \in \mathcal{N}_\alpha$, it follows from $R_{\alpha\bar{\alpha}\zeta\bar{\zeta}} = 0$ that $\alpha \otimes \bar{\zeta}$ is a zero–eigenvector of Q. Conjugation on $\mathfrak{m}^+$ gives rise to a conjugation κ: $Gr(\mathfrak{m}^+,n(X)) \longrightarrow Gr(\mathfrak{m}^-,n(X))$, which is a conjugate biholomorphism. Denote by $\bar{\nu}$: $\mathcal{S}_o \longrightarrow Gr(\mathfrak{m}^-,n(X))$ the composite map of ν and the conjugation κ. To prove (*) it is equivalent to show that $\bar{\nu}$ is holomorphic. Since the mapping λ: $\mathbb{P}(\mathfrak{m}^+) \times Gr(\mathfrak{m}^-,n(X)) \longrightarrow Gr(\mathfrak{m}^+\otimes\mathfrak{m}^-,n(X))$ defined by $\lambda([\xi],E) = \xi \otimes E$ is a holomorphic embedding it suffices to show that $\mu([\alpha]) := \lambda([\alpha],\bar{\nu}([\alpha]))$ is holomorphic. As the zero eigenspace $\mathcal{N}(Q)$ of Q is a complex linear subspace, the map μ: $\mathcal{S}_o \longrightarrow Gr(\mathfrak{m}^+\otimes\mathfrak{m}^-,n(X))$ is given by $\mu([\alpha]) = \{\alpha \otimes \bar{\zeta} \in \mathfrak{m}^+ \otimes \mathfrak{m}^-: \alpha \otimes \bar{\zeta} \in \mathcal{N}(Q)\} = (\alpha \otimes \mathfrak{m}^-) \cap \mathcal{N}(Q)$, implying clearly that μ is holomorphic as $\mathcal{N}_\alpha(Q) \subset \mathfrak{m}^+ \otimes \mathfrak{m}^-$ is a complex vector subspace, proving (*).

To deduce Prop.3 from (*) we polarize the vectors α and ζ. Fix a characteristic vector α^o at o and fix $\zeta^o \in \mathcal{N}_{\alpha^o}$. Let $\{\alpha(z)\}_{z\in\Delta}$ be a holomorphic 1–parameter family of characteristic vectors in $\mathfrak{m}^+$ such that $\alpha(o) = \alpha$. From (*) it follows that there exists a family $\{\zeta(z)\}_{z\in\Delta}$ conjugate holomorphic in z such that $\zeta(o) = \zeta^o$ and $\zeta(z) \in \mathcal{N}_{\alpha(z)}$. As $\tau(z) := \alpha(z) \otimes \beta(z) \in W$ for any $z \in \Delta$ and $W \subset \mathfrak{m}^+ \otimes \mathfrak{m}^-$ is a complex vector subspace it follows that all partial derivatives of τ at 0 are also in W. Since $\zeta(z)$ is anti–holomorphic in z we have

$$\partial_z^k\tau(0) = \partial_z^k\alpha(0) \otimes \zeta^o \in W. \tag{1}$$

As the holomorphic family $\{\alpha(z)\}_{z\in\Delta}$ is arbitrary it follows from (1) that W contains $E \otimes \zeta^o$, where $E \subset \mathfrak{m}^+$ is the subspace generated by all possible partial derivatives $\{\partial_z^k\alpha(0): k \in \mathbb{Z}, k \geq 0\}$. $E \subset \mathfrak{m}^+$ is equivalently the complex vector subspace generated by the set of all α such that $[\alpha]$ lies on some coordinate open neighborhood U of $[\alpha^o]$ in $\mathcal{S}_o$. By the identity theorem on real–analytic functions it follows that a vector ξ is orthogonal to $E_o = \Sigma_{[\alpha]\in U} \mathbb{C}\alpha$ if and only if it is orthogonal to $E = \Sigma_{[\alpha]\in \mathcal{S}_o} \mathbb{C}\alpha$. As E is invariant under K and K acts irreducibly on $\mathfrak{m}^+$ we have $E = \mathfrak{m}^+$, so that

$$\mathfrak{m}^+ \otimes \zeta^o \subset W \quad \text{for any } \zeta^o \in \mathcal{N}_{\alpha^o}. \tag{2}$$

As $[\alpha^o] \in \mathcal{S}_o$ is arbitrary by varying α^o and hence ζ^o it follows that $\{\zeta^o\}$ generate again a K–invariant complex vector subspace of $\mathfrak{m}^+$, so that in fact

$$\mathfrak{m}^+ \otimes \mathfrak{m}^+ \subset W, \tag{3}$$

as desired. The proof of Prop.2 and hence the Hermitian metric rigidity theorem [(1.1), Thm.1] is completed. ▌

§3 An Alternative Proof Using Moore's Ergodicity Theorem

(3.1) In this section we are going to give another derivation of the Hermitian metric rigidity theorem [(1.1), Thm.1] based on a slightly different integral formula on the characteristic bundle $\mathcal{S}$ and Moore's Ergodicity Theorem. This proof applies equally to the irreducible, locally reducible case, which we will formulate later on.

We start with a modified integral formula on the characteristic bundle. We use the notations of §2. Recall in particular that $\pi\colon \mathbb{P}T_X \to X$ is the base projection and that $\nu = -c_1(L,\hat{g}) + \pi^*\omega$ is a Kähler form on $\mathbb{P}T_X$. We have the following integral formula on $\mathcal{S}$, whose proof is the same as that of [(2.1), Prop.1].

PROPOSITION 1

Let (X,g) be a compact quotient of an irreducible bounded symmetric domain Ω of dimension n and of rank ≥ 2. Write $q = n(X)$ for the null dimension of X. Then, on the characteristic bundle $\mathcal{S}$ over X we have

$$\int_{S} (-c_1(L,\hat{h})) \wedge (-c_1(L,\hat{g}))^{2n-2q-1} \wedge \nu^{q-1}$$
$$= \int_{S} (-c_1(L,\hat{g}))^{2n-2q} \wedge \nu^{q-1} = 0.$$

In particular, if $(T_X,\bar{h})$ is of seminegative curvature, we have

$$(-c_1(L,\hat{h})) \wedge (-c_1(L,\hat{g}))^{2n-2q-1} \equiv 0 \quad \text{on } S.$$

Consider now on X a Hermitian metric h of seminegative curvature. Write $\hat{g} + \hat{h} = e^u \hat{g}$ on $\mathbb{P}T_X$. We have

$$c_1(L,\hat{g}+\hat{h})) = c_1(L,\hat{g}) - \frac{\sqrt{-1}}{2\pi}\,\partial\bar{\partial}u. \tag{1}$$

The identity in Prop.1 applied to the metrics g and g + h implies immdediately that

$$-\sqrt{-1}\,\partial\bar{\partial}u \wedge (-c_1(L,\hat{g}))^{2n-2q-1} \wedge \nu^{q-1} \equiv 0 \quad \text{on } S. \tag{2}$$

Multiplying (2) by u and integrating by parts over S we obtain immediately

$$-\sqrt{-1}\,\partial u \wedge \bar{\partial}u \wedge (-c_1(L,\hat{g}))^{2n-2q-1} \wedge \nu^{q-1} \equiv 0 \quad \text{on } S. \tag{3}$$

We are going to deduce from (3) that u is constant on S and that hence h = cg for some global constant c on X. To do this we consider parallel transport on the unit sphere bundle of X along geodesics.

(3.2) We start with stating Moore's Ergodicity Theorem

THEOREM 1 (cf. ZIMMER [ZIM, Thm.(2.2.6), p.19])
Let G be a simple Lie group and Γ be a lattice on G, i.e., $\Gamma\backslash G$ is of finite volume in the left invariant Haar measure. Suppose $H \subset G$ is a closed subgroup. Consider the action of H on $\Gamma\backslash G$ by multiplication on the right. Then, H acts ergodically if and only if H is non–compact.

The statement that H acts ergodically on $\Gamma\backslash G$ means that the only subsets

$E \subset \Gamma\backslash G$ invariant under the action of H is either of zero or of full measure. We return now to the situation of the bounded symmetric domain $\Omega \cong G/K$ and a compact quotient $X = \Gamma\backslash G/K$. (We write Γ on the left only in this section.) We exploit Thm.1 by considering the geodesic flow on the unit sphere bundle S over X. Let $S(\mathcal{S}) \subset S$ be the subset corresponding to unit characteristic vectors. Here $\xi \in T_X$ is identified with the real tangent vector $\sqrt{2}Re\xi$. From the invariance of $\mathcal{S}$ under G it follows that $S(\mathcal{S})$ is invariant under the induced action of G. Given any two vectors $v, w \in S$ at $x \in X$ consider the parallel transport $w_{\gamma(t)}$ of w along the unique geodesic on $\gamma = \{\gamma(t): -\infty < t < \infty\}$ on X such that γ passes through x and v is tangent to γ. Write γ^* for the curve $\{w_{\gamma(t)}: -\infty < t < \infty\}$ in S lying above γ. γ^* lies on $S(\mathcal{S})$ because $\mathcal{S}$ and hence $S(\mathcal{S})$ is invariant under parallel transport. Fix a unit characteristic vector α. Let $L \subset G$ be the subgroup fixing the point α on $S(\mathcal{S})$. Then, $S(\mathcal{S}) \cong \Gamma\backslash G/L$ as a locally homogeneous space. Without loss of generality assume that x can be lifted to the origin $o \in \Omega$ and write $\tilde{\gamma}$, $\tilde{\gamma}^*$ for the lifting of γ and γ^* to Ω and $S(\mathcal{S}(\Omega))$ (similarly defined) resp. Recall from [Ch.3, (1.1) & App.I] that $\tilde{\gamma}^*$ is the orbit of a non–compact one–parameter family H of isometries, called transvections, obtained by composing the symmetries $\sigma_{\tilde{\gamma}(t)}$ along $\tilde{\gamma}$. (Here we use the fact that the transvections H induce parallel transport along γ.) For any $\varphi \in G$, $\varphi(\tilde{\gamma}) = \varphi H(o) = \varphi H\ (e\ mod\ K)$; $\varphi(\tilde{\gamma}^*) = \varphi H(\alpha) = \varphi H\ (e\ mod\ L)$. Since G acts as isometries it follows that $\varphi(\tilde{\gamma})$ is a geodesic on (Ω,g) and $\varphi(\tilde{\gamma}^*)$ is the lifting of $\varphi(\tilde{\gamma})$ to $S(\mathcal{S}(\Omega))$. By Moore's Ergodicity Theorem it follows that $\Gamma\backslash G$ and hence $\Gamma\backslash G/L$ is ergodic under the action of the non–compact group H. To give an alternative proof of the Hermitian metric rigidity theorem [(1.1), Thm.1] we proceed to prove

(i) The function $u|_{\mathcal{S}}$ is invariant under the action of H, so that $h_{\alpha\bar{\alpha}}(x) = cg_{\alpha\bar{\alpha}}(x)$ for any $x \in X$ and for a global constant c.

(ii) If h is a Hermitian metric on X such that $h_{\alpha\bar{\alpha}}(x) = cg_{\alpha\bar{\alpha}}(x)$ for any characteristic vector α, then $h = cg$ on X.

We will deduce (i) from the integral formula [(3.1), Prop.1]. First of all we are going to prove (ii). Fix $x \in X$ and consider the Hermitian matrix $s_{i\bar{j}}(x) = h_{i\bar{j}}(x) -$

$cg_{i\bar{j}}(x)$ with respect to any orthonormal frame at x. We have $s_{\alpha\bar{\alpha}}(x) = 0$ for any characteristic vector α at x. Consequently, for any two vectors $\mu, \nu \in T_x(X)$, $s_{\mu\bar{\nu}}(x) = 0$ if $\mu \otimes \bar{\nu}$ is in the complex linear span W of $\alpha \otimes \bar{\alpha} \in m^+ \otimes m^-$. As α varies holomorphically in m^+, $\bar{\alpha}$ varies anti-holomorphically in m^-. It follows therefore from the same polarization argument as in [(2.1), proof of Prop.3] that $W = m^+ \otimes m^-$, implying therefore that $s_{\mu\bar{\nu}} = 0$ for any $\mu, \nu \in T_x(X)$. In other words, we have proved that $h = cg$ everywhere on X.

It remains now to prove (i) using the integral formula and Moore's Ergodicity Theorem. Recall that we have by (3.1)

$$-\sqrt{-1}\,\partial u \wedge \bar{\partial} u \wedge (-c_1(L,\hat{g}))^{2n-2q-1} \wedge \nu^{q-1} \equiv 0 \text{ on } \mathcal{S}. \tag{1}$$

Since ν is a Kähler form we have in fact

$$-\sqrt{-1}\,\partial u \wedge \bar{\partial} u \wedge (-c_1(L,\hat{g}))^{2n-2q-1} \equiv 0 \text{ on } \mathcal{S}. \tag{2}$$

We lift to the origin of the bounded symmetric domain Ω and assume $\alpha = \partial/\partial z_1$. Recall in the notations of [(2.1), proof of Prop.1] the formula

$$\begin{aligned} &2\pi\, c_1(L,\hat{g})([\alpha]) \\ &= -\sqrt{-1}\sum_{1\le k\le n-1} dw^k \wedge d\overline{w}^k + \sqrt{-1}\sum_{1\le i,j\le n} R_{1\bar{1}i\bar{j}}(g)\, dz^i \wedge d\bar{z}^j \end{aligned} \tag{3}$$

in terms of coordinates on $\mathbb{P}T_X$ arising from the bounded symmetric domain Ω. Identify as usual $T_{[\alpha]}(\mathcal{S}) \cong T_0(\mathcal{S}_0 \times \Omega) \cong T_0(\mathcal{S}_0) \oplus m^+$. With this identification the identity (3) implies clearly that the nonnegative (1,1)–form $\sqrt{-1}\,\partial u \wedge \bar{\partial} u$ vanishes on $\mathcal{N}_\alpha$. Since u is real this implies

$$du|\mathcal{N}_\alpha \equiv 0. \tag{4}$$

In what follows recall that we sometimes identify $\xi \in T_x(X)$ with the real tangent vector $\sqrt{2}Re\xi$. Let now $\zeta \in \mathcal{N}_\alpha$ and γ be the geodesic on X determined by ζ and γ^* be the curve on $S(\mathcal{S})$ lying above γ obtained by parallel transport of α along γ. It follows from (4) that in fact u is constant along $\tilde{\gamma}^*$. Recall that there is a one–parameter non–compact subgroup H of G such that for the liftings $\tilde{\gamma}$, $\tilde{\gamma}^*$ to Ω; $\tilde{\gamma}^*$ is the orbit under H of α. Moreover, by the same argument as in (4) for $\varphi \in G$ and $\varphi(\alpha) = \varphi \bmod L \in G/L \cong S(\mathcal{S})$, u is constant on the H–orbit $\varphi H \bmod L = \varphi(\tilde{\gamma}^*)$. In other words the function $\hat{u}$ on $\Gamma\backslash G$ induced by u is invariant under the action of H. If $\hat{u}$ is non–constant on $\Gamma\backslash G$ then for some real numbers $a < b$ the open set $\{y \in \Gamma\backslash G: a < \hat{u}(y) < b\} \subset \Gamma\backslash G$ is a non–empty open subset which is not of full measure on $\Gamma\backslash G$ (equipped with the left–invariant Haar measure), contradicting with Moore's Ergodicity Theorem [(3.1), Thm.1]. We have therefore established that u is in fact constant on $\mathcal{S}$. By definition we have $g_{\alpha\bar{\alpha}} + h_{\alpha\bar{\alpha}} = u([\alpha])g_{\alpha\bar{\alpha}}$ for any characteristic vector α on X, so that $h_{\alpha\bar{\alpha}} = cg_{\alpha\bar{\alpha}}$ for a global constant c on X, as asserted in (i). With this we have completed the alternative proof of the Hermitian metric rigidity theorem [(1.1), Thm.1]. ▌

§4 The Case of Irreducible and Locally Reducible Compact Quotients

(4.1) We are now in a position to formulate a metric rigidity theorem for the case of irreducible, locally reducible quotients. Any such X is of rank ≥ 2 since it is locally reducible. The most interesting case is that of irreducible compact quotients of the polydisc.

To start with we define the notion of irreducibility for quotients of bounded symmetric domains.

DEFINITION 1

Let $X = \Omega/\Gamma$ be a quotient of finite volume of a bounded symmetric domain Ω. We say that X is reducible if there exists a subgroup $\Gamma_0 \subset \Gamma$ of finite index, a decomposition $\Omega \cong \Omega' \times \Omega''$ into a product of bounded symmetric domains Ω' and Ω'' such that $\Gamma_0 \subset \mathrm{Aut}(\Omega') \times \mathrm{Aut}(\Omega'')$. Otherwise, X is said to be irreducible.

Geometrically, $X = \Omega/\Gamma$ is reducible if and only if some finite covering $\tilde{X}$ of X can be decomposed isometrically. Suppose Ω is reducible, $\Omega = \Omega_1 \times \ldots \times \Omega_k$, $k \geq 2$, is the decomposition of Ω into irreducible components and $X = \Omega/\Gamma$

is an irreducible quotient of finite volume. From a classical theorem of Cartan (cf. NARASIMHAN [NA3, Ch.5, p.56ff]) all automorphisms of Ω are given by automorphisms of individual irreducible factors and by permutation of isomorphic factors. Write $G = \mathrm{Aut}(\Omega)$ and $\mathrm{Aut}_o(\Omega)$ for the identity component of G. It follows that for any lattice Γ in $G = \mathrm{Aut}(\Omega)$ there is a subgroup $\Gamma_o \subset \Gamma$ of finite index such that $\Gamma_o \subset \mathrm{Aut}_o(\Omega) = \mathrm{Aut}_o(\Omega_1) \times \ldots \times \mathrm{Aut}_o(\Omega_k)$. For the formulation of the Hermitian metric rigidity theorem on X we can always pass to the finite covering Ω/Γ_o. In other words, we need to consider only the case of $\Gamma \subset \mathrm{Aut}_o(\Omega)$. For such quotients $X = \Omega/\Gamma$ we have the important

LEMMA 1 (special case of RAGHUNATHAN [RA, Cor.(5.21), p.86]
Let $I = (i(1), \ldots, i(p))$, $1 \leq i(1) < \ldots < i(p) \leq k$, be a multi-index and pr_I: $\mathrm{Aut}_o(\Omega) \longrightarrow \mathrm{Aut}_o(\Omega_{i(1)}) \times \ldots \times \mathrm{Aut}_o(\Omega_{i(p)})$ be the canonical projection. Then, $pr_I(\Gamma)$ is dense in $\mathrm{Aut}_o(\Omega_{i(1)} \times \ldots \times \mathrm{Aut}_o(\Omega_{i(p)})$ whenever $p < k$.

Lemma 1 is a consequence of a lemma of BOREL [BO1].

We now formulate the Hermitian metric rigidity theorem for irreducible compact quotients of bounded symmetric domains of rank ≥ 2. We have a canonical decomposition $(T_X,g) \cong (T_1,g_1) \oplus \ldots \oplus (T_k,g_k)$ of the Hermitian holomorphic tangent bundle of (X,g). We have

THEOREM 1
Let $X = \Omega/\Gamma$ be a compact quotient of a reducible bounded symmetric domain $\Omega = \Omega_1 \times \ldots \times \Omega_k$ such that $\Gamma \subset \mathrm{Aut}_o(\Omega)$. Write g for a canonical metric on Ω and $(T_X,g) \cong (T_1(X),g_1) \oplus \ldots \oplus (T_k(X),g_k)$ for the canonical decomposition of the Hermitian holomorphic tangent bundle (T_X,g). Suppose h_i is a Hermitian metric of seminegative curvature on the holomorphic vector bundle $T_i(X)$. Then, $h_i = cg_i$ for some global constant c on X.

We remark that there may exist Hermitian metrics h of seminegative curvature on X which are not equal to the canonical metric g up to normalizing constants. We will discuss such examples in (4.2). We also remark that the individual factors Ω_i may be of rank 1, as is in the case of irreducible quotients of the polydisc.

Proof:

The proof of Thm.1 follows from the alternative proof given in §3 of the Hermitian metric rigidity theorem for compact quotients in the locally irreducible case [(3.1), Thm.1] and Lemma 1 above. In place of using the characteristic bundle $\mathcal{S}$ we use the projective bundles $\mathbb{P}T_i(X)$, embedded in $\mathbb{P}T_X$ using the decomposition $T_X \cong T_1(X) \oplus \ldots \oplus T_k(X)$. We consider the tautological line bundle L over X equipped with the Hermitian metrics $\hat{g}_i$, $\hat{h}_i$ and $\hat{g}_i + \hat{h}_i$ of seminegative curvature. Write $\hat{g}_i$, $\hat{h}_i + \hat{g}_i = e^u \hat{g}_i$. We use the analogue of the integral formula [(3.1), Prop.1], giving

$$-\sqrt{-1}\,\partial u \wedge \bar{\partial} u \wedge (-c_1(L,\hat{g}_i))^{2n-2q-1} \equiv 0 \text{ on } \mathbb{P}T_i(X) \qquad (1)$$

As before we lift to the origin o in Ω and make the identification $T_o(\mathbb{P}T_i(X)) \cong T_{[\xi]}(\mathbb{P}T_{o,i}(X)) \oplus T_o(\Omega)$. Write $m^+ = m_1^+ \oplus \ldots \oplus m_k^+$ for $m^+ \cong T_o(\Omega)$. The null space of the seminegative (1,1)-form $c_1(L,\hat{g}_i)$ contains in particular the vector space $m_1^+ \oplus \ldots \oplus \hat{m}_i^+ \oplus \ldots \oplus m_k^+ \subset m^+ \subset T_o(\mathbb{P}T_i(X))$, where ^ denotes omission. For $a_i \in \Omega_i$ define $\Omega_i^*(a) = \Omega_1 \times \ldots \times \Omega_{i-1} \times \{a_i\} \times \Omega_{i+1} \times \ldots \times \Omega_k$ and denote by τ: $\Omega \to X$ the covering projection. It follows that the function u is constant on $\tau(\Omega_i^*(a_i))$ for any $a_i \in \Omega_i$. For any such a_i, $\tau(\Omega_i^*(a_i))$ is in fact dense in X by Lemma 1. Consequently, the function u is constant on X, from which one deduces that $h_i(x) = cg_i$ for a global constant c on X, proving Thm.1. ∎

(4.2) We discuss in this section Hermitian rigidity phenomena for the entire tangent bundle in the irreducible and locally reducible case. We have

PROPOSITION 1

Let $X = \Delta^2/\Gamma$ be an irreducible compact quotient of the bidisc Δ^2 such that $\Gamma \subset (\mathrm{Aut}(\Delta))^2$. Then, there exists on X a Hermitian metric h of seminegative curvature such that (X,h) is not a Hermitian locally symmetric space. Moreover, the space of all Hemitian metrics h of seminegative curvature on X can be parametrized by a convex set on a finite dimensional real Euclidean space. Any such metric h is Kähler.

REMARKS

Prop.1 should be contrasted with the Riemannian metric rigidity phenomenon which implies in the present case that for any Riemannian metric h of seminegative Riemannian sectional curvature on X, (X,h) must be Riemannian symmetric. (In the case of compact quotients of the polydiscs this follows from EBERLEIN [EBE]. For the general case, see BALLMANN [BAL]).

Proof of Prop.1:

We are in fact going to describe all possible Hermitian metrics h of seminegative curvature on $X = \Delta^2/\Gamma$. By the curvature–decreasing property for Hermitian holomorphic subbundles for $i = 1, 2$, $h|T_i(X)$ defines a Hermitian metric of seminegative curvature. From Thm.1 it follows that $h|T_i(X)$ is a canonical metric. Lifting h to the covering bidisc Δ^2 and retain the same symbol. We have

$$h = 2Re\left[\frac{c_1 dz^1 \otimes d\bar{z}^1}{(1-|z_1|^2)^2} + \frac{c_2 dz^2 \otimes d\bar{z}^2}{(1-|z_2|^2)^2} + 2h_{1\bar{2}}(z_1,z_2)\, dz^1 \otimes d\bar{z}^2\right]. \quad (1)$$

Write $(h^{i\bar{j}})$ for the conjugate inverse of $(h_{i\bar{j}})$. We have

$$\begin{aligned} 0 \geq R_{1\bar{1}2\bar{2}}(h) &= -\frac{\partial^2 h_{1\bar{1}}}{\partial z_2\, \partial \bar{z}_2} + \sum_{p,q} h^{p\bar{q}} \frac{\partial h_{1\bar{q}}}{\partial z_2} \frac{\partial h_{p\bar{1}}}{\partial \bar{z}_2} \\ &= h^{2\bar{2}} \frac{\partial h_{1\bar{2}}}{\partial z_2} \frac{\partial h_{2\bar{1}}}{\partial \bar{z}_2} \geq 0. \end{aligned} \quad (2)$$

It follows that $R_{1\bar{1}2\bar{2}}(h) = 0$ and that $\partial h_{1\bar{2}}/\partial z_2 \equiv 0$, i.e, $h_{1\bar{2}}$ is anti–holomrophic in z_2. Interchanging z_1 and z_2 we have $R_{2\bar{2}1\bar{1}}(h) = 0$ and $\partial h_{2\bar{1}}/\partial z_1 \equiv 0$. It follows therefore that $h_{1\bar{2}}$ is holomorphic in z_1 and anti–holomorphic in z_2. A straightforward calculation as in (2) together with the new structure equations $\partial h_{1\bar{2}}/\partial z_2 \equiv 0$ and $\partial h_{2\bar{1}}/\partial z_1 \equiv 0$ yields immediately the formulas

$$R_{1\bar{1}1\bar{2}}(h) \equiv R_{2\bar{2}2\bar{1}}(h) \equiv 0. \quad (3)$$

Thus, the only possibly non–vanishing curvature terms are given by $R_{1\bar{1}1\bar{1}}(h)$, $R_{2\bar{2}2\bar{2}}(h)$ and $R_{1\bar{2}1\bar{2}}(h)$ and its conjuate $R_{2\bar{1}2\bar{1}}(h)$. Let $\xi = \xi^i(\partial/\partial z_i)$ and $\eta =$

$\eta^i(\partial/\partial z_i)$ be tangent vectors of type (1,0) at some $x \in \Delta^2$. We have

$$R_{\xi\bar{\xi}\eta\bar{\eta}} = |\xi^1|^2|\eta^1|^2 R_{1\bar{1}1\bar{1}}(h) + |\xi^2|^2|\eta^2|^2 R_{2\bar{2}2\bar{2}}(h) + 2Re(\xi^1\bar{\xi}^2\eta^1\bar{\eta}^2 R_{1\bar{2}1\bar{2}}(h)). \tag{4}$$

It follows from the fact that (X,h) is of seminegative curvature that

$$(*)\, |R_{1\bar{2}1\bar{2}}(h)|^2 \leq R_{1\bar{1}1\bar{1}}(h) R_{2\bar{2}2\bar{2}}(h) \text{ at every point } x \in \Delta^2.$$

Conversely suppose h is a Hermitian metric on Δ^2 of the form given in (1) invariant under the action of Γ such that $h_{1\bar{2}}$ is holomorphic in z_1 and anti–holomorphic in z_2 and such that h verifies the inequality (*) everywhere on Δ^2, it follows readily from the Cauchy–Schwarz inequality (and the negativity of $R_{1\bar{1}1\bar{1}}(h)$ and $R_{2\bar{2}2\bar{2}}(h)$) that the quotient metric, denoted also by h, is of seminegative curvature on X. To find a non–standard Hermitian metric h of seminegative curvature on X the problem is to find a function $h_{1\bar{2}}$ holomorphic in z_1 and anti–holomorphic in z_2 such that $\eta = h_{1\bar{2}}\, dz^1 \wedge d\bar{z}^2$ is Γ–invariant. Given any such a function $h_{1\bar{2}}$ one can consider the family of symmetric 2–tensors $\{h_t = g + 2tRe(\eta)\colon t \in \mathbf{R}\}$. Since X is compact, for t sufficiently small h defines a Hermitian metric on X such that (*) is satisfied. (As $t \to 0$ we have $R_{1\bar{2}1\bar{2}}(h_t) \to 0$.) To construct $h_{1\bar{2}}$ consider the change of variable $(z_1,z_2) \to (z_1,\bar{z}_2)$. The (1,1)–form η is now transformed to the (2,0)–from $\nu(z_1,z_2) = h_{1\bar{2}}(z_1,\bar{z}_2)\, dz^1 \wedge dz^2 = \nu_{12}\, dz^1 \wedge dz^2$. The condition that $h_{1\bar{2}}$ is holomorphic in z_1 and anti–holomorphic in z_2 now translates to the condition that ν_{12} is holomorphic and that hence ν is a holomorphic (2,0)–form. Define by Γ' the subgroup of $Aut(\Delta) \times Aut(\Delta)$. $\Gamma' = \{(\gamma_1,\overline{\gamma_2})\colon (\gamma_1,\gamma_2) \in \Gamma\}$. Clearly Γ' acts properly discontinously on Δ^2 without fixed points such that $X' := \Delta^2/\Gamma'$ is compact (and diffeomorphic to X). The Γ–invariance of η is then equivalent to the Γ'–invariance of ν, so that it descends to a holomorphic 2–form, to be denoted by the same symbol, on X'. The advantage of working with ν on X' is that tools such as the Riemann–Roch Theorem or Selberg's trace formula are sometimes applicable for determining the dimension of holomorphic sections of a line bundle over a compact complex manifold. ν is equivalently a holomorphic section of $K_{X'}$, the canonical line bundle of X'. The dimension of the space of such sections $\Gamma(X',K_{X'})$ were already determined by SHIMIZU [SHI, Thm.11] using Selberg's trace formula, which showed in particular that it is non–zero. Let $\{\nu_1, ..., \nu_r\}$ be a basis of $\Gamma(X',K_{X'})$, corresponding to the (1,1)–forms $\{h_{1\bar{2}}^{(1)}, ..., h_{1\bar{2}}^{(r)}\}$ on Δ^2. The space of Hermitian metrics of seminegative curvature h on X are then given by all those h expressed as in (1) with $h_{1\bar{2}} = \Sigma a_i h_{1\bar{2}}^{(i)}$ for which h defines Hermitian metric on X and for which the extra inequality (*) is satisfied. From the compactness of X it is immediate that such h is parametrized by a convex set in a real Euclidean space.

§5 Applications of the Hermitian Metric Rigidity Theorem and Its Proofs

(5.1) In this section we give two applications of the Hermitian metric rigidity theorem [(1.1), Thm.1] and its proofs. We study first of all holomorphic mappings of the quotients $X = \Omega/\Gamma$ into Hermitian manifolds of seminegative curvature, proving in particular [Ch.1, (2.1), Thm.4]. We will also use the integral formula of §2 and §3 to study Hemitian metrics on the covering bounded symmetric domains Ω, proving that if Ω admits a Hermitian metric of strictly negative curvature (in the sense of Griffiths) bounded between two negative constants and of bounded torsion, then Ω is necessarily of rank 1. The latter result is a generalization of a theorem of YANG [YAN] in the Kähler case.

We now turn to rigidity phenomena for holomorphic mappings. In the compact case we have [Ch.1, (2.1), Thm.4], which we recall here.

THEOREM 1

Let (X,g) be a Hermitian locally symmetric manifold of finite volume uniformized by an irreducible bounded symmetric domain Ω of rank ≥ 2. Suppose (N,h) is any Hermitian locally symmetric manifold of non–compact type. Then, any holomorphic mapping $f: X \to N$ is necessarily an isometric immersion up to a scaling constant. If (N,h) is in addition Kähler, then f is in fact a totally geodesic isometric immersion.

Proof:

Except for the last statement, Thm.1 was already proved in [Ch.1, (2.1)] as a deduction of the Hermitian metric rigidity theorem given there as [Ch.1, (2.1),

Thm.1], as a result simply of the fact that $(X,g+f^*h)$ is of seminegative curvature. To prove the last statement it suffices therefore to prove that the second fundamental form σ of the isometric immersion (rescaling the Kähler metric h if necessary) $f: (X,g) \longrightarrow (Y,h)$ vanishes. Since (Y,h) is Kähler the Hermitian connection agrees with the Riemannian connection, so that it agrees with the second fundamental form of the induced bundle map $(X,T_X) \longrightarrow (Y,T_Y)$. As the computation is local we will identify (X,g) with $(f(X),h|_{f(X)})$. Let $x \in X$ and $\xi, \eta \in T_x(X)$. Write

$$R_{\xi\bar{\xi}\eta\bar{\eta}}(X,g) = R_{\xi\bar{\xi}\eta\bar{\eta}}(Y,h) - S_{\xi\bar{\xi}\eta\bar{\eta}} \tag{1}$$

where $S_{\xi\bar{\xi}\eta\bar{\eta}} \geq 0$. If $\sigma : T_X \otimes T_X \longrightarrow N_{X|Y}$ denotes the second fundamental form of f then we have $S_{\xi\bar{\xi}\eta\bar{\eta}} = \|\sigma(\xi,\eta)\|^2$ in terms of canonically induced metrics. In the special case when $\xi = \alpha$ is a characteristic vector and $\eta = \zeta \in \mathcal{N}_\alpha$, we know from the integral formula [(2.1), proof of Prop.2] that in fact

$$R_{\alpha\bar{\alpha}\zeta\bar{\zeta}}(X,g) = R_{\alpha\bar{\alpha}\zeta\bar{\zeta}}(Y,h) = S_{\alpha\bar{\alpha}\zeta\bar{\zeta}} = 0. \tag{2}$$

It follows therefore that $\sigma(\alpha,\zeta) = 0$. Applying now the polarization argument of [(2.1), proof of Prop.3] we conclude that in fact $\sigma \equiv 0$ on X, proving therefore that f is in fact a totally geodesic isometric immersion. ∎

Thm.1 applies in particular to the special case when (Y,h) is itself a quotient of a bounded symmetric domain (Y not necessarily compact or of rank ≥ 2). In this special case we will actually prove a rigidity theorem in the more general situation where the domain manifold is of finite volume. This relies on a generalization of the integral formulas of §2 and §3 to the case of finite volume and the Ahlfors–Schwarz lemma in a special case. The most general statement for such rigidity phenomena will be proved in Chapter 9 using some deep structure theory on quotients of bounded symmetric domains of finite volume. The case covered here can nonetheless be proved by elementary means. We state

THEOREM 2

Let (X,g) be as in Thm.1. Suppose (Y,h) is a Hermitian locally symmetric manifold of non–compact type and $f: X \longrightarrow Y$ is not totally degenerate. Then, f is up to a scaling factor a totally geodesic isometric immersion.

For the proof of Thm.2 we will first prove a first extension of the Hermitian metric rigidity theorem [(1.1), Thm.1] to the case of finite volume. We state

THEOREM 3

Let X be a quotient of an irreducible bounded symmetric domain of rank ≥ 2. Let h be a Hermitian metric of seminegative curvature in the sense of GRIFFITHS [GRI1] such that $h \lesssim Cg$ for a canonical metric g on X and for some global constant C. Then, $h \equiv cg$ for some global constant c on X.

We start with

Proof of Thm.3:

We will give a proof based on the alternative proof of [(1.1), Thm.1], which relies on the integral formula [(3.1), Prop.1] and Moore's Ergodicity Theorem [(3.1), Thm.1]. As the latter is valid on non–uniform lattices $\Gamma \subset G$ (i.e., $\Gamma\backslash G$ being non–compact and of finite volume with respect to a left invariant Haar measure) and the rest of the derivation is completely local, it remains to establish the integral formula [(2.1), Prop.1] in case X is only assumed to be of finite volume under the condition that $h \lesssim Cg$. We recall that ω is the Kähler form of (X,g), π: $\mathbb{P}T_X \longrightarrow X$ is the canonical base projection, $\mathcal{S}$ is the characteristic bundle over X, $\hat{g}$ and $\hat{h}$ are the induced Hermitian metrics on the tautological line bundle $L \longrightarrow \mathbb{P}T_X$ and that $\nu = (-c_1(L,\hat{g})) + \pi^*\omega$ is a Kähler form on $\mathbb{P}T_X$. We need to establish the integral formula for $q = n(X)$

$$(*) \qquad \int_{\mathcal{S}} (-c_1(L,\hat{g}+\hat{h})) \wedge (-c_1(L,\hat{g}))^{2n-2q-1} \wedge \nu^{q-1}$$
$$= \int_{\mathcal{S}} (-c_1(L,\hat{g}))^{2n-2q} \wedge \nu^{q-1} = 0.$$

To do this on the possibly non–compact manifold $\mathcal{S}$ it suffices to justify one single integration by parts. Writing $\hat{g}+\hat{h} = e^u\hat{g}$ on $\mathcal{S}$ it suffices to show

$$(*)' \qquad \int_{\mathcal{S}} \partial\bar{\partial}u \wedge (-c_1(L,\hat{g}))^{2n-2q-1} \wedge \nu^{q-1} = 0.$$

We observe first of all that by fiber integration the Kähler manifold $(\mathbb{P}T_X,\nu)$ is of finite volume, that the (1,1)–form $c_1(L,\hat{g})$ is bounded in $(\mathbb{P}T_X,\nu)$ and that $u \geq 0$.

The assumption $h \lesssim Cg$ implies that u is bounded on $\mathcal{S}$. Fix a base point $x_0 \in X$ and denote by $B(r)$ the geodesic ball with center x_0 and radius r. In what follows all norms on X and $\mathcal{S}$ resp. will be measured in terms of the Kähler form ω and ν resp. Denote by ρ_R a smooth function on X such that

$$\begin{cases} \rho_R \equiv 1 & \text{on } B(R), \\ \quad\;\; \equiv 0 & \text{on } X - B(2R), \\ \|\nabla\rho_R\| \lesssim 2/R & \text{on } B(2R) - B(R). \end{cases} \tag{1}$$

We will denote the function $\pi^*\rho_R$ on $\mathcal{S}$ also by the same symbol. Since ρ_R is of compact support we have by Stokes' Theorem certainly

$$\int_{\mathcal{S}} d(\rho_R\bar\partial u) \wedge (-c_1(L,\hat g))^{2n-2q-1} \wedge \nu^{q-1} = 0. \tag{2}$$

Writing $d(\rho_R\bar\partial u) = \rho_R\partial\bar\partial u + \partial\rho_R \wedge \bar\partial u$ and using (1) it is immediate that (*)′ would follow if we can show that $\|\bar\partial u\|^2$ is integrable on $(\mathcal{S},\nu|_{\mathcal{S}})$, i.e., writing $k = 2n - q - 1 = \dim_{\mathbb{C}}\mathcal{S}$ we have to justify

$$(\#) \qquad \int_{\mathcal{S}} \sqrt{-1}\,\partial u \wedge \bar\partial u \wedge \nu^{k-1} < \infty.$$

To this end we use the integral formula

$$\int_{\mathcal{S}} \sqrt{-1}\, d(\rho_R^2 u\,\bar\partial u) \wedge \nu^{k-1} = 0. \tag{3}$$

Write $d(\rho_R^2 u\,\bar\partial u) = \rho_R^2\,\partial u \wedge \bar\partial u + \rho_R^2 u\,\partial\bar\partial u + 2\rho_R\,u\,\partial\rho_R \wedge \bar\partial u$. Observe that

$$\frac{\sqrt{-1}}{2\pi}\partial\bar\partial u = (-c_1(L,\hat g+\hat h)) + c_1(L,\hat g) \geq c_1(L,\hat g) \geq -\nu. \tag{4}$$

Since $u \geq 0$ we have $\sqrt{-1}\,\rho_R^2 u\,\partial\bar\partial u \geq -2\pi\,\rho_R^2 u\,\nu$, so that

$$\sqrt{-1}\,d(\rho_R^2 u\,\bar\partial u) \geq \sqrt{-1}\,\rho_R^2\,\partial u \wedge \bar\partial u + \sqrt{-1}\,2\rho_R u\,\partial\rho_R \wedge \bar\partial u - 2\pi\,\rho_R^2 u\,\nu. \tag{5}$$

It follows then from (3) that in fact

$$\int_{\mathcal{S}} \sqrt{-1}\,\rho_R^2\,\partial u \wedge \bar\partial u \wedge \nu^{k-1} \leq \left|\int_{\mathcal{S}} 2\sqrt{-1}\,\rho_R\,\partial\rho_R \wedge \bar\partial u \wedge \nu^{k-1}\right| + \int_{\mathcal{S}} \rho_R^2 u\,\nu^k. \tag{6}$$

Since by assumption u is bounded the last integral is bounded independent of R. Furthermore by the Cauchy–Schwarz inequality

$$2\sqrt{-1}\,\rho_R u\,\partial\rho_R \wedge \bar\partial u \le \frac{1}{2}\sqrt{-1}\,\rho_R^2\,\partial u \wedge \bar\partial u + 2\sqrt{-1}u^2\,\partial\rho_R \wedge \bar\partial\rho_R \tag{7}$$

as (1,1)–forms. It follows then readily from (6) and (7) that

$$\int_{\mathcal{S}} \sqrt{-1}\,\rho_R^2\,\partial u \wedge \bar\partial u \wedge \nu^{k-1} \le 2\left|\int_{\mathcal{S}} \sqrt{-1}\,u^2\,\partial\rho_R \wedge \bar\partial\rho_R \wedge \nu^{k-1}\right| + 2\int_{\mathcal{S}} \rho_R^2 u\,\nu^k, \tag{8}$$

which clearly implies by $\|\nabla\rho_R\| \le 2/R$ that the integral on the integral on the left hand side is bounded independent of R. Passing to limit as $R \longrightarrow \infty$ we have established that ∇u is square–integrable on $(\mathcal{S},\nu|_{\mathcal{S}})$, proving (#) and thus establishing the integral formula (*), from which we deduce Thm.3, as desired. ∎

REMARKS

As is clear from the proof it suffices to assume that $h(x) \le C(x)g(x)$ over X such that C(x) is square–integrable on (X,g).

We are now ready to prove the rigidity theorem on holomorphic mappings.

<u>Proof of Thm.2</u>:

Recall that (X,g) and (Y,h) are Hermitian locally symmetric manifold of non-compact type such that (X,g) is of finite volume, locally irreducible and of rank $\ge$ 2. To prove that a non–constant holomorphic mapping $f: X \longrightarrow Y$ is a totally geodesic isometric immersion by the proof of Thms.1 & 3 it suffices to show that $f^*h \le Cg$ for some global constant C. But the latter follows immediately from the Ahlfors–Schwarz lemma on bounded domains, as given in [Ch.4, (1.3) Prop.3].

The Ahlfors–Schwarz lemma is true in a much more general context. We state here a general form of it for complete Kähler manifolds. As given in [Ch.5, (2.1)] any Hermitian symmetric manifold of non–compact type is of negative holomorphic sectional curvature (bounded between two negative constants by homogeneity). The Ahlfors–Schwarz lemma that we used is therefore a special case of (cf. REMARKS for references)

THEOREM 4

Let (M,ω_M) be a complete Hermitian manifold of bounded torsion and of holomorphic sectional curvature bounded from below by a negative constant $-K_1$ such that the holomorphic bisectional curvatures are also bounded from below. Let (N,ω_N) be a Hermitian manifold with holomorphic sectional curvature bounded from above by a negative constant $-K_2$. Then, there exists a constant C depending only on K_1, K_2 and the dimension of M such that for any holomorphic mapping $f: M \to N$ we have

$$f^*\omega_N \leq C\omega_M.$$

REMARKS

Thm.4 was first proved in case (M,ω_M) is Kähler and the holomorphic bisectional curvatures of (N,ω_N) is bounded from above by a negative constant by YAU [YAU], in which case one only needs that the Ricci curvatures of (M,ω_M) is bounded form below. In the form stated but assuming that (M,ω_M) is Kähler, the theorem is due to CHEN–CHENG–LU [CCL] and ROYDEN [ROY]. The improvement to Hermitian manifolds (M,ω_M) is due to CHEN–YANG [CY].

(5.2) We apply now the methods of proof of Hermitian metric rigidity theorems to study Hermitian metrics on bounded symmetric domains. We are going to prove

THEOREM 1 (MOK [MOK3, (5.1), Prop.(5.1.2), p.145])

Let Ω be a bounded symmetric domain. Suppose Ω admits a complete Hermitian metric of bounded torsion and of negative (bisectional) curvature bounded between two negative constants. Then, Ω is of rank 1.

Thm.1 in the special case of Kähler metrics was first proved by YANG [YAN] by a different method. Our presentation of the proof of Thm.1, unlike all results obtained so far, is less self–contained. It relies in part on the Ahlfors–Schwarz lemma [(5.1), Thm.4], some theorems in Several Complex Variables and Borel's Theorem on the existence of uniform lattices. First, we need the solution of the Levi Problem in Several Complex Variables in a general form. To formulate it, we say that an upper–semicontinuous real–valued function φ on a complex manifold M is strictly plurisubharmonic at $x \in M$ if there exists a neigborhood U of x and a smooth strictly plurisubharmonic function ψ such that $\varphi - \psi$ is plurisubharmonic. We say that φ is an exhaustion function if and only if the sublevel set $\{\varphi < c\}$ is relatively compact in M for any real number $c < \sup_M \varphi$. We have

DEFINITION 1

A complex manifold M is said to be strongly pseudoconvex if there exists a smooth exhaustion function φ on M such that for some compact subset K of M, φ is strictly plurisubharmonic on $M - K$. If the smoothness assumption on φ is removed, we say that M is a strongly pseudoconvex manifold in the generalized sense.

THEOREM 2 (GRAUERT [GRA1], FORNÆSS–NARASIMHAN [FN])

Let M be a strongly pseudoconvex manifold in the generalized sense. Then, there exists a compact subset K of M, a normal Stein space M′, a proper holomorphic mapping $\Phi: M \longrightarrow M'$ such that $\Phi|_{M-K}$ is a biholomorphism onto its image.

In case the exhaustion function φ is assumed to be smooth, Thm.2 in Grauert's famous solution to the Levi problem. The generalization to arbitrary (upper–semicontinuous) φ is due to FORNÆSS–NARASIMHAN [FN]. The set E on which Φ fails to be a local biholomorphism is a compact subvariety on M. As holomorphic functions separate points on the Stein space M′ by the maximum principle on holomorphic functions any positive–dimensional compact subvariety on M must be mapped to isolated points on M′. Clearly E is the union of all positive–dimensional compact subvarieties on M. E is called the maximal compact subvariety of M. As a consequence of Thm.2 and the argument of GRAUERT [GRA1], we have

THEOREM 3

Let M be a strongly pseudoconvex manifold in the generalized sense. Suppose the maximal compact subvariety E is non–singular and of codimension one. Then, the holomorphic normal bundle $N_{E|M}$ admits a Hermitian metric of negative curvature.

By Kodaira's Embedding Theorem the assertion in Thm.3 is equivalent to the fact that the dual bundle $N^*_{E|M}$ is ample. In case E is possibly singular and of arbitrary positive dimension Thm.3 remains valid with a proper definition of the normal space $N_{E|M}$ as a linear space over E and of negativity of such spaces (cf. GRAUERT [GRA2]).

Finally we need the following existence theorem on cocompact lattices.

THEOREM 4 (BOREL [BO2], cf. also RAGHUNATHAN [RA, Ch.XIV, p.215ff.]) Let G be a semisimple Lie group. Then, there exists a cocompact lattice Ψ in G. As a consequence, for any bounded symmetric domain Ω there exists a torsion–free discrete subgroup $\Gamma \subset G = \mathrm{Aut}_0(\Omega)$ such that $X = \Omega/\Gamma$ is compact.

We remark that any lattice Ψ in G admits a torsion–free subgroup Γ of finite index in Ψ (cf. SATAKE [SA2, Ch.IV, Lemma 7.2, p.196]).

Proof of Thm.1

Write g for a canonical metric on Ω. First by the Ahlfors–Schwarz Lemma [(5.1), Thm.4] the identity mappings $(\Omega,g) \to (\Omega,h)$ and its inverse are both distance decreasing up to a constant, so that g and h must be uniformly equivalent to each other. Writing $\hat{g}$ and $\hat{h}$ for the induced Hermitian metrics on the tautological line bundle $L \to \mathbb{P}T_\Omega$ this means that $\hat{h} = e^{2\pi u}\hat{g}$ on $\mathbb{P}T_\Omega$ for a bounded function u. Let $X = \Omega/\Gamma$ be a compact quotient. From now on we will often use the same notations for entities defined over X and their lifting to Ω. Associated to g is the homogeneous Kähler form ν on $\mathbb{P}T_X$ defined as in [(2.1), Lemma 1]. The idea is to make h descend to X to obtain a Hermitian metric μ of strictly negative curvature on X, which would then contradict the integral formulas in §4 since $c_1(L,\hat{\mu})$ would be negative definite on $\mathbb{P}T_X$. The difficulty is that h is not invariant under Γ. However, we have just seen that h is uniformly equivalent to g, which is Γ–invariant. This will allow us to obtain instead an upper semi–continuous complex Finsler metric μ on X uniformly equivalent to g of strictly negative curvature in a generalized sense. By this we mean equivalently an upper–semicontinuous Hermitian metric $\hat{\mu}$ on L such that the generalized first Chern form $c_1(L,\hat{\mu}) \leq -C\nu$ on $\mathbb{P}T_X$. Here $c_1(L,\hat{\mu})$ is defined in the same way as in the case of smooth Hermitian metrics using the same local formula, except that it is in general only a distribution. The inequality $c_1(L,\hat{\mu}) \leq -C\nu$ is interpreted in the sense of distribution. It suffices to construct on Ω an upper–semicontinuous function v invariant under Γ and define $\hat{\mu} = e^{2\pi v}\hat{g}$ such that $-\frac{\sqrt{-1}}{2\pi}\partial\bar{\partial}v + c_1(L,\hat{g}) \leq -C\nu$ for some positive constant C. From the curvature assumption on the Hermitian metric h and the fact that h is uniformly equivalent to the canonical metric g on Ω it is clear that over Ω we have $c_1(L,\hat{h}) \leq -C\nu$ on Ω. Define on $\mathbb{P}T_\Omega$

$$v_o(z) = \sup\nolimits^*_{\gamma\in\Gamma} u(\gamma z). \tag{1}$$

Since u is bounded on Ω, the function v_o is well–defined, finite and non–zero over Ω. Let v be the upper–semicontinuous regularization of v_o, i.e., $v(z) = \limsup_{w\to z} v_o(w)$. We write $v(z) = \sup^*_{\gamma\in\Gamma} u(\gamma z)$ to denote this. It is clear that v_o and hence v is Γ–invariant. We argue that the Hermitian metric

$$\hat{\mu} = e^{2\pi v}\hat{g}. \tag{2}$$

has the desired curvature property. For an arbitrary $z \in \mathbb{P}T_\Omega$. Let U be an open coordinate neighborhood of z on Ω such that the Γ–orbits of U are disjoint from each other and such that there exists a smooth function η on U satisfying $C_1\nu \leq \sqrt{-1}\partial\bar{\partial}\eta \leq C_2\nu$ for some positive constants C_1 and C_2 on U. Transport the function η to the disjoint orbits γU and denote by the same notation η the function thus defined on $\bigcup_{\gamma\in\Gamma}\gamma U$. We may assume that in fact $-c_1(L,\hat{h}) \geq \sqrt{-1}\partial\bar{\partial}\eta$. Assume also that $-c_1(L,\hat{g}) = \sqrt{-1}\,\partial\bar{\partial}\lambda$ for some Γ–invariant function λ on $\bigcup_{\gamma\in\Gamma}\gamma U$. We have by (1) and (2)

$$\begin{aligned}-c_1(L,\hat{\mu})(z) &= \sqrt{-1}\partial\bar{\partial}v(z) - c_1(L,\hat{g})(z) = \sqrt{-1}\partial\bar{\partial}\sup\nolimits^*_{\gamma\in\Gamma}(u(\gamma z) + \lambda(\gamma z)) \\ &= \sqrt{-1}\partial\bar{\partial}\sup\nolimits^*_{\gamma\in\Gamma}(u(\gamma z) + \lambda(\gamma z) - \eta(\gamma z)) + \sqrt{-1}\partial\bar{\partial}\eta(z)\end{aligned} \tag{3}$$

Define $\beta(z) = u(z) + \lambda(z) - \eta(z)$ on $\bigcup_{\gamma\in\Gamma}\gamma U$. Then $\sqrt{-1}\partial\bar{\partial}\beta = -c_1(L,\hat{h}) - \sqrt{-1}\partial\bar{\partial}\eta \geq 0$, so that β is plurisubharmonic. As the upper–semicontinuous regularization of the supremum of a family of plurisubharmonic functions is again plurisubharmonic it follows therefore from (3) that in fact

$$-c_1(L,\hat{\mu})(z) = \sqrt{-1}\partial\bar{\partial} \sup^*_{\gamma\in\Gamma}\beta(\gamma z) + \sqrt{-1}\partial\bar{\partial}\eta(z) \geq C_1\nu. \tag{4}$$

Since $\hat{\mu}$ is by definition Γ–invariant we have obtained an upper–semicontinuous Hermitian metric of strictly negative curvature on the compact quotient X.

Let $B \subset \mathbb{P}T_X$ be the unit ball bundle on with respect to $\hat{\mu}$, i.e., B consists of all vectors on $L \longrightarrow \mathbb{P}T_X$ of length < 1, i.e., $B = \{v \in \mathbb{P}T_X: \hat{\mu}(v,v) < 1\}$. B is open since $\varphi(v) = \hat{\mu}(v,v)$ is upper–semicontinuous on $\mathbb{P}T_X$. φ is an exhaustion function on B. We claim that φ is plurisubharmonic on B and strictly plurisubharmonic outside the zero section X. To see this let e be a holomorphic basis of L over some open set U of $\mathbb{P}T_X$. Write $L|_U \cong U \times \mathbb{C}$ accordingly. Write $(z;w)$, $z = (z_1,\ldots,z_{2n-1})$ for the base and fiber coordinates resp. Suppose $\hat{\mu}(e,e)(z) = \eta(z)$ as a function of $z \in U$. Then, $\varphi(z;w) = \eta(z)|w|^2$ and we have

$$\begin{aligned}\sqrt{-1}\partial\bar{\partial}\varphi(z,w) &= (\sqrt{-1}\partial_z\bar{\partial}_z\eta(z))\,|w|^2 + \eta(z)\sqrt{-1}\,dw\wedge d\bar{w}.\\ &\geq |w|^2\eta(z)(\sqrt{-1}\partial_z\bar{\partial}_z\log\eta(z)) + \eta(z)\sqrt{-1}\,dw\wedge d\bar{w}.\end{aligned} \tag{5}$$

Denote by $p: L \longrightarrow \mathbb{P}T_X$ the canonical base projection. As $c_1(L,\hat{\mu}) \leq -C\nu$ for some positive constant C it follows from (5) that on $L|_U$ we have

$$\sqrt{-1}\partial\bar{\partial}\varphi(z,w) \geq c|w|^2\eta(z)p^*(\nu) + \eta(z)\sqrt{-1}\,dw\wedge d\bar{w} \tag{6}$$

for some positive constant c, showing clearly that φ is plurisubharmonic on $L|_U$ and strictly plurisubharmonic whenever $w \neq 0$, i.e., outside the zero section X of L, proving our assertion. Thus, B is a strictly pseudoconvex manifold in the generalized sense whose maximal compact subvariety is the zero section of X. It follows then from Thm.3 that the normal bundle $N_{X|B}$ of X in B is negative. By writing down transition functions it is easy to see that $N_{X|B}$ is precisely the line bundle L over $\mathbb{P}T_X$. Thus by Thm.3 L admits a smooth Hermitian metric θ of negative curvature, i.e., $-c_1(L,\theta)$ is a strictly positive (1,1)–form on X. To prove Thm.2 we are going to argue by contradiction. Suppose Ω is irreducible and of rank ≥ 2. Let $\mathcal{S}$ be the characteristic bundle over X. Write $s = \dim_{\mathbb{C}}\mathcal{S}$. Then, as implied by the integral formula [(3.1), Prop.1] we have $(-c_1(L,\hat{g}))^s \equiv 0$

on X. This contradicts immediately with $\int_{\mathcal{S}}(-c_1(L,\theta))^s > 0$. If Ω is reducible we may assume that $\Gamma \subset \mathrm{Aut}_0(\Omega)$ and replace $\mathcal{S}$ by $\mathbb{P}T_i(X)$ for some canonical direct factor $T_i(X)$ of T_X. A slight modification of [(2.1), Prop.1] gives again the identity $(-c_1(L,\hat{g}))^s \equiv 0$ on $\mathbb{P}T_i(X)$, of complex dimension s, yielding again a contradiction with $-c_1(L,\theta) > 0$. The proof of Thm.2 is completed. ▌

REMARKS

Instead of replacing the upper-semicontinuous metric $\hat{\mu}$ by the smooth metric θ one can also argue by contradiction directly with the metric $\hat{\mu}$ by justifying the equation $\int_{\Sigma}(-c_1(L,\hat{\mu}))^s = \int_{\Sigma}(-c_1(L,\hat{g}))^s = 0$ and the inequality $\int_{\Sigma}(-c_1(L,\hat{\mu}))^s > 0$ for $\Sigma = \mathcal{S}$ or $\mathbb{P}T_i(X)$ by defining $(-c_1(L,\hat{\mu}))^s$ in the sense of distribution. This can be done using the definition of the Monge–Ampère operator $(\sqrt{-1}\partial\bar{\partial}\psi)^s$ of a bounded plurisubharmonic function ψ in s–dimensional manifold S due to Bedford–Taylor [BT]. For this line of argument see [Ch.8, (2.2), proof of Prop.(1.2)].

(5.3) On the bounded symmetric domains Ω one can formulate problems of rigidity of holomorphic mappings between such domains. One possible direction is to replace compactness by properness. We formulate here the following

CONJECTURE 1

Let Ω and Ω' be two bounded symmetric domains. Suppose Ω is irreducible and of rank ≥ 2, $\mathrm{rank}(\Omega') \leq \mathrm{rank}(\Omega)$ and $f: \Omega \longrightarrow \Omega'$ is a proper holomorphic map. Then, f is a totally geodesic isometric embedding up to a scaling constant.

For the case of compact quotients of bounded symmetric domains Ω we obtained rigidity theorems for holomorphic mappings by using Hermitian metric rigidity theorems on such quotients. It is plausible that to solve Conjecture 1 one can make use of another type of rigidity phenomena, *viz.* the rigidity theorem on bounded convex realizations of irreducible bounded symmetric domains of rank ≥ 2, as stated in [Ch.5, (2.3)].

CHAPTER 7 THE KÄHLER METRIC RIGIDITY THEOREM IN THE SEMIPOSITIVE CASE

§1 Hermitian Symmetric Manifolds of Compact Type

(1.1) In this chapter we will prove a metric rigidity theorem for Hermitian symmetric manifolds of compact type [Ch.1, (2.1), Thm.2], which is in some sense dual to the Hermitian metric rigidity theorem on quotients of bounded symmetric domain. Before we recall the theorem we start with some preliminary discussion of Hermitian symmetric manifolds of compact type.

From duality and the list given in Chapter 5 of irreducible Hermitian symmetric manifolds of noncompact type we have a corresponding list for those of compact type. The classical series, already given in [Ch.4, §2] are (I) $G(q,p)$, $p, q \geq 1$, the Grassmannian of complex q–planes in $\mathbb{C}^{p+q}$; (II) $G^{II}(n,n) \subset G(n,n)$, $n \geq 2$, the submanifold of isotropic n–planes with respect to a non–degenerate alternating form J; (III) $G^{III}(n,n)$, $n \geq 2$, the submanifold of isotropic n–planes with respect to a non–degenerate complex symmetric bilinear form Σ; and (IV) the hyperquadrics $Q^n \subset \mathbb{P}^{n+1}$, $n \geq 3$. All these manifolds have been discussed in relation to the Borel Embedding Theorem. We studied the hyperquadrics Q^n in some detail in [Ch.4, (3.1)] in order to compute the curvature tensor of the dual domains D_n^{IV}. The standard inclusion $Q^n \subset \mathbb{P}^{n+1}$ yields in fact a Hermitian symmetric structure on Q^n by restricting a Fubini–Study metric from $\mathbb{P}^{n+1}$ to Q^n. To provide further examples where one can see the symmetric structure we discuss in some details the case of the Grassmannian $G(q,p)$.

We define a holomorphic mapping from $G(q,p)$ into some projective space $\mathbb{P}^N$. Fix a complex linear space $W \cong \mathbb{C}^{p+q}$ and write $G(q,p) = \{[V]: V \subset W$ is a complex vector subspace of dimension $q.\}$. We use the notation $[V]$ when we think of V as an element of $G(q,p)$. $G(q,p)$ has a covering by open subsets $U \cong \mathbb{C}^{pq}$. With respect to some choice of basis on W elements of U are described by matrices $Z \in M(p,q;\mathbb{C})$ so that each element $[V] \in G(q,p)$ corresponds to the matrix $\begin{bmatrix} Z \\ I_q \end{bmatrix}$. Let now $[V] \in G(q,p)$ be arbitrary. We write $V = \Sigma_{1 \leq i \leq q} \mathbb{C}v_i$. Define $\tau: G(q,p) \to \mathbb{P}(\Lambda^q W)$ by $\tau([V]) = [v_1 \wedge \ldots \wedge v_p] \in \mathbb{P}(\Lambda^q W)$. Obviously τ is independent of the choice of the basis. Fix such an open set U and use the matrix representation $[V] = \begin{bmatrix} Z \\ I_q \end{bmatrix}$. Write $\{e_i\}$ for the corresponding basis on W, V_o for the q–plane $\Sigma_{1\leq i\leq q} \mathbb{C}e_{p+i}$ corresponding to the origin in U and $(\zeta_i)_{1\leq i\leq q}$ for the q vectors in $\Sigma_{1\leq i\leq p} \mathbb{C}e_i$ represented by the column vectors of Z. Then,

$$\begin{aligned}\tau([V]) &= [(e_{p+1} + \zeta_1) \wedge \dots \wedge (e_{p+q} + \zeta_q)] \\ &= [(e_{p+1} \wedge \dots \wedge e_{p+q}) + \zeta_1 \wedge (e_{p+2} \wedge \dots \wedge e_{p+q}) + \dots + \\ &\quad + (e_{p+1} \wedge \dots \wedge e_{p+q-1}) \wedge \zeta_q + R],\end{aligned}$$

where the remainder terms in R are at least quadratic in (ζ_i), i.e., in Z. It follows readily from this description that τ is holomorphic. Moreover the tangent map at $[V_o]$ can be represented by $(\zeta_1, \dots, \zeta_q) \to \zeta_1 \wedge (e_{p+2} \wedge \dots \wedge e_{p+q}) + \dots + (e_{p+1} \wedge \dots \wedge e_{p+q-1}) \wedge \zeta_q$, which is visibly injective, so that τ is a local biholomorphism. Since $G(q,p)$ is simply connected it follows readily that τ: $G(q,p) \to \mathbb{P}(\Lambda^q W)$ is a holomorphic embedding. τ is called the Plücker embedding. From now on we identify $G(q,p)$ with its image under τ.

Fix a Euclidean metric on W so that $\{e_i: 1 \leq i \leq p+q\}$ is an orthonormal basis. This induces a Euclidean metric on $\Lambda^q W$ and hence a Fubini–Study metric g on $\mathbb{P}(\Lambda^q W)$. We endow $G(q,p)$ with the restriction of g and use the same symbol g for $g|_{G(q,p)}$. We assert that $(G(q,p),g)$ is a Hermitian symmetric manifold. The unitary group $U(p+q)$ on W induces a group of isometries on $\mathbb{P}(\Lambda^q W)$ preserving $G(q,p)$ and acting transitively on the Grassmannian. Thus, to show that $(G(q,p),g)$ is Hermitian symmetric it suffices to exhibit a symmetry at $[V_o]$. Write $V_o^\perp$ for the orthogonal complement of V with respect to the Euclidean metric on W. Then, $\zeta_i \in V_o^\perp$ for $1 \leq i \leq q$. The complex linear isometry L on $W = V_o \oplus V_o^\perp$ given by $L|V_o = \mathrm{id}$ and $L|V_o^\perp = -\mathrm{id}$ induces a holomorphic isometry σ on $\mathbb{P}(\Lambda^q W)$ which yields the map $Z \to -Z$ on the affine open subset U of $G(q,p)$. Thus, σ preserves $G(q,p)$ and induces at $[V_o]$ a symmetry on $(G(q,p),g)$, proving that the latter is Hermitian symmetric.

By using the Euclidean coordinates on U and the Plücker embeddings τ

one can furthermore write down a potential for g on U. Recall that on $\mathbb{C}^N \subset \mathbb{P}^N$ the standard Fubini–Study metric is given by the potential $\log\|z\|^2$, where $z = (z_1,\dots,z_n)$ are the Euclidean coordinates and $\|.\|$ is the Euclidean norm. It follows that a potential for $(U,g|_U)$ is given by $\Phi(Z) = \log \|(e_{p+1} + \zeta_1) \wedge \dots \wedge (e_{p+q} + \zeta_q)\|^2$, where $Z = [\zeta_1,\dots,\zeta_q]$. To compute this consider the mapping μ: $G(q,p) \to G(p,q)$ defined by $\mu([V]) = [V^\perp]$. μ is a conjugate–biholomorphic map, which is an isometry with respect to the Kähler metrics induced by Plücker embeddings. In terms of matrices we write $\mu\begin{bmatrix} Z \\ I_q \end{bmatrix} = \begin{bmatrix} I_p \\ -Z^t \end{bmatrix}$. Denote by $\xi_i \in W$, $1 \le i \le p$, the vectors in W represented by the p columns of $-Z^t$. We have

$$\begin{aligned} \Phi(Z) &= \log \|(e_{p+1} + \zeta_1) \wedge \dots \wedge (e_{p+q} + \zeta_q)\|^2 \\ &= \log \|(e_1 + \xi_1) \wedge \dots \wedge (e_p + \xi_p)\|^2 \\ &= \log \|(e_1 + \xi_1) \wedge \dots \wedge (e_p + \xi_p) \wedge (e_{p+1} + \zeta_1) \wedge \dots \wedge (e_{p+q} + \zeta_q)\| \\ &= \log\ \det \begin{bmatrix} I_p & Z \\ -Z^t & I_q \end{bmatrix} = \log\det(I_p + ZZ^t). \end{aligned}$$

Here we use the fact that $[V]$ and $\mu([V])$ represent orthogonal planes. Writing $g|_U = \sqrt{-1}\,\partial\bar\partial \log\det(I_p + ZZ^t)$ we can compute the curvature directly, showing that it has semipositive holomorphic bisectional curvatures. The computation is of course the same as could be obtained from the computation on $D^I_{p,q}$ and duality.

(1.2) The standard inclusions $Q^n \subset \mathbb{P}^{n+1}$ and the Plücker embeddings are special cases of the following general embedding theorems on Hermitian symmetric manifolds of compact type, which we formulate only for the irreducible case.

THEOREM 1 (NAKAGAWA–TAKAGI [NT])

Let (X_c,g_c) be an irreducible Hermitian symmetric manifold of compact type. Then, for every positive integer ν there exist isometric embeddings σ_ν: $(X_c,g_c) \to (\mathbb{P}^{N(\nu)},g_\nu)$, unique up to projective–linear transformations, where g_ν is a Fubini–Study metric on the projective space $\mathbb{P}^{N(\nu)}$ such that if the maximum holomorphic sectional curvature of (X_c, g_c) is m, then $(\mathbb{P}^{N(\nu)}, g_\nu)$ is of constant holomorphic sectional curvature νm.

The isometric embedding $\sigma_\nu: (X_c, g_c) \to (\mathbb{P}^{N(\nu)}, g_\nu)$ is called the ν-th canonical embedding. It is easy to see that an isometric embedding $\sigma: (X_c, g_c) \to (\mathbb{P}^N, g)$ is the first canonical embedding if there exists some rational line C in the image of σ, as can be seen by considering curvatures on C. Examples of the first canonical embedding are given by (i) the identification $\mathbb{P}^N \equiv \mathbb{P}^N$, (ii) the standard embedding of the hyperquadric $Q^n \hookrightarrow \mathbb{P}^{n+1}$, and (iii) the Plücker embedding of the Grassmannian $G(q,p) \hookrightarrow \mathbb{P}(\Lambda^p \mathbb{C}^{p+q})$. (ii) and (iii) can be seen to be the first canonical embedding by finding some rational line in Q^n and $G(q,p)$ resp. In case of Q^n, $n \geq 3$, with the standard equation $\Sigma z_i^2 = 0$ in homogeneous coordinates an example of rational line is given by $C = \{[\zeta, i\zeta, 1, i..., 0] : \zeta \in \mathbb{C}\} \cup \{[1, i, 0, ..., 0]\}$. In case of the Grassmannian $G(q,p) = \{$complex q–planes in $W \cong \mathbb{C}^{p+q}\}$ one can consider a $(q-1)$–plane $E \subset W$ and a 2–plane $F \subset W$ such that $E \cap F = \{0\}$. Then, $C = \{[E + \mathbb{C}\xi] : 0 \neq \xi \in F\}$ is a rational line on $G(q,p) \hookrightarrow \mathbb{P}(\Lambda^q W)$, as can be easily seen from the definition of the Plücker embedding.

The submanifolds $G^{II}(n,n) \subset G(n,n)$ and $G^{III}(n,n) \subset G(n,n)$, with induced metrics obtained by restriction of properly chosen canonical metrics g on $G(n,n)$, are totally geodesic Hermitian symmetric submanifolds, by the same argument as in the dual cases [Ch.4, (3.1)]. The embedding $G^{III}(n,n) \subset G(n,n) \hookrightarrow \mathbb{P}^N$ given by the Plücker embedding $\tau: G(n,n) \hookrightarrow \mathbb{P}^N$ is in fact the first canonical embedding. This can be seen from the curvature computation (on their dual bounded symmetric domains) in [Ch.4, (3.1)], which shows that they have the same maximal holomorphic sectional curvature. On the other hand, the corresponding embedding $G^{II}(n,n) \subset G(n,n) \hookrightarrow \mathbb{P}^N$ gives on the contrary only the second canonical embedding on $G^{II}(n,n) \subset G(n,n) \hookrightarrow \mathbb{P}^N$, by the computations in [Ch.4, (3.1)]. Another example of the second canonical embedding is given by the Veronese map $\mathbb{P}^n \hookrightarrow \mathbb{P}^N$, where $N = \frac{1}{2}(n^2+3n)$ (cf. [App.III.3]).

For our proof of metric rigidity theorems on (X_c, g_c) we will need to analyze a class of rational curves which are totally geodesic with respect to any choice of canonical metric g_c. For this purpose it will be convenient, although not strictly necessary, to use the embedding theorem of NAKAGAWA–TAKAGI [NT]. We refer the reader to [NT] for a proof, which is obtained from embedding $G_c = \mathrm{Aut}_0(X_c, g_c)$ into a projective linear group using representation theory.

Regarding the topology of Hermitian symmetric manifolds of compact type we have

PROPOSITION 1
Let (X_c, g_c) be an irreducible Hermitian symmetric manifold of compact type. Then, $H_2(X,\mathbb{Z}) \cong \mathbb{Z}$.

In case of the Grassmannians $G(q,p)$, Prop.1 can be seen using the decomposition of $G(q,p)$ into Schubert cells (cf. Griffiths–Harris [GH]), which are complex manifolds (biholomorphic to complex Euclidean spaces). In particular, all cells are real even–dimensional. Consequently, all odd–dimensional homology and cohomology groups vanish and $H_{2k}(G(q,p),\mathbb{Z})$ is a free abelian group of rank equal to the number of Schubert cells of complex k–dimensions. There is only one Schubert cell of complex dimension 1.

§2 The Dual Characteristic Bundle $\mathcal{S}^*$ and an Integral Formula

(2.1) We recall the statement of the metric rigidity theorem for Hermitian symmetric manifolds of compact type, which we call the Kähler metric rigidity theorem for the semipositive case.

THEOREM 1 (MOK [MOK3,1987])
Let (X_c, g_c) be an irreducible Hermitian symmetric manifold of compact type and of rank ≥ 2. Suppose h is a Kähler metric of semipositive holomorphic bisectional curvature on X_c. Then, there exists a biholomorphism Ψ of X such that $h = c\Psi^* g_c$ for some constant c.

There are two major differences between Thm.1 and the the Hermitian metric rigidity theorem for the seminegative case [Ch.6, (2.1), Thm.1]. First of all, the analogue of Thm.1 for Hermitian metrics is not valid. Moreover, we have in Thm.1 uniqueness only up to biholomorphisms and a normalizing constant. For the sake of simplicity we will drop the suffix c when referring to Hermitian symmetric manifolds of compact type.

A Hermitian metric on a vector bundle is of seminegative curvature if and only if the induced Hermitian metric θ on the associated tautological line bundle is of seminegative curvature. This is the case if θ can be expressed locally as a sum of norms of holomorphic functions f_i, using the formula for $\sqrt{-1}\,\partial\bar{\partial}\,(\log \Sigma|f_i|^2)$ as given in [Ch.4, (1.1), proof of Prop.1]. Let W be a holomorphic vector field on X_c. It follows that $Re(W \otimes \overline{W})$ can be regarded as a degenerate Hermitian metric of seminegative curvature on T^*_X. Write g^* for the Hermitian metric on T^*_X defined by g. Then, $g^* + Re(W \otimes \overline{W})$ defines a *bona fide* Hermitian metric on T^*_X of seminegative curvature, as could be seen from [Ch.2, (3.2), Prop.2], which holds true even though one of the metrics is degenerate. The dual metric $h = (g^* + Re W \otimes \overline{W})^*$ defines then a Hermitian metric of semipositive curvature on T^*_X which is in general different from the canonical metrics. Regarding Hermitian metrics of semipositive curvature on X we will nonetheless prove a partial rigidity result later on [(4.3), Thm.1].

Let now h be a Hermitian metric of semipositive holomorphic bisectional curvature on X. In a rather obvious analogy to the dual case, we can work with the induced Hermitian metric h^* of seminegative curvature on the holomorphic cotangent bundle T^*_X and obtain an integral formula on a submanifold $\mathcal{S}^*(X) \subset \mathbb{P}T^*_X$ called the dual characteristic bundle, to obtain a partial vanishing theorem for the ∇h^*, where ∇ denotes the connection on (X,g) for some fixed canonical metric g on X. As the canonical metrics on X are only unique up to biholomorphism and scaling factors one cannot expect as in the dual case to deduce from the integral formula that h^* is parallel on (X,g) even if h is Kähler. In fact the integral formula does not distinguish between Kähler metrics and Hermitian metrics, so that the partial vanishing result obtained is even valid for the non–Kähler metrics of semipositive curvature described above.

To prove the Kähler rigidity theorem for the semipositive case one has to find first of all algebro–geometric properties that are shared by all the possible choices of canonical metrics. It follows from a theorem of MATSUSHIMA [MA1] that the canonical metrics on X are characterized as the set of all possible Kähler–Einstein metrics on X. On the geometric side we find that there is a class of rational curves C on X which are totally geodesic w.r.t any choice of canonical metric g. These are precisely the minimal rational curves, which are rational curves representing the positive generator of $H_2(X,\mathbf{Z}) \cong \mathbf{Z}$. Our strategy is to show that the minimal rational curves C are also totally geodesic w.r.t. any Kähler metric h of semipositive bisectional curvature on X and then to deduce that (X,h) is Kähler–Einstein. For the first statement the problem is reduced to a vanishing theorem on second fundamental forms. We will obtain this vanishing theorem from the partial vanishing obtained from the integral formula and from the Kähler property of h. This entails some finer study of the characteristic bundle $\mathcal{S}$. The deduction of the Einstein property of (X,h) will be obtained by algebraic means.

In this section we will start with the dual integral formula. In §3 we will study the minimal rational curves on X using the Polysphere Theorem and the theorem of NAKAGAWA–TAKAGI [NT] on isometric embeddings of X into projective space [(1.2), Thm.1]. In §4 we will give a proof of Thm.1.

(2.2) Let X be a Hermitian symmetric space of compact type. Let Ω be a bounded symmetric domain dual to X and write $\Omega \subset\subset \mathbf{C}^N \subset X$ for the Harish–Chandra and the Borel embeddings. Write $\Omega = G_0/K$ and $X = G_c/K \cong G^{\mathbf{C}}/P$ as usual. We define the notion of characteristic vectors on X in the same way as we did on Ω as highest weights of the isotropy representations on the tangent spaces of type (1,0) (cf. [Ch.6, (1.2)]). Write $\mathcal{S}(X) \subset \mathbf{P}T_X$ for the characteristic bundle on X thus defined. Fix a characteristic vector α at $o \in \Omega$. Since the isotropy representations for Ω and X are the same at o, α is also a characteristic vector for X. As proved in [Ch.6, (1.2), Prop.2], $\mathcal{S}(\Omega)$ is the part of the $G^{\mathbf{C}}$–orbit of $[\alpha]$ that lies above Ω. It follows from the proof given that $\mathcal{S}(X)$ is the G_c–orbit of $[\alpha]$ and hence equal to the $G^{\mathbf{C}}$–orbit of $[\alpha]$. In other words, we have

PROPOSITION 1

Let π: $\mathbb{P}T_X \to X$ be the canonical base projection and $\Omega \subset\subset \mathbb{C}^N \subset X$ be the Harish–Chandra and Borel embeddings. Identifying in the usual way $\mathbb{P}T(\mathbb{C}^N)$ with $\mathbb{C}^N \times \mathbb{P}^{N-1}$ and writing $\mathcal{S}_0$ for $\mathcal{S}_0(X) = \mathcal{S}_0(\Omega)$, we have

$$\mathcal{S}(\Omega) = \mathcal{S}(X) \cap \pi^{-1}(X);\ \mathcal{S}(X) \cap \pi^{-1}(\mathbb{C}^N) = \mathbb{C}^N \times \mathcal{S}_0.$$

Write in local coordinates $g = 2Re\Sigma g_{i\bar{j}} dz^i \wedge d\bar{z}^j$. For $x \in X$ define the lifting operator Φ_x: $T_x(X) \to T_x^*(X)$ by $\Phi_x(\Sigma a^i \frac{\partial}{\partial z_i}) = \Sigma g_{i\bar{j}} \bar{a}^j dz^i$. Φ_x is conjugate complex–linear. Denote by the same symbol the induced maps $\mathbb{P}T_x(X) \to \mathbb{P}T_x^*(X)$, yielding a fiber–by–fiber conjugate–holomorphic map Φ: $\mathbb{P}T_X \to \mathbb{P}T_X^*$. We define the dual characteristic bundle $\mathcal{S}^*(X) \subset \mathbb{P}T_X^*$ by $\mathcal{S}^*(X) = \Phi(\mathcal{S}(X))$. By a dual characteristic vector at $x \in X$ we mean the lifting $\alpha^* = \Phi_x(\alpha)$ of a characteristic vector. $G^{\mathbb{C}}$ acts on $\mathbb{P}T_X^*$ in the usual way dual to the action of $G^{\mathbb{C}}$ on $\mathbb{P}T_X$, i.e., for $\gamma \in G^{\mathbb{C}}$ and $\gamma(x) = y$, we have for $\eta^* \in T_x^*(X)$ and $\xi \in T_y(X)$, $\gamma(\eta^*)(\xi) = \eta(\gamma^{-1}\xi)$. It follows that at the origin $M^- = \exp(\mathfrak{m}^-)$ acts trivially on $T_0^*(X)$. $\mathcal{S}^*(X)$ is the G_c–orbit of a single $[\alpha^*] \in \mathcal{S}_0^*$, as in the proof of [Ch.6, (1.2), Prop.2]. It follows that $\mathcal{S}^*(X)$ is invariant under $G^{\mathbb{C}}$. Hence, we have

PROPOSITION 2

In the notations of Prop.1 and writing $\mathcal{S}_0^*$ for $\mathcal{S}_0^*(X)$, we have

$$\mathcal{S}^*(X) \cap \pi^{-1}(\mathbb{C}^N) = \mathbb{C}^N \times \mathcal{S}_0^*.$$

Based on Prop.2 and the proof of the integral formula in the dual case [Ch.6, (2.1), Prop.1] we obtain immediately the following integral formula on $\mathcal{S}^*(X)$.

PROPOSITION 3

Let (X,g) be an irreducible Hermitian symmetric manifold of dimension N of rank ≥ 2. Let ω be the Kähler form of (X,g), π: $\mathbb{P}T_X^* \to X$ the canonical base projection, and μ be any Hermitian metric on T_X^*. Let $\hat{\mu}$ denote the induced Hermitian metric on the tautological line bundle $\Lambda \to \mathbb{P}T_X^*$ and $c_1(\Lambda,\hat{\mu})$ be the first Chern form of Λ defined by $\hat{\mu}$. Write $q = n(X)$ for the dimension of nullity of X. Then, on the dual characteristic bundle $\mathcal{S}^* := \mathcal{S}^*(X)$ over X we have

$$\int_{\mathcal{S}^*} [-c_1(\Lambda,\hat{\mu})]^{2N-2q} \wedge (\pi^*\omega)^{q-1}$$

$$= \int_{\mathcal{S}^*} [-c_1(\Lambda,\hat{g}^*)]^{2N-2q} \wedge (\pi^*\omega)^{q-1} = 0.$$

In particular, if (T_X^*,μ) is of seminegative curvature, we have

$$[-c_1(\Lambda,\hat{\mu})]^{2N-2q} \wedge (\pi^*\omega)^{q-1} \equiv 0 \quad \text{on } X.$$

PROPOSITION 4

In the notations of Prop.3 let h be a Hermitian metric on X such that (T_X,h) is of semipositive curvature. Write h^* for the dual Hermitian metric on the cotangent bundle and regard h^* also as a contravariant (1,1)–tensor on X. Denote by ∇ the connection on (X,g). Then, for any $[\alpha] \in \mathcal{S}_x$, $\alpha^* = \Phi(\alpha)$, $\zeta \in \mathcal{N}_\alpha$ and $\eta^* \in T_x^*(X)$ we have

$$\nabla_\zeta h^{\alpha^*\overline{\eta}^*} = 0.$$

In the proof of the total geodesy of minimal rational curves in (X,h) we will need a slightly different partial vanishing statement on ∇h^* which can nonetheless be deduced from Prop.4. We state first

PROPOSITION 5

In the notations of Prop.4 and writing ζ^* for $\Phi(\zeta)$ we have

$$\nabla_\alpha h^{\zeta^*\overline{\eta}^*} = 0.$$

Consider the isotropy action of K on $\mathfrak{m}^+ \otimes \mathfrak{m}^- \cong T_o(X) \otimes \overline{T_o(X)}$. Recall the Hermitian bilinear form Q on $\mathfrak{m}^+ \otimes \mathfrak{m}^-$ defined by $Q(\alpha\otimes\overline{\beta};\nu\otimes\overline{\mu}) = R_{\alpha\overline{\beta}\mu\overline{\nu}}(X,g)$ and extended by $\mathbb{C}$–linearity. By [Ch.3, (2.2), Prop.1] Q is positive semidefinite. Since K preserves the curvature tensor it follows that the zero eigenspace $\mathcal{N}(Q)$ of Q is invariant under K. Prop.5 can now be deduced from

LEMMA 1 (BOREL [BO4])

K acts irreducibly on $\mathcal{N}(Q) \subset \mathfrak{m}^+ \otimes \mathfrak{m}^-$.

Proof of Prop.5

We use Euclidean coordinates at $o \in \mathbf{C}^n \subset X$ and identify $\overline{T_o(X)}$ with $T_o^*(X)$ via contraction of the metric tensor g with $g_{i\bar{j}}(o) = \delta_{ij}$. Thus, for $\xi \in T_o(X)$, $\bar{\xi}$ is identified with $\Phi(\xi) = \xi^*$. Define $M = \Sigma\, \mathbf{C}(\zeta \otimes \alpha^*) \subset T_o(X) \otimes \overline{T_o(X)}$. For any $A = \Sigma\, a^{i\bar{j}} \frac{\partial}{\partial z_i} \otimes \frac{\partial}{\partial \bar{z}_j} \in M$ we deduce from Prop.4 that $\Sigma\, \nabla_i h^{j\bar{k}} = 0$ for $1 \le k \le n$. Clearly M is K–invariant. As $R_{\bar{\zeta}\alpha\bar{\alpha}\zeta}(X,g) = 0$ it follows that $\zeta \otimes \alpha^* \in \mathcal{N}(Q)$. From Lemma 1 we conclude that $M = \mathcal{N}(Q)$. As $R_{\alpha\bar{\zeta}\zeta\bar{\alpha}}(X,g) = 0$ we also have $\alpha \otimes \zeta^* \in \mathcal{N}(Q) = M$. Consequently, $\nabla_\alpha h^{\zeta^* \bar{\eta}^*} = 0$, as asserted. ▌

REMARKS

One can avoid using the lemma of Borel by using a different integral formula. For $1 \le k < \mathrm{rank}(X) = r$ we define the k–th characteristic bundle $\mathcal{S}_k = \mathcal{S}_k(X)$ in the same way as $\mathcal{S}_k(\Omega)$ in the dual case (cf, [App.III.4]. Define $\mathcal{S}_k^* = \Phi(\mathcal{S}_k)$ and call it the k–th dual characteristic bundle. By taking $k = r-1$ we have an analogous integral formula

$$\int_{\mathcal{S}^*_{r-1}} [-c_1(\Lambda,\hat{\mu})]^s \wedge (\pi^*\omega)^t$$
$$= \int_{\mathcal{S}^*_{r-1}} [-c_1(\Lambda,\hat{g}^*)]^s \wedge (\pi^*\omega)^t = 0.$$

for appropriate choices of exponents s and t such that $s + t = \dim_{\mathbf{C}} \mathcal{S}^*_{r-1}$. Given $[\alpha] \in \mathcal{S}$ and $\zeta \in \mathcal{N}_\alpha$ we have from $\Theta_{\zeta^*\bar{\zeta}^*\alpha\bar{\alpha}} = -R_{\zeta\bar{\zeta}\alpha\bar{\alpha}} = 0$ that $[\zeta^*] \in \mathcal{S}_{r-1}$. From the preceding integral formula we obtain by the same reasoning as in Props. 3 & 4 that $\nabla_\alpha h^{\zeta^*\bar{\eta}^*} = 0$ for any $\eta^* \in T^*_x$, where $x = \pi(\alpha)$, as desired.

§3 The Characteristic Bundle and Minimal Rational Curves

(3.1) In order to discuss the minimal rational curves it is convenient to use the first canonical embedding of X into projective spaces using the embedding theorem of NAKAGAWA–TAKAGI [NT]. We will use exclusively the first canonical embedding. Sometimes we write $\mathbb{P}^N(k)$ to indicate that $\mathbb{P}^N$ is equipped with a Fubini–Study metric of constant holomorphic sectional curvature k. We will normalize the canonical metric g on X so that the maximum of holomorphic sec–

tional curvatures on (X,g) is 2. Recall that we have $H_2(X,\mathbf{Z}) \cong \mathbf{Z}$ for every irreducible Hermitian symmetric manifold X of compact type (cf. [(1.2), Prop.1]). By a minimal rational curve C on X we will mean a rational curve representing a generator of $H_2(X,\mathbf{Z}) \cong \mathbf{Z}$. We have

PROPOSITION 1

For any $x \in X$ and any characteristic vector α at x there exists a unique minimal rational curve C passing through x such that $T_x(C) = \mathbb{C}\alpha$. Furthermore, all minimal rational curves are obtained this way.

Proof:

We use the first canonical embedding $\sigma: (X,g) \hookrightarrow (\mathbb{P}^N(2),g_1)$. By the Polysphere Theorem [Ch.5, (1.1), Thm.1] given $x \in X$ and $[\alpha] \in \mathcal{S}_x$ there exists a compact Riemann surface $C \subset X$ totally geodesic in (X,g) such that $C \cong \mathbb{P}^1$. Moreover, by [App.III.1, Prop.1] a unit characteristic vector α is equivalently a unit vector realizing the maximum of holomorphic sectional curvatures. Consequently,

$$R_{\alpha\bar{\alpha}\alpha\bar{\alpha}}(C,g|_C) = R_{\alpha\bar{\alpha}\alpha\bar{\alpha}}(X,g) = R_{\alpha\bar{\alpha}\alpha\bar{\alpha}}(\mathbb{P}^N(2),g_1) = 2.$$

In particular, C is totally geodesic in $(\mathbb{P}^N(2),g_1)$. Since any totally geodesic algebraic curve in $\mathbb{P}^N$ is linear C must itself be a rational line in $\mathbb{P}^N$. From the existence of one minimal rational curve C it follows that the embedding $\sigma: X \hookrightarrow \mathbb{P}^N$ induces an isomorphism $\mathbf{Z} \cong H_2(X,\mathbf{Z}) \cong H_2(\mathbb{P}^N,\mathbf{Z}) \cong \mathbf{Z}$. In particular, C must represent a generator of $H_2(X,\mathbf{Z})$, as claimed. From the isomorphism $H_2(X,\mathbf{Z}) \cong H_2(\mathbb{P}^N,\mathbf{Z}) \cong \mathbf{Z}$ it follows that a minimal rational curve D on $X \subset \mathbb{P}^N$ must also represent a generator in $H_2(\mathbb{P}^N,\mathbf{Z})$ and hence be a rational line. For every unit vector α of type (1,0) tangent to D at $x \in D$ from

$$2 = R_{\alpha\bar{\alpha}\alpha\bar{\alpha}}(D,g|_D) \leq R_{\alpha\bar{\alpha}\alpha\bar{\alpha}}(X,g) \leq R_{\alpha\bar{\alpha}\alpha\bar{\alpha}}(\mathbb{P}^N(2),g_1) = 2$$

it follows that $R_{\alpha\bar{\alpha}\alpha\bar{\alpha}}(X,g) = 2$ so that α is a characteristic vector. From the Polysphere Theorem and the preceding argument there exists again a minimal rational curve C such that $T_x(C) = \mathbb{C}\alpha = T_x(D)$. Since both D and C are rational lines we must have $C = D$. The proof of Prop.1 is completed. ∎

The space of minimal rational curves $\mathscr{C} = \mathscr{C}_X$ on X only depends on the complex structure of X. On the other hand, as shown in the proof of Prop.1, a minimal rational curve C is equivalently a totally geodesic Riemann sphere in (X,g) representing a generator of $H_2(X,\mathbf{Z}) \cong \mathbf{Z}$. In other words, a minimal rational curve C has the remarkable property of being totally geodesic in (X,g) with respect to any possible choice of canonical metrics g on X. Based on this observation, our strategy in proving the Kähler rigidity theorem in the semipositive case [(2.1), Thm.1] is to show first of all that every $C \in \mathscr{C}$ is totally geodesic in (X,h) for any Kähler metric h of semipositive curvature on X. For this purpose we will need some finer structure of the characteristic bundle $\mathcal{S}$.

We use the same conventions and notations Δ, Δ^+, Δ_M^+ for the space of roots, positive roots and positive non–compact roots, etc. as in [Ch.5, (1.1)]. We need first of all

LEMMA 1

Let α be a unit characteristic vector on (X,g) at x and write H_α for the Hermitian bilinear form on $T_x(X)$ defined by $H_\alpha(\xi,\eta) = R_{\alpha\bar{\alpha}\xi\bar{\eta}}(X,g)$. Then, the eigenspace decomposition of $T_x(X)$ with respect to H_α is given by

$$T_x(X) = \mathbf{C}\alpha + \mathscr{H}_\alpha + \mathscr{N}_\alpha,$$

corresponding to the eigenvalues 2, 1 and 0 resp.

Proof:

Write $X = G_c/K$ as usual. We may take $x = o = eK$ and identify $T_o(X)$ with $\mathfrak{m}^+$. Write $\alpha = e_\mu$ for a highest root μ with respect to some choice of Cartan subalgebra $\mathfrak{h}$ of $\mathfrak{k}$. Let now $\mathfrak{m}^+ = \Sigma\, \mathbf{C}e_\varphi$ be the decomposition of $\mathfrak{m}^+$ into root spaces correponding to non–compact positive roots φ. We have by [Ch.3, (1.3), Prop.1] the curvature formula

$$R(\alpha,\bar{\alpha};e_\varphi,\overline{e_\psi}) = R(e_\mu,\overline{e_\mu};e_\varphi,\overline{e_\psi}) = c\,([e_\mu,e_{-\varphi}];[e_\mu,e_{-\psi}]) \qquad (1)$$

where (.,.) denotes $-B(.,\bar{.})$ for the Killing form B of $\mathfrak{g}^{\mathbf{C}}$; and c is some positive constant. Norms given by (.,.) will be denoted by $\|.\|$. From [App.I.3, Prop.3] we have

$$\begin{cases} [e_\mu, e_{-\mu}] = H_\mu \in \mathfrak{h}^{\mathbf{C}}; \\ [e_\mu, e_{-\varphi}] = N_{\mu,-\varphi} e_{\mu-\varphi} & \text{if } \mu - \varphi \in \Delta; \\ [e_\mu, e_{-\varphi}] = 0 & \text{if } \varphi \neq \mu \text{ and } \mu - \varphi \notin \Delta. \end{cases} \tag{2}$$

Moreover, if $\mu - \varphi \in \Delta$ and $\{\varphi + k\mu \in \Delta : k \in \mathbf{Z} \text{ and } -a \leq k \leq b\}$ is the maximal μ–string of roots attached to φ, we have the fomula

$$N^2_{\mu,-\varphi} = \frac{a(b+1)}{2} \|H_\mu\|^2. \tag{3}$$

In the present case since μ is a dominant root $\varphi + \mu$ is not a root, so that $b = 0$. We assert that furthermore $\varphi - 2\mu$ cannot be a root, so that $a = 1$. Suppose otherwise. As $e_{\varphi-\mu} \in \mathfrak{k}^{\mathbf{C}}$ and $[\mathfrak{k}^{\mathbf{C}}, \mathfrak{m}^-] \subset \mathfrak{m}^-$ we would have $\varphi - 2\mu \in \Delta^-_M$, so that $2\mu - \varphi = \psi \in \Delta^+_M$. But then $\varphi + \psi = 2\mu$ and so one of φ and ψ would have to dominate μ, contradicting with the assumption that μ is the highest root. It follows hence from (1), (2) and (3) that for any positive non–compact root $\varphi \neq \mu$ either

$$\begin{aligned} R(\alpha,\overline{\alpha};e_\varphi,\overline{e_\varphi}) &= \tfrac{1}{2} R_{\alpha\overline{\alpha}\alpha\overline{\alpha}} = 1; \text{ or} \\ R(\alpha,\overline{\alpha};e_\varphi,\overline{e_\varphi}) &= 0. \end{aligned} \tag{4}$$

On the other hand it is clear that for distinct positive non–compact roots λ, ν

$$R(\alpha,\overline{\alpha};e_\lambda,\overline{e_\nu}) = 0, \tag{5}$$

as the vectors $H_\mu \in \mathfrak{h}^{\mathbf{C}}$ and $\{e_{\mu-\varphi} : \mu - \varphi \text{ is a root}\}$ are all orthogonal to each other with respect to $(.,.)$. The proof of the lemma is completed. ▌

The eigenspace decomposition in Lemma 1 will be called the (α,g)–decomposition of $T_o(X)$.

REMARKS

By using the fact that a unit vector α is characteristic if and only if it is maximal, i.e., it realizes the maximum of holomorphic sectional curvatures on (X,g) ([App.III.1, Prop.1]), one can also prove Lemma 1 by using the maximum principle on tensors (cf. [MOK3]).

Consider now the characteristic projective subvariety $\mathcal{S}_o \subset \mathbb{P}(\mathfrak{m}^+)$ at $o \in X$. We make the identification $T_{[\alpha]}(\mathbb{P}(\mathfrak{m}^+))$ with $\mathfrak{m}^+/\mathbb{C}\alpha$ as usual. By [Ch.6, (1.2), proof of Prop.1] the holomorphic tangent space $T_o(\mathcal{S}_o)$ can be identified with $(\mathbb{C}\alpha+V)/\mathbb{C}\alpha$, where $V = \Sigma\{\mathbb{C}e_\varphi: \varphi \in \Delta_M^+$ and $\mu - \varphi \in \Delta\}$. From Lemma 1 it follows that $V = \mathcal{H}_\alpha$ and hence

PROPOSITION 2

Identifying $T_{[\alpha]}(\mathbb{P}(\mathfrak{m}^+))$ with $\mathfrak{m}^+/\mathbb{C}\alpha$, we have $T_{[\alpha]}(\mathcal{S}_o) \cong (\mathbb{C}\alpha+\mathcal{H}_\alpha)/\mathbb{C}\alpha$.

In order to exploit the partial vanishing result [(2.2), Prop.4] we will need to study the variation of ζ as $[\alpha]$ varies in $\mathcal{S}_o$. By Prop.2 and the fact that $\mathcal{S}_o = K[\alpha]$ we have $\mathcal{H}_\alpha \subset ad(\mathfrak{k})(\mathbb{C}\alpha) \subset \mathcal{H}_\alpha+\mathbb{C}\alpha$. It is more convenient to express this in the form $ad(\mathfrak{k}^{\mathbb{C}})(\mathbb{C}\alpha) = \mathcal{H}_\alpha+\mathbb{C}\alpha$, recalling that $[z,\alpha] = \sqrt{-1}\alpha$ for some central element of $\mathfrak{k}$ (cf [Ch.4, Prop.]). We assert

PROPOSITION 3

$ad(\mathfrak{k}^{\mathbb{C}})(\mathcal{N}_\alpha) = \mathcal{N}_\alpha + \mathcal{H}_\alpha$.

Proof:

For any root $\varphi \in \Delta$ let e_φ denote a corresponding unit root vector. Define $N = \{\nu \in \Delta_M^+ : \mu - \nu \notin \Delta\}$. Then $\mathcal{N}_\alpha = \Sigma_{\nu \in N} \mathbb{C}e_\nu$. Since $[\mathfrak{k}^{\mathbb{C}},\mathfrak{m}^+] \subset \mathfrak{m}^+$ for $\nu \in N$, we have for compact roots γ either $[e_\gamma,e_\nu] = 0$ or $\gamma + \nu \in \Delta_M^+$ and $[e_\gamma,e_\nu] = N_{\gamma,\nu}e_{\gamma+\nu}$. If $\gamma + \nu = \mu$ we would have $\mu - \nu = \gamma \in \Delta$, contradicting with the definition of N. Consequently, $\mathcal{N}_\alpha \subset ad(\mathfrak{k}^{\mathbb{C}})(\mathcal{N}_\alpha) \subset \mathcal{N}_\alpha + \mathcal{H}_\alpha$. Write $ad(\mathfrak{k}^{\mathbb{C}})(\mathcal{N}_\alpha) = \mathcal{N}_\alpha + E$ for some complex vector subspace $E \subset \mathcal{H}_\alpha$. Let $K_s = [K,K]$ be the semisimple part of K. We have $\mathcal{S}_o = K_s[\alpha]$. Let $L \subset K_s$ be the isotropy subgroup at $[\alpha]$. As a homogeneous space $\mathcal{S}_o \cong K_s/L$. Since L fixes $\mathbb{C}\alpha$ and preserves the curvature tensor of (X,g) it also fixes the null space $\mathcal{N}_\alpha$. Thus L acts on $\mathcal{H}_\alpha$. From $ad(\mathfrak{k}^{\mathbb{C}})(\mathcal{N}_\alpha) = \mathcal{N}_\alpha + E$ it follows that E is invariant under L. To prove Prop.3 it would suffice to show

(1) $E \neq 0$, and

(2) L acts irreducibly on $\mathcal{H}_\alpha$.

We are going to prove (1) by contradiction. If $E = 0$ we would have $ad(\mathfrak{k}^{\mathbb{C}})(\mathcal{N}_\alpha) = (\mathcal{N}_\alpha)$. In other words, $\mathcal{N}_\alpha$ is infinitesimally invariant under $\mathfrak{k}^{\mathbb{C}}$. As $[\alpha] \in \mathcal{S}_o$ is arbitrary this implies that $\mathcal{N}_\alpha$ is invariant under $\mathfrak{k}^{\mathbb{C}}$, contradicting with the fact that $\mathfrak{k}$ acts irreducibly on $\mathfrak{m}^+$. (2) is however not always true. By [App.-III.3.] the Kähler submanifold $\mathcal{S}_o \subset \mathbb{P}(\mathfrak{m}^+)$ (with a Fubini–Study metric induced by the Hermitian metric $g|\mathfrak{m}^+$) is a Hermitian symmetric manifold of rank ≤ 2, which is irreducible except in the case when X is of type I and of rank ≥ 2. In this case X is a Grassmannian $G(q,p)$, $p, q \geq 2$, and we have $K_s = SU(p)\times SU(q)$ and $L = U(p-1)\times U(q-1)$, which gives $\mathcal{S}_o \cong \mathbb{P}^{p-1} \times \mathbb{P}^{q-1}$. In place of (2) we actually have the decomposition $\mathcal{H}_\alpha \cong T_{[\alpha]}(\mathcal{S}_o) = V_1 \oplus V_2$, where V_1 and V_2 come from the two direct factors of $\mathcal{S}_o$. It remains to prove Prop.3 for this case.

It suffices to show that there exists $\xi_i \in V_i$ such that $\xi_i \in E$. This can be checked by direct computation. Use the identification $T_o(X) \cong M(p,q;\mathbb{C})$ given in [Ch.4, (2.2)]. By [Ch.4, (3.2), Prop.1] $[\alpha] \in \mathcal{S}_o$ if and only if α is a matrix of rank 1 under this identification. Define the $p\times q$ matrix E_{ij}, $1 \leq i \leq p$, $1 \leq j \leq q$, by $(E_{ij})_{st} = \delta_{is}\delta_{jt}$. Then, $[E_{11}] \in \mathcal{S}_o$. $\mathcal{N}_{[E_{11}]} = \left\{\begin{bmatrix} 0 & 0 \\ 0 & A \end{bmatrix} : A \in M(p-1,q-1;\mathbb{C})\right\}$. For any real number s, $[\alpha_s] := [E_{11}+sE_{12}] \in \mathcal{S}_o$. From the curvature formula of [Ch.4, (3.1)] it follows that $\zeta_s \in \mathcal{N}_{[\alpha_s]}$ if and only if $\alpha_s\bar{\zeta}_s^t = 0$ and $\alpha_s^t\bar{\zeta}_s = 0$ as matrices, where A^t denotes the transpose of A. It is clear then that one can take $\zeta_s = E_{22} - sE_{21}$. As $\mathcal{S}_o = K[E_{11}]$ it follows that $E_{21} \in ad(\mathfrak{k}^{\mathbb{C}})(\mathcal{N}_\alpha)$. By a similar argument we also have $E_{12} \in ad(\mathfrak{k}^{\mathbb{C}})(\mathcal{N}_\alpha)$. We have hence shown the existence of elements $\xi_i \in V_i$, $i = 1, 2$, as desired. The proof of Prop.2 is completed. ▌

§4 Proof of the Metric Rigidity Theorem

(4.1) Recall that (X,g) is an irreducible Hermitian symmetric manifold of compact type and of rank ≥ 2 and that h is a Kähler metric of semipositive curvature on X. To prove the metric rigidity theorem [(2.1), Thm.1] we are going to use

THEOREM 1 (MATSUSHIMA [MA1], special case)
Let (M,s) be a Hermitian symmetric manifold of compact type and t be a Kähler–Einstein metric on M. Then, there exists a biholomorphism Φ of M such that t and Φ^*s are equal up to normalizing constants (on the individual de Rham factors of (M,s)).

REMARKS
Thm.1 was also proved for Kähler metrics of constant positive scalar curvature on M. The result of MATSUSHIMA [MA1] concerns more generally the space of such metrics on an arbitrary compact complex manifold. In another direction BANDO–MABUCHI [BM] proved that in fact on any compact complex manifold Z admitting a Kähler–Einstein metric of constant Ricci curvature +1, the space of such metrics is homogeneous under the identity component $\mathrm{Aut}_0(Z)$ of the group Aut(Z) of holomorphic automorphisms of Z, which implies in particular Thm.1.

To prove the Kähler rigidity theorem for the semipositive case [(1.1), Thm.1] it suffices therefore to show that (X,h) is Einstein for any Kähler metric h of semipositive curvature h on X. To prove this we will first of all exploit the space of minimal rational curves on X and show

PROPOSITION 1
Let (X,g) be an irreducible Hermitian symmetric manifold of compact type and of rank ≥ 2. Then, for any Kähler metric h of semipositive bisectional curvature on X and any minimal rational curve $C \in \mathscr{C}_X$ on X, C is totally geodesic in (X,h).

We use the Harish–Chandra coordinates on an affine part $\mathbf{C}^N$ of X (cf. [Ch.5, (2.1), Remarks]). To prove Prop.1 we fix g and h and may assume that C passes through $o \in \mathbf{C}^N$. Suppose $T_0(C) = \mathbf{C}\alpha^o$ for some unit vector α at o. By a unitary transformation on $\mathbf{C}^N$ we may assume that $\alpha^o = \partial/\partial z_1$. We specify furthermore the choice of Euclidean coordinates on $\mathbf{C}^N$ in such a way that $\mathscr{H}_{\alpha^o} = \Sigma_{2\leq i\leq s+1}\mathbf{C}(\partial/\partial z_i)$ and $\mathscr{N}_{\alpha^o} = \Sigma_{s+2\leq i\leq N}\mathbf{C}(\partial/\partial z_i)$, where $s = \dim_{\mathbf{C}}\mathscr{H}_{\alpha^o}$ and $N - s - 1 = n(X)$. We will use indexes p and q for $\mathscr{H}_{\alpha^o}$ and $\mathscr{N}_{\alpha^o}$ resp. To prove that C is totally geodesic in h it is equivalent to show that the second fundamental from of the inclusion $(C,h|_C) \hookrightarrow (X,h)$ vanishes. Since any point on

C can be regarded as the origin and C is arbitrary it suffices to show that

(#) For the Riemann–Christoffel symbols $(\Gamma^k_{ij})_{1\leq i,j,k\leq N}$ at o we have
$$\Gamma^k_{11} = 0 \quad \text{for } 2 \leq k \leq N.$$

We write $(\#)_k$ for the statment for a particular index k. Recall the formula for Riemann–Christoffel symbols on Kähler manifolds (cf. [Ch.2, (2.1)])

$$\Gamma^k_{ij}(z) = h^{k\bar{r}} \frac{\partial}{\partial z_i} h_{j\bar{r}}(z) \tag{1}$$

Denote by ∇ the Riemannian connection of (X,g). Since the Euclidean coordinates (z_i) are complex geodesic coordinates at o we have at o

$$\Gamma^k_{ij} = h^{k\bar{r}} \nabla_i h_{j\bar{r}}(o). \tag{2}$$

Recall the partial vanishing result [(2.2), Prop.5] obtained from an integral formula on the dual characteristic bundle $\mathcal{S}^*$. We had $\nabla_1 h^{\zeta^*\bar{k}}(o) = 0$ for any $\zeta^* = \Phi(\zeta)$ with $\zeta \in \mathcal{N}_\alpha$. (From now on we write η^* for $\Phi(\eta)$.) Thus,

$$\nabla_1 h^{q\bar{r}}(o) = 0 \qquad \text{for } s+2 \leq q \leq N. \tag{3}$$

From $h^{k\bar{r}} h_{j\bar{r}} = \delta_{jk}$ it follows that $h^{k\bar{r}}\, \nabla_i h_{j\bar{r}}(o) = -\, h_{j\bar{r}}\, \nabla_i h^{k\bar{r}}(o)$. Consequently we have by (3)

$$\Gamma^q_{11} = h^{q\bar{r}}\, \nabla_i h_{j\bar{r}}(o) = h_{1\bar{r}}\, \nabla_1 h^{q\bar{r}}(o) = 0 \qquad \text{for } s+2 \leq k \leq N. \tag{4}$$

We note that (4) is obtained without using at all the Kähler property of h. To complete the proof of Prop.2 it remains to show $(\#)_p$, i.e., $\Gamma^p_{11} = 0$, for $2 \leq p \leq s+1$. Let $W \subset T_o(X) \otimes T^*_o(X)$ be the complex vector subspace generated $\alpha \otimes \zeta^*$ for $[\alpha] \in \mathcal{S}_o$ and $\zeta^* \in \Phi(\mathcal{N}_\alpha)$. For $A = \Sigma\, a^i_j \frac{\partial}{\partial z_i} \otimes dz^j \in W$ we have from the partial vanishing result [(2.2), Prop.5]

$$\sum_{i,j} a^i_j\, \nabla_i h^{j\bar{k}}(o) = 0. \tag{5}$$

We are going to prove $(\#)_p$ for $2 \leq p \leq s+1$ by using (5) and the Kähler property of h. To extract elements $A \in W$ we consider $[\alpha^o] \in \mathcal{S}_o$ and real curves $\{[\alpha(t)]: -\epsilon < t < \epsilon\}$ in $\mathcal{S}_o$ with $\alpha(o) = \alpha^o$. By [(3.1), Prop.2] we may take $\frac{d\alpha}{dt}(o)$ to be an arbitrary vector $\xi \in \mathcal{H}_{\alpha^o}$. Let $\zeta(t) \in \mathcal{N}_{\alpha(t)}$ be chosen such that $t \longrightarrow \zeta(t)$ is a smooth function. By [(3.1), Prop.3] we know that one always have $\eta := \frac{d\zeta}{dt}(o) \in \mathcal{H}_{\alpha^o}$ and that moreover by varying the choice of $[\alpha(t)]$ we can make $\eta \in \mathcal{H}_{\alpha^o}$ arbitrary. We have now $\alpha(t) = \alpha^o + t\xi + O(t^2)$, $\zeta(t) = \zeta(o) + t\eta + O(t^2)$, so that

$$\alpha(t) \otimes \zeta^*(t) = \{\alpha^o + t\xi + O(t^2)\} \otimes \{\zeta^*(o) + t\eta^* + O(t^2)\} \in W. \tag{6}$$

Expanding in Taylor series, all the coefficients of t^k are elements of W. In particular for $k = 1$ we have

$$\alpha^o \otimes \eta^* + \xi \otimes \zeta^*(o) \in W. \tag{7}$$

As η is arbitrary it follows therefore from (5) that for any p with $2 \leq p \leq s+1$ we can find $\xi \in \mathcal{H}_{\alpha^o}$ and $\zeta \in \mathcal{N}_{\alpha^o}$ such that for $\zeta^* = \Phi(\zeta)$ we have for $1 \leq k \leq n$

$$\nabla_1 h^{p\bar{k}}(o) + \nabla_\xi h^{\zeta^*\bar{k}}(o) = 0. \tag{8}$$

We now compute

$$\begin{aligned}\Gamma^p_{11} &= h^{p\bar{r}}\nabla_1 h_{1\bar{r}}(o) = -h_{1\bar{r}}\nabla_1 h^{p\bar{r}}(o) \\ &= h_{1\bar{r}}\nabla_\xi h^{\zeta^*\bar{r}}(o) = -h^{\zeta^*\bar{r}}\nabla_\xi h_{1\bar{r}}(o).\end{aligned} \tag{9}$$

Since (X,h) is Kähler and $\nabla_i h_{j\bar{k}}(o) = \partial_i h_{j\bar{k}}(o)$ we have

$$\nabla_\xi h_{1\bar{r}}(o) = \nabla_1 h_{\xi\bar{r}}(o). \tag{10}$$

Consequently, by (9) and (4)

$$\begin{aligned}\Gamma^p_{11} &= -h^{\zeta^*\bar{r}}\nabla_\xi h_{1\bar{r}}(o) = -h^{\zeta^*\bar{r}}\nabla_1 h_{\xi\bar{r}}(o) \\ &= h_{\xi\bar{r}}\nabla_1 h^{\zeta^*\bar{r}}(o) = 0,\end{aligned} \tag{11}$$

proving $(\#)_p$ for $2 \leq p \leq s+1$ and hence (*). The proof of Prop.2 is completed. ∎

(4.2) We now deduce from the total geodesy of minimal rational curves C in (X,h) that the latter is Einstein. We work as usual at the point $o \in X$. To begin with we introduce an intermediate algebraic property for the canonical metrics g on X. Recall that P is the Hermitian bilinear form on $S^2\mathfrak{m}^+$ defined by $P(\alpha\circ\beta,\mu\circ\nu) = R_{\alpha\bar{\mu}\beta\bar{\nu}}(X,g)$ and extended by Hermitian bilinearity. We assert

LEMMA 1

For every characteristic vector $\alpha \in \mathfrak{m}^+$, $\alpha\circ\alpha \in S^2\mathfrak{m}^+$ is an eigenvector for the Hermitian bilinear form P.

Proof:

We may assume that α is of unit length. For some choice of Cartan subalgebra $\mathfrak{h}$ of $\mathfrak{k}$ we can write $\alpha = e_\mu$ for the highest noncompact root μ with respect to $\mathfrak{h}^C$. Let φ, $\psi \in \Delta_M^+$, not necessarily distinct, be such that at least one of them is different from μ. From the curvature formula we have

$$R(\alpha,e_\varphi;\alpha,e_\psi) = c\,([e_\mu,e_{-\varphi}];[e_\psi,e_{-\mu}]), \qquad (1)$$

which can be non–zero only if $\mu - \varphi$ and $\psi - \mu$ are equal roots, by [AI.3, Prop.3]. This can only happen if $2\mu = \varphi + \psi$, which contradicts with the assumptions $\mu \geq \varphi$, ψ and $(\varphi,\psi) \neq (\mu,\mu)$. The proof of Lemma 1 is completed. ▌

Given a Kähler metric h of semipositive bisectional curvature, we can define similarly a Hermitian bilinear form P^h on $S^2\mathfrak{m}^+$. We are going to prove

PROPOSITION 1

For any characteristic vector $\alpha \in \mathfrak{m}^+$, $\alpha\circ\alpha$ is an eigenvector of the Hermitian bilinear form P^h on $S^2\mathfrak{m}^+$ induced by h, i.e.,

$$P^h(\alpha\circ\alpha,\xi\circ\eta) = 0 \text{ whenever } h(\alpha,\xi) = 0 \text{ or } h(\alpha,\eta) = 0.$$

Proof:

Quantities defined using the Kähler metric h will be indicated by an h as a subscript or a superscript. We also write $u \perp_h v$ to mean that u and v are orthogonal with respect to h. Since every minimal rational curve C is totally geodesic in

(X,h) we have $R^h(\alpha,\alpha)\alpha = \text{Const}.\alpha$ for any characteristic vector $\alpha \in \mathfrak{m}^+$. Consequently, for $\eta \perp_h \alpha$, $\eta \in \mathfrak{m}^+$, we have $R^h_{\alpha\bar{\alpha}\alpha\bar{\eta}} = 0$. As $\mathcal{S}_o$ is complex–analytic we may let α vary holomorphically on some parameters while $\bar{\alpha}$ varies anti-holomorphically. By polarization and the fact that K acts irreducibly on $\mathfrak{m}^+$ it follows that $R^h_{\alpha\bar{\xi}\alpha\bar{\eta}} = 0$ whenever $\xi \in \mathfrak{m}^+$ and $\eta \perp_h \alpha$, proving Prop.1. ▌

We now deduce

PROPOSITION 2
The Kähler metric h on X is Einstein.

Proof:
We will present a simpler proof than that given in [MOK3]. First observe that as $\mathcal{S}_o$ is connected it follows from Prop.1 that the eigenvalues of P^h associated to any $[\alpha] \in \mathcal{S}_o$ are the same, so that we have $R^h_{\alpha\bar{\alpha}\alpha\bar{\alpha}} = \text{Const}.\|\alpha\|_h^4$. We are going to deduce that

$$(*)\quad R_{\alpha\bar{\alpha}} = c\|\alpha\|_h^2 \text{ for some (positive) constant independent of } \alpha \text{ at } o.$$

First of all we deduce Prop.1 from (*). As in [Ch.6, (2.1), proof of Prop.3] using the fact that $\mathcal{S}_o$ is complex–analytic one deduces by polarization that $R^h_{i\bar{j}}(o) = ch_{i\bar{j}}(o)$ for some constant c. As any point $x \in X$ can be regarded as the origin we have $\rho_h := \text{Ric}(X,h) = c(x)\omega_h$ for the Kähler form ω_h of (X,h). It then follows readily from $0 \equiv d\rho_h = dc \wedge \omega_h$ that the constants c(x) are in fact indepedent of x, proving that (X,h) is Kähler–Einstein.

It remains to prove (*). To this end we imitate the (α,g)–decomposition of $\mathfrak{m}^+$ as follows. Let H^h_α be the h–orthogonal complement of $\mathbb{C}\alpha$ in $\mathbb{C}\alpha + \mathcal{H}_\alpha$ and N^h_α be the h–orthogonal complement of $\mathbb{C}\alpha + \mathcal{H}_\alpha$ in $\mathfrak{m}^+$. We have the h–orthogonal decomposition

$$(\#)\quad \mathfrak{m}^+ = \mathbb{C}\alpha + H^h_\alpha + N^h_\alpha.$$

(Once we have proved that h is a canonical metric, (#) will be seen to be the (α,h)–decomposition of m^+.) Let λ be the common values of $R^h_{\alpha\bar{\alpha}\alpha\bar{\alpha}}$. We are going to show

(**) The decomposition (#) is the eigenspace decomposition of $H^h_\alpha(\xi,\eta) = R^h_{\alpha\bar{\alpha}\xi\bar{\eta}}$ into eigenspaces corresponding to the eigenvalues λ, $\lambda/2$ and 0.

Clearly (**) implies (*) and consequently Prop.2. First from the proof of the integral formula on $\mathcal{S}^*$ and [(1.1) Prop.] (or just from the integral formula on $\mathcal{S}^*_{r-1}$, $r = \mathrm{rank}(X)$, cf. [(2.2), Remarks]), we have $R^h_{\alpha\bar{\alpha}\zeta^*\bar{\zeta}^*} = 0$ for $\zeta \in \mathcal{N}_\alpha$ and $\zeta^* = \Phi(\zeta)$. Write $\Phi_h\colon m^+ \to (m^+)^*$ for the lifting operator using h. We have thus $R^h_{\alpha\bar{\alpha}\sigma\bar{\sigma}} = 0$ for $\sigma = \Phi_h^{-1}(\zeta^*)$. By definition $\mathcal{N}_\alpha$ is the g–orthogonal complement of $\mathbb{C}\alpha + \mathcal{H}_\alpha$, so that $\Phi(\mathcal{N}_\alpha)$ is the annihilator of $\mathbb{C}\alpha + \mathcal{H}_\alpha$ with respect to the natural pairing beteen m^+ and its dual $(m^+)^*$ (defined without using metrics). Thus, $\Phi_h^{-1}\Phi(\mathcal{N}_\alpha)$ is the h–orthogonal complement of $\mathbb{C}\alpha + \mathcal{H}_\alpha$ in m^+, i.e., $\Phi_h^{-1}\Phi(\mathcal{N}_\alpha) = N^h_\alpha$ in the definition of (#) and we have $R^h_{\alpha\bar{\alpha}\sigma\bar{\sigma}} = 0$ for $\sigma = \Phi_h^{-1}(\zeta^*) \in N^h_\alpha$. As (X,h) is of semipositive curvature, N^h_α is in fact in the zero–eienspace of H^h_α. It remains to show that $R^h_{\alpha\bar{\alpha}\alpha\bar{\xi}} = 0$ for $\xi \in H^h_\alpha$ and $R^h_{\alpha\bar{\alpha}\xi\bar{\xi}} = \lambda/2$. From the idenification $T_o(\mathcal{S}_o) \cong (\mathbb{C}\alpha + \mathcal{H}_\alpha) / \mathbb{C}\alpha$ given any $\xi \in H^h_\alpha$ with $\|\xi\|_h = 1$ we can find a smooth curve $\{[\alpha(t)]: -\epsilon < t < \epsilon\}$ such that $\alpha(o) = \alpha$ and $\frac{d\alpha}{dt}(o) = \xi$. Write $\alpha(t) = \alpha + t\xi + O(t^2)$ as an h–orthogonal decomposition. From $R^h_{\beta\bar{\beta}\beta\bar{\beta}} = \lambda\|\beta\|_h^4$ for any $[\beta] \in \mathcal{S}_o$ it follows that

$$R^h(\alpha(t),\overline{\alpha(t)};\alpha(t),\overline{\alpha(t)}) = \lambda\,(1 + 2t^2 + O(t^4)). \tag{1}$$

On the other hand

$$R^h(\alpha(t),\overline{\alpha(t)};\alpha(t),\overline{\alpha(t)})$$
$$= R^h_{\alpha\bar{\alpha}\alpha\bar{\alpha}} + (4\mathit{Re}R^h_{\alpha\bar{\alpha}\alpha\bar{\xi}})t + (4R^h_{\alpha\bar{\alpha}\xi\bar{\xi}})t^2 + O(t^3). \tag{2}$$

Comparing (1) and (2) we see immediately that

$$ReR^{h}_{\alpha\bar{\alpha}\alpha\bar{\xi}} = 0, \text{ and} \tag{3}$$
$$R^{h}_{\alpha\bar{\alpha}\xi\bar{\xi}} = \lambda/2. \tag{4}$$

By changing ξ to $\sqrt{-1}\,\xi$ (3) actually implies $R^{h}_{\alpha\bar{\alpha}\alpha\bar{\xi}} = 0$ for $\xi \in H^{h}_{\alpha}$. Together with (4) we have completed the proof of (**), proving hence (*) and Prop.2. From Prop.2 and the theorem of Matsushima [(2.1), Thm.1] we have completed the proof that any Kähler metric h of semipositive bisectional curvature on X is in fact a canonical metric, proving the Kähler metric rigidity theorem in the semipositive case [(1.1), Thm.1].

(4.3) We mention in this section a partial rigidity theorem for Hermitian metrics of semipositive curvature. For a Hermitian metric s on T^{*}_{X} write $\Theta(s)$ for the curvature on (T^{*}_{X},s). For any Hermitian metric h on X we can define the Hermitian bilinear form Q^{h}_{x} at x on $T_{x} \otimes T^{*}_{x}$ by $Q^{h}_{x}(\beta\otimes\xi^{*},\gamma\otimes\eta^{*}) = -\Theta_{\xi^{*}\bar{\eta}^{*}\beta\bar{\gamma}}(h^{*})$ for $\beta, \gamma \in T_{x}$ and $\xi^{*}, \eta^{*} \in T^{*}_{x}$ and extended by Hermitian bilinearity. Q^{g}_{x} is positive semidefinite for a canoncial metric g. (Q^{g}_{0} is the same as Q defined in [(2.2), after Prop.5] if we identify $(m^{+})^{*}$ with m^{-} via contraction with the metric tensor g.) Denote by Z_{x} the zero–eigenspace of Q^{g}_{x}. Let K_{x} be the isotropy group of (X,g) at $x \in X$. From [(2.2), Lemma 1] Z_{x} is irreducible as a K_{x}–representation space. We have

Theorem 1

Let (X,g) be an irreducible Hermitian symmetric manifold of rank ≥ 2. Let h be a Hermitian metric of semipositive curvature on X. Then, for any $x \in X$ the Hermitian bilinear form Q^{h}_{x} vanishes identically on Z_{x}.

<u>Proof</u>:

For $[\alpha] \in \mathcal{S}_{x}$, $\zeta \in \mathcal{N}_{\alpha}$ and $\alpha^{*} = \Phi_{x}(\alpha)$ the integral formula of [(2.1), Prop.3] implies that $\Theta_{\alpha^{*}\bar{\alpha}^{*}\zeta\bar{\zeta}}(h^{*}) = 0$. Since (T^{*}_{X},h^{*}) is of seminegative curvature we have also $\Theta_{\alpha^{*}\bar{\eta}^{*}\zeta\bar{\zeta}}(h^{*}) = 0$ for any $\eta^{*} \in T^{*}_{x}(X)$. Recall that $\mathcal{S}_{x}$ is complex–analytic and that as α varies holomorphically, ζ can be made to vary anti–holomorphically. Since Φ_x is conjugate complex–linear we can make ζ vary holomorphically while $\zeta \otimes \alpha^*$ varies anti–holomorphically, so that by polarization we obtain from $\Theta_{\alpha^*\overline{\eta}^*\zeta\overline{\zeta}}(h^*) = 0$ actually $\Theta_{\alpha^*\overline{\eta}^*\zeta\overline{\beta}}(h^*) = 0$ for any $\eta^* \in T_x^*$ and any $\beta \in T_x$. In other words, we have $Q_x^h(\zeta\otimes\alpha^*, \beta\otimes\eta^*) = 0$. Since Z_x is irreducible as a K_x–representation space we have $\Sigma\, \mathbb{C}(\zeta\otimes\alpha^*) = Z_x$, so that in particular $Q_x^h | Z_x \equiv 0$, as asserted in Thm.1. ▌

PART II

FURTHER DEVELOPMENT

CHAPTER 8 THE HERMITIAN METRIC RIGIDITY THEOREM FOR QUOTIENTS OF FINITE VOLUME

§1 Compactifications of Arithmetic Varieties and an Integral Formula

(1.1) The Hermitian metric rigidity theorem for the seminegative case [Ch.6, (1.1), Thm.1] was recently generalized by To [To] to the case of finite volume

THEOREM 1 (To [To,1988])

Let X be a quotient of an irreducible bounded symmetric domain Ω of rank ≥ 2 of finite volume in the canonical metric. Let h be a Hermitian metric on X of seminegative curvature in the sense of Griffiths. Then, h is necessarily a constant multiple of the canonical metric.

Similar theorems in the irreducible, locally reducible case [Ch.6, (4.1), Thm.1] remain true in the general case of quotients of finite volume. The strength of Thm.1 lies in the fact that absolutely no growth conditions are imposed on the Hermitian metric h. As will be seen, the proof relies rather heavily on precise knowledge of compactifications $\overline{X}$ of X and the asymptotic behavior of the canonical metrics near $\overline{X} - X$. In this chapter we will prove Thm.1 based on such results, which we will collect in the present section. For applications, the Kähler case is of particular interest. We will also give another proof of Thm.1 in the special case of Kähler metrics h. The alternative proof in this special case has the advantage of being more elementary and is based on the fact that such manifolds X admit compactifications $\overline{X}$ such that $\overline{X} - X$ is of codimension ≥ 2 in $\overline{X}$.

We recall that in the proof of the rigidity theorem for holomorphic mappings [Ch.6, (5.1), Thm.1] we proved the special case of Thm.1 under the assumption that $h \leq Cg$ for some constant C. The proof there actually works under the weaker assumption that $h(x) \leq C(x)g(x)$ on X such that $\log C(x)$ is square–integrable with respect to (X,g). To prove Thm.1 To showed that the latter condition is automatically satisfied. He used the singular minimal compactification of SATAKE–BAILY–BOREL ([SA1] and [BB]) and non–singular toroidal compactifications of ASH–MUMFORD–RAPOPORT–TAI [AMRT] for certain arithmetic varieties. (The prototype is the case of $\Omega = \Delta \cong H$, the upper half–plane, with $\Gamma \subset SL(2,\mathbb{Z}) \subset SL(2,\mathbb{R})$ a torsion–free subgroup in $SL(2,\mathbb{Z})$ of finite index under this identification). Here by a compactification of X we mean a compact complex space $\overline{X}$ together with a holomorphic embedding $\sigma: X \hookrightarrow \overline{X}$ such that $\overline{X} - \sigma(X)$ is a complex–analytic subvariety of $\overline{X}$. We will always identify X with its image $\sigma(X)$. Regarding such compactifications of arithmetic varieties we quote

THEOREM 2 (SATAKE [SA1] & BAILY–BOREL [BB])
Let X be an n–dimensional, $n \geq 2$, irreducible quotient of a bounded symmetric domain by a (torsion–free) arithmetic subgroup Γ of $\mathrm{Aut}(\Omega)$. Then, there exists a compactification $\overline{X}_{min}$ of X such that (i) $\overline{X}_{min}$ is a normal projective algebraic variety and (ii) the complex–analytic variety $\overline{X}_{min} - X$ is of complex codimension ≥ 2 in $\overline{X}_{min}$.

We call the Satake–Baily–Borel compactification the minimal compactification (hence the notation $\overline{X}_{min}$) because of the following theorem of Borel [BO5].

THEOREM 3 (BOREL [BO5])
Let $\overline{X}$ be any compactification of X. Then, there exists a (unique) holomorphic map $\tau: \overline{X} \longrightarrow \overline{X}_{min}$ inducing the identity map on X.

The minimal compactifications are in general highly singular. On the other hand we have the toroidal compactifications of ASH–MUMFORD–RAPOPORT–TAI [AMRT] with mild singularities. We refer the reader to [AMRT] for definitions and properties of such compactifications. Our main concern is the behavior of the canonical metric of X on non–singular toroidal compactifications $\overline{X}$ of X (if one exists). Let M be any non–singular compactification of X such the $D := M - X$ is a union of smooth divisors on M intersecting at worst at normal crossings. Let L be a holomorphic line bundle on M and h be a Hermitian metric on $L|_X$. Denote by [D] the divisor line bundle and $\|.\|$ the norm on [D] arising from some Hermitian metric on [D]. Let s be the canonical section on M of [D]. We have

DEFINITION 1
The Hermitian metric h on $L|_X$ is said to be of logarithmic growth with respect to M if and only if for any local basis e of L over an open subset U_0 of M and for any open $U \subset\subset U_0$ there exist nonnegative numbers α and β such that

$$\mathrm{Const.}|\log\|s\||^{-\beta} \leq h(e,e) \leq \mathrm{Const.}|\log \|s\||^{\alpha}.$$

Denote by g a canonical metric on X and by ω its Kähler form. For a complex manifold N we denote by K_N the canonical line bundle. We have

THEOREM 4 (ASH–MUMFORD–RAPOPORT–TAI [AMRT], MUMFORD [MUM1]) Replacing Γ by a subgroup of finite index if necessary we may always assume that X admits a non–singular toroidal compactification M such that on X the volume form ω^n of the canonical metric defines a Hermitian metric on $K_X \otimes [D]|_X \cong (K_M \otimes [D])|_X$ of logarithmic growth with respect to M.

For a sharper upper bound of the volume form ω^n we have the following Ahlfors–Schwarz lemma for volume forms proved in YAU [YAU].

THEOREM 5 (YAU [YAU]) Let (M,μ) be a complete Kähler manifold such that $\mathrm{Ricci}(M,\mu) \geq -1$ and (N,ν) be a Kähler manifold of the same dimension such that $\mathrm{Ricci}(N,\nu) \leq -1$. Let $f: M \to N$ be a holomorphic mapping. Then, we have everywhere on M

$$\frac{f^*(\nu^n)}{\mu^n} \leq 1.$$

It follows from Thm.3 and the Ahlfors–Schwarz lemma for volume forms that ω^n of (X,g) satisfies stronger estimates. Write $D = \cup_i D_i$ for the decomposition of D into irreducible components. We have correspondingly $[D] = [D_1] \otimes \ldots \otimes [D_k]$. Write s_i for the canonical section of $[D_i]$. Equip $[D_i]$ with Hermitian metrics and denote also by $\|.\|$ norms arising from such metrics. Use the norm on $[D]$ such that $\|s\| = \Pi_i\|s_i\|$. We have

$$\mathrm{Const.}|\log\|s\||^{-\beta} \leq h(e,e) \leq \mathrm{Const.}|\Pi_i \log\|s_i\||^{-2} \qquad (1)$$

for some real number β. The lower bound is given by Thm.4. The upper bound follows from Thm.5 by considering at points b on $\overline{X} - X$ neigborhoods U' on $\overline{X}$ biholomorphic to the polydisc Δ^n such that $U' \cap X$ can be identified with $(\Delta^*)^k \times \Delta^{n-k}$ (called a punctured polydisc if $k \geq 1$). Let θ be the Kähler form of the Poincaré metric on $U' \cap X$. Then, applying the Ahlfors–Schwarz lemma to the inclusion map $(U\cap X,\theta|_{U\cap X}) \hookrightarrow (X,\omega)$ for any $U \subset\subset U'$ one obtains immediately the upper bound in (1).

Crucial to the proof of Thm.1 of To is the following relationship between the toroidal and the minimal compactifications of arithmetic varieties.

THEOREM 6 (cf. MUMFORD [MUM1])
Let X be an arithmetic variety admitting a non–singular toroidal compactification $\overline{X}$. Denote by X_{min} the minimal compactification. Let $\tau: \overline{X} \longrightarrow X_{min}$ be the unique holomorphic map extending the identity map on X. Denote by D the divisor $\overline{X} - X$ and by $K_{\overline{X}}$ the canonical line bundle on $\overline{X}$. Then, there exists a holomorphic line bundle Λ on X_{min} such that $K_{\overline{X}} \otimes [D] \cong \tau^*(\Lambda)$.

Finally we will need the following lemma on plurisubharmonic functions.

LEMMA 1
Let M be a connected complex–analytic variety and $V \subset M$ be a complex–analytic subvariety. Let φ be any plurisubharmonic function on $M - V$. Then,

(a) If M is smooth and φ is uniformly bounded from above on compact subsets of M, then φ extends to a plurisubharmonic function on M.

(b) If V is of codimension ≥ 2 in M (not necessarily smooth), then φ is uniformly bounded on compact subsets of M.

The extension $\tilde{\varphi}$ of φ is defined by $\tilde{\varphi}(y) = \lim\sup_{x \to y, x \in U-V} \varphi(x)$. (b) is simply a consequence of the maximum principle for plurisubharmonic functions. (One restricts φ to analytic discs.).

(1.2) We give here the proof of [(1.1), Thm.1] of To. Let Ω be an irreducible bounded symmetric domain of rank ≥ 2. By the Arithmeticity Theorem of Margulis (cf. ZIMMER [ZIM]) any lattice Γ in $\mathrm{Aut}(\Omega)$ is arithmetic. By [(1.1), Thm.4] it suffices to consider those Γ for which $X = \Omega/\Gamma$ admits a non–singular toroidal compactification $\overline{X}$. By [(1.1), Thm.6] of Mumford $K_{\overline{X}} \otimes [D] \cong \tau^*(L)$ for some holomorphic line bundle L on X_{min}. Cover now X_{min} by open sets $\mathcal{U} = \{U_\alpha\}$ such that $L|U_\alpha$ is holomorphically trivial. Write $\mathcal{W} = \{W_\alpha\}$ for the corresponding covering on $\overline{X}$ defined by $W_\alpha = \tau^{-1}(U_\alpha)$. We assert

PROPOSITION 1 (TO [TO])

Let s be a canonical section of the divisor line bundle [D]; $\|.\|$ be a norm arising from any smooth Hermitian metric on [D] over $\overline{X}$; and σ be any positive (1,1)–form. Then, there exists a function ψ on $\overline{X}$ bounded from above such that on X

$$\theta = e^{\psi}\,\|s\|^{-2}\sigma^n.$$

Proof:

Fix $U = U_\alpha$ and let e be a holomorphic basis of $L|_U$ on U. Thus $\epsilon = \tau^*e$ is a holomorphic basis of $K_{\overline{X}} \otimes [D]$ over $W = W_\alpha = \tau^{-1}(U)$. $\sqrt{-1}\,\epsilon \wedge \overline{\epsilon}$ is a positive (n,n)–form on W_α. Write $\theta = e^{\eta}(\sqrt{-1}\,\epsilon \wedge \overline{\epsilon})$, where η is a smooth function on $W - D$. Since ϵ is holomorphic and nowhere–vanishing on $W - D$, we have $2\pi c_1(K_X^{-1}|_{W-D}, \theta) = -\sqrt{-1}\,\partial\overline{\partial}\eta$. As (K_X^{-1}, θ) is of seminegative curvature, η is a plurisubharmonic function on $W - D$. Write S for $X_{min} - X$. Write φ for the function on $U - S$ such that $\tau^*(\varphi) = \eta$. By Lemma 1 φ is a plurisubharmonic function uniformly bounded from above on compact subsets of U. Since W is smooth, again by Lemma 1 $\eta = \tau^*(\varphi)$ extends to a plurisubharmonic function on $W = W_\alpha$. As $\theta = e^{\eta}(\sqrt{-1}\,\epsilon \wedge \overline{\epsilon})$ on $W = W_\alpha$ and ϵ is a basis of $K_{\overline{X}} \otimes [D]$ over W we have $(\sqrt{-1}\,\epsilon \wedge \overline{\epsilon}) = \gamma\,\|s\|^{-2}\sigma^n$ on W, where γ is a smooth function on W. Hence,

$$\theta = e^{\eta}\,\gamma\,\|s\|^{-2}\sigma^n.$$

Since η extends to a plurisubharmonic function on W, the function $\psi := \eta + \log(\gamma)$ is uniformly bounded from above on compact subsets of W. On the other hand ψ is globally defined by the equation $\theta = e^{\psi}\,\|s\|^{-2}\sigma^n$; i.e. ψ is the same as the function defined in the proposition. The proof of Prop. 1 is thus completed. ▌

Proof of [(1.1), Thm.1]:

The Hermitian metric rigidity theorem [(1.1), Thm.1] follows now from Prop.1 and a justification of the integral formula in [Ch.6, (5.1), Prop.2]. Recall that ω is the Kähler form of the canonical metric g on X and h is some Hermitian metric of seminegative curvature on X. Replacing h by g + h we may assume that $h \geq g$ on X. Write $\hat{g}$ resp. $\hat{h}$ for the induced Hermitian metric on the tautological line bundle L over $\mathbb{P}T_X$. Write $\hat{h} = e^u\hat{g}$ on X. Since $\hat{h} \geq \hat{g}$ we have $u \geq 0$. By [Ch.6, (5.1), Remarks] it suffices to prove that u is square–integrable on the char–

acteristic bundle $\pi: \mathcal{S} \to X$, $\mathcal{S} \subset \mathbb{P}T_X$ w.r.t. the Kähler form $\nu = -c_1(L,\hat{g}) + \pi^*\omega$ on $\mathcal{S}$. Write η for the Hermitian form of h. Since $\eta \geq \omega$ we have $\eta^n = e^v\omega^n$ with $v \geq 0$. For the same reason, given any $x \in X$, it is clear that $v(x) \geq u([\xi])$ for $[\xi] \in \mathbb{P}T_x(X)$. It follows that for $p = \dim_{\mathbb{C}}\mathcal{S}_x$, $n = \dim_{\mathbb{C}}X$

$$\int_{\mathcal{S}} u^2\,\nu^{n+p} = \text{Const.}\int_X u^2\left[\int_{\mathcal{S}_x}(-c_1(L,\hat{g}))^p\right]\omega^n$$

$$\leq \text{Const.}\int_X v^2\,\omega^n \tag{1}$$

since the integral $\int_{\mathcal{S}_x}(-c_1(L.\hat{g}))^p$ is independent of $x \in X$ and since $u([\xi]) \leq v(x)$ for $[\xi] \in \mathbb{P}T_x(X)$. The proof of [(1.1), Thm.1] is now reduced to showing that v is square–integrable on (X,ω). We use Prop.1 and [(1.1), Thms. 4 & 5] to deduce the latter statement. In the notations of the proof of Prop.1 we have by [(1.1), Thms. 4 & 5] and Prop.1

$$\text{Const. } \|s\|^{-2}|\log\|s\||^{-\beta}\sigma^n \leq \omega^n \leq \text{Const. } \|s\|^{-2}|\Pi_i\log\|s_i\||^{-2}\sigma^n \tag{2}$$

$$\omega^n \leq \eta^n \leq \text{Const. } \|s\|^{-2}\sigma^n. \tag{3}$$

Consequently, we have $\eta^n = e^v\omega^n$ with

$$v \leq \text{Const.} -\beta\log(|\log\|s\||). \tag{4}$$

From (3) and (1) we deduce

$$\int_X v^2\,\omega^n \leq \text{Const.}\int_X (\log|\log\|s\||)^\beta\,\omega^n$$

$$\leq \text{Const.}\int_X \frac{(\log|\log\|s\||)^{2\beta}}{\|s\|^2\,|\Pi_i\log\|s_i\||^2}\,\sigma^n \tag{5}$$

for some nonnegative real number β. Consider in local coordinates $W = W_\alpha \cong \Delta^n$ and $W - D \cong (\Delta^*)^k \times \Delta^{n-k}$. Write $(z_1,\ldots,z_n)$ for the Euclidean coordinates on Δ^n and denote by $d\lambda$ the Euclidean volume form. To justify the finiteness of the last integral it suffices to observe that for $1 \leq k \leq n$ and $\delta > 0$

$$\int_{\Delta^n(1/2)} \frac{1}{|z_1|^2\ldots|z_k|^2\,[(-\log|z_1|)\ldots(-\log|z_k|)]^{1+\delta}}\,d\lambda$$

$$\leq \text{ Const.} \left[\iint_{\Delta(1/2)} \frac{1}{|z|^2 \, (-\log|z|)^{1+\delta}} \sqrt{-1}\, dz \wedge d\bar{z}\right]^k < \infty. \tag{6}$$

We have thus established

$$\int_S u^2 \nu^{n+p} < \text{Const.} \int_X v^2 \omega^n < \infty, \tag{7}$$

proving by the integral formula [Ch.6, (5.1), proof of Thm.3] that indeed $h = cg$, implying the Hermitian metric rigidity theorem [(1.1), Thm.1] of To's.

As an immediate consequence of [(1.1), Thm.1] and the proof of [Ch.6, (5.1), Thm.3] we have the following rigidity theorem on holomorphic mappings.

THEOREM 1 (TO [TO,1988])
Let X be a quotient of an irreducible bounded symmetric domain Ω of rank ≥ 2 of finite volume in the canonical metric. Let (N,h) be any Kähler manifold of seminegative bisectional curvature and $f: X \longrightarrow N$ be a holomorphic mapping which is not totally degenerate. Then, f is necessarily a totally geodesic isometric immersion up to a normalizing constant.

§2 An Alternative Proof in the Kähler Case

(2.1) The proof of the Hermitian metric rigidity theorem [(1.1), Thm.1] requires very precise information from the theory of compactification of arithmetic varieties. For pedagogical reasons we include in this section an alternative proof in the Kähler case, which implies the rigidity theorem [(1.2), Thm.1] for holomorphic mappings, based on less precise information on X. We will only use the fact that $X = \Omega/\Gamma$ admits compactifications M such that $M - X$ is of codimension ≥ 2.

Recall that [(1.1), Thm.1] was first proved under some growth conditions on the Hermitian metric h on X of seminegative curvature. A reason to expect [(1.1), Thm.1] to hold is the so-called Koecher's principle (cf. BOREL [BO3]): one expects growth conditions for holomorphic objects to be automatic on X because X is pseudoconcave. Here we are concerned instead with Hermitian metrics of seminegative curvature. From the maximum principle for plurisubharmonic functions it is indeed not difficult to show that in some sense h does not have "essential singularities" along D. Nonetheless, it is not possible to justify the integral formula based on pseudoconcavity alone. In the Kähler case one is dealing with Kähler forms, which are closed positive (1,1)–forms. It is reasonable to expect Koecher's principle to hold for closed positive currents, as most extension theorems on holomorphic functions are valid for closed positive currents. Our formulation of the alternative in the Kähler case is based on this perspective. We will justify the integral formula by showing that any Kähler metric on X is of finite volume. We do this by proving an extension theorem of the Kähler form as a closed positive (1,1)–current to any smooth compactification of X.

Our principal tool is the Bishop Extension Theorem for closed positive currents. To formulate this we start with some definitions. Let $D \subset \mathbb{C}^n$ be a domain. A (p,p)–current T on D is said to be positive if for any smooth (1,0)–forms $\alpha_1, ..., \alpha_{n-p}$ of compact support the distribution $T \wedge \sqrt{-1}\, \alpha_1 \wedge \overline{\alpha}_1 \wedge \dots \wedge \sqrt{-1}\, \alpha_{n-p} \wedge \overline{\alpha}_{n-p}$ is a nonnegative measure. Let $\beta = \frac{\sqrt{-1}}{2} \partial\overline{\partial}|z|^2$ be the Kähler form of the Euclidean metric. T is said to be closed if it is a closed current in the sense of de Rham. Given any positive (p,p)–current T on D we define $\sigma_T := T \wedge \frac{1}{(n-p)!} \beta^{n-p}$, which is a nonnegative measure. The mass of T on a σ_T–measurable subset E of D is defined to be $\mathrm{Mass}(T;E) := \sigma_T(E)$.

The prototype of a closed positive (p,p)–current is given by a complex–analytic subvariety $V \subset D$ of pure complex codimension p. Associated to V is the (p,p)–current [V] defined by $[V](\varphi) = \int_{\mathrm{Reg}(V)} \varphi$ for an (n–p,n–p) form φ of compact support. It was proved in LELONG [LE] that [V] is defined even in the singular case. The definition of the mass of closed positive currents agrees with that of volumes of complex–analytic varieties; i.e., $\mathrm{Mass}([V];E) = \mathrm{Volume}(V \cap E;\beta)$. For a closed positive (p,p)–current T, LELONG [LE] defined a density number n(T,x) at a point $x \in \mathrm{Supp}(T)$ which measures the concentration of mass in small neighborhoods of x. The Lelong number n(T,x) is a generalization of multiplicities of complex–analytic subvarieties in the sense that for V a complex–analytic subvariety of D of pure codimension p and [V] the integral closed (p,p)–current defined by V, we have $n([V],x) = \mathrm{mult}(V;x)$ by a theorem of THIE [THI].

The Bishop Extension Theorem for complex analytic varieties is given by

THEOREM 1 (cf. STOLZENBERG [STO])
Let $D \subset \mathbb{C}^n$ be a domain and $A \subset D$ be a complex–analytic subvariety. Let $V \subset D - A$ be a pure–dimensional complex–analytic subvariety of finite volume with respect to the Euclidean metric. Then, the topological closure $\overline{V}$ of V is a pure–dimensional complex–analytic subvariety of D.

The Bishop Extension Theorem for closed positive currents was proved by SKODA [SKO] and later generalized to the case of closed pluripolar sets A by EL MIR [EL]. ($A \subset D$ is said to be pluripolar if it is locally contained in the $(-\infty)$–set of some plurisubharmonic function ψ.) Obviously a complex–analytic subvariety is pluripolar. More recently, a simpler proof was given by SIBONY [SI]. We have

THEOREM 2 (SKODA [SKO2] & EL MIR [EL])
Let $D \subset \mathbb{C}^n$ be a domain, $A \subset D$ be a closed pluripolar subset; and V be a closed (p,p)–current defined on $D - A$ of finite mass with respect to the Euclidean metric. Then, T can be trivially extended to a closed positive (p,p)–current $\tilde{T}$ on D.

Let $\{\chi_i\}$ be a sequence of smooth functions with compact support on $D - A$. The trivial extension $\tilde{T}$ of T is defined by $\tilde{T}(\varphi) = \lim_{i\to\infty} T(\chi_i\varphi)$ for a smooth $(n-p,n-p)$–form φ of compact support on D. From the assumption that T is positive and of finite mass it follows readily that the coefficients of T are integrable. Consequently $\tilde{T}$ can be defined and is independent of the choice of $\{\chi_i\}$. The crux of Thm.2 is to show that $\tilde{T}$ is closed.

Let $T = [V]$ be the integral closed (p,p)–current defined by V on $D - A$. Let $\tilde{T}$ be the closed trivial extension of T to D as guaranteed by Thm.2. By a theorem of SIU [SIU1], the set $E_1([V]) = \{x \in D: n(\tilde{T},x) \geq 1\}$ is a complex–analytic subvariety of D. By THIE [THI], $E_1([V])$ contains $\overline{V}$. We have necessarily $E_1([V]) = \overline{V}$ since $\tilde{T}$ is a trivial extension of T, which shows that indeed Thm.2 is a generalization of the classical Bishop Extension Theorem (Thm.1).

We will also need the Riemann Extension Theorem for closed positive (1,1)–currents, which is a very easy case of a Thullen Extension Theorem of SIU [SIU1].

THEOREM 3 (SIU [SiU1])
Let $D \subset \mathbb{C}^n$ be a domain and $A \subset D$ be a complex–analytic subvariety of codimension ≥ 2. Let T be a closed positive (1,1)–current defined on $D - A$. Then, T extends trivially to a closed positive (1,1)–current on D.

(2.2) We return now to the proof of the Kähler case of To's Theorem [(1.1), Thm.1]. Recall that X is a complex–analytic manifold admitting a projective–algebraic compactification Z such that $S = Z - X$ is of codimension ≥ 2 in Z. In the notations of §1, we formulate

LEMMA 1
Let (X,h) be as in [(1,1), Thm.1] with Hermitian form η. Let g, with Kähler form ω, denote a canonical metric on X. Suppose $g + h$ is of finite volume, then $h = cg$ for some constant c.

Proof:
Write $(\eta + \omega)^n = e^v \omega^n$. We have $v \geq 0$. It follows readily from $\text{Volume}(X,\eta+\omega) < \infty$ that in fact v is square–integrable with respect to ω^n.

The key point of the alternative proof in the Kähler case is the following proposition, which is of some independent interest.

PROPOSITION 1
Let X be a complex–analytic manifold admitting a projective–algebraic compactification Z such that $S = Z - X$ is of codimension ≥ 2 in Z. Then, any Kähler metric on X is of finite volume.

We remark that Prop.1 remains true if M is only assumed to be a compact complex space.

Let now M be a smooth projective–algebraic (hence Kähler) compactification of X obtained by desingularizing Z. Denote by $\rho: M \longrightarrow Z$ the holomorphic map thus obtained extending the identity map on X. We break up the proof of Prop.1 into

PROPOSITION 1.1

Let η be a closed positive (1,1)–form on $X \subset M$. Then, η can be trivially extended as a closed positive (1,1)–current to M.

PROPOSITION 1.2

Let T be a closed positive (1,1)–current on the n–dimensional compact manifold M such that $T|_X = \eta$ is smooth. Then, $\int_X \eta^n < \infty$.

Proof of Prop. 1.1:

Fix a Kähler metric on M with Kähler form ν. For a positive current T and E a Borel subset of M we define $\text{Mass}(T,\nu;E)$ in the same way as in (2.1) by replacing β by ν. Denote the infinity $M - X$ by D. By the Bishop Extension Theorem for closed positive currents [(2.2), Thm.2] it suffices to show that $\text{Mass}(\eta,\nu;X) < \infty$. To prove this we make use of the compactification $X \hookrightarrow Z$ such that $S = Z - X$ is of codimension ≥ 2 in Z. Cover Z by coordinate open sets U_α and write $W_\alpha = \rho^{-1}(U_\alpha)$. It suffices to show that for any choice of $U = U_\alpha$; $W = W_\alpha$, we have $\text{Mass}(\eta,\nu;W) < \infty$. We are going to define a closed positive current P on W such that on $W - D$, $P \geq \eta$ as currents and such that P extends trivially to a closed positive current on W. Shrinking U and hence W if necessary this shows that $\text{Mass}(P,\nu;W) < \infty$ and hence $\text{Mass}(\eta,\nu;W) \leq \text{Mass}(P,\nu;W) < \infty$, which implies again by [(2.2), Thm.2] that η can be trivially extended to a closed positive (1,1)–current on M (by using the covering $\{W_\alpha\}$ of M).

To construct P we make use of the Riemann extension for plurisubharmonic functions [(1.1), Lemma 1]. By the Weierstrass Preparation Theorem one can take a coordinate projection $\pi: U \longrightarrow G$ to a domain G on some $\mathbf{C}^n$ such that $G \cong \Delta^n$ and such that π is a finite proper map. The composition $\pi \circ \rho|_W: W \longrightarrow G$ is a proper map. $\theta := \pi \circ \rho|_{W-D}: W - D \longrightarrow G - \pi(S)$ is a branched covering. Note that $\pi(S)$ is of codimension ≥ 2 in G. One can define the direct image $\theta_*(\eta)$ by using the projection formula $\theta_*(\eta)(\varphi) = \eta(\theta^*(\varphi))$ for a smooth (n–1,n–1)–form φ of compact support on $G - \pi(S)$. By the Thullen Extension Theorem [(2,2), Thm.3] $\theta_*(\eta)$ extends trivially to a closed positive (1,1)–current Q on G. On $G \cong \Delta^n$ one can solve the Poincaré–Lelong equation $\sqrt{-1}\partial\bar\partial\psi = Q$. The function $\theta^*(\psi)$ is a plurisubharmonic function on W. We define now $P = \sqrt{-1}\partial\bar\partial(\theta^*\psi)$ on

W. Clearly $P = \theta^*\theta_*(\eta) \geq \eta$ on W-D. P is therefore the closed positive (1,1)–current we looked for on W. The proof of Prop. 1 is completed. ∎

Proof of Prop. 1.2:
Let T be a closed positive (1,1)–current on M such that $T|_X = \eta$ is smooth. As (M,ν) is Kähler there is a harmonic (1,1)–form σ such that $T = \sigma + \sqrt{-1}\partial\overline{\partial}\mu$ for some integrable function μ on M. Since $T \geq 0$ and $\sigma \leq$ -Const.ν as (1,1)–currents it follows that $\sqrt{-1}\partial\overline{\partial}\eta \geq$ –Const.ν on M. In particular μ is bounded from above on M. We want to show that $\int_X \eta^n = \int_X (\sigma + \sqrt{-1}\partial\overline{\partial}\mu)^n < \infty$. Define $\mu_c = \max(\mu,c)$. Suppose U is an open subset of M and λ is a smooth plurisubharmonic function such that $\mu + \lambda$ is plurisubharmonic on U. Then, $\mu_c = \max(\mu+\lambda,c+\lambda) - \lambda$, so that $\mu_c + \lambda$ is a plurisubharmonic function. It follows that there exists a positive constant C independent of c such that $\sqrt{-1}\partial\overline{\partial}\mu_c \geq -C\nu$ everywhere on M. Define $T_c := \sigma + \sqrt{-1}\partial\overline{\partial}\mu_c$. On any $K \subset\subset X \subset M$, $T_c \equiv T$ for $c \leq c(K)$. The closed (1,1)–current $R_c := T_c + C\nu$ is positive. We are going to define R_c^n as a measure in such a way that $\int_M R_c^n = \int_M (\sigma + C\nu)^n$, hence independent of c. We use the definition of the Monge–Ampère operator on bounded plurisubharmonic functions as given in BEDFORD–TAYLOR [BT].

Locally $R_c = \sqrt{-1}\partial\overline{\partial}\xi$ for some bounded plurisubharmonic function ξ. Define $(\sqrt{-1}\partial\overline{\partial}\xi)^2 := \sqrt{-1}\,\partial\overline{\partial}(\xi\sqrt{-1}\partial\overline{\partial}\xi)$. $(\sqrt{-1}\partial\overline{\partial}\xi)^2$ is defined as a closed (2,2)–current because ξ is bounded. Clearly if ξ is replaced by $\xi + \eta$ for a pluriharmonic function η, we obtain the same current. Thus, $R_c^2 = (\sqrt{-1}\partial\overline{\partial}\xi)^2$ is well–defined. Shrinking the domain of definition of ξ if necessary we can mollify ξ to get smooth plurisubharmonic functions $\{\xi_\epsilon\}_{\epsilon>0}$ decreasing to ξ. It follows that $R_c^2 = (\sqrt{-1}\partial\overline{\partial}\xi)^2$ is the limit of the positive closed currents $(\sqrt{-1}\partial\overline{\partial}\xi_\epsilon)^2$, so that R_c^2 is a closed positive (2,2)–current. From this it is clear how one can proceed inductively to define R_c^i, $1 \leq i \leq n$, to arrive at the positive (n,n)–current $(\sqrt{-1}\partial\overline{\partial}\xi)^n = R_c^n$. We have $R_c = \sigma + C\nu + \sqrt{-1}\partial\overline{\partial}\mu_c$. We are going to show

$$\int_M R_c^n = \int_M (\sigma + C\nu + \sqrt{-1}\partial\overline{\partial}\mu_c)^n = \int_M (\sigma + C\nu)^n. \qquad (1)$$

By expressing μ_c locally as a difference $(\mu_c+\lambda) - \lambda$ where $\mu_c+\lambda$ is plurisubharmonic and λ is smooth it is clear how the closed (p,p)–forms $(\sqrt{-1}\partial\bar\partial\mu_c)^p$, $1 \leq p \leq n$ are defined. It remains true that $(\sqrt{-1}\partial\bar\partial\mu_c)^2 = \sqrt{-1}\partial\bar\partial(\mu_c\sqrt{-1}\partial\bar\partial\mu_c)$ for the function μ_c bounded from above, etc. It is then immediate that one can justify inductively integration by parts for the forms $(\sqrt{-1}\partial\bar\partial\mu_c)^p$ to arrive at the equation (1), as if μ_c were a smooth function. Fix any compact subset K of X and choose $c \geq c(K)$ so that $T_c \equiv T \equiv \eta$ on K. It follows now from (1) and the fact that $R_c^n \geq 0$ that

$$\int_K \mu^n \leq \int_M T_c^n \leq \int_M (\sigma + C\nu)^n < \infty. \tag{2}$$

Since the compact subset $K \subset X$ is arbitrary we have established Prop.1.2. ∎

Let $X = \Omega/\Gamma$ be a quotient of finite volume (in the canonical metric) of an irreducible bounded symmetric domain of rank ≥ 2. By the Arithmeticity Theorem of Margulis (cf. ZIMMER [ZIM]), X is necessarily an arithmetic variety. It follows that there exists by [(1.1), Thm.2] a compactification $X \hookrightarrow X_{min}$ such that $S := X_{min} - X$ is of codimension ≥ 2 on X_{min}. The alternative proof of the metric rigidity theorem of To [(1.1), Thm.1] in the Kähler case is now completed.

(2.3) The fact that $X = \Omega/\Gamma$ (irreducible and of dimension ≥ 2) satisfies the hypothesis of Prop.1 is based on the compactification theorem of SATAKE–BAILY–BOREL ([SA1] & [BB]). To give a proof of the Kähler case of [(1.1), Thm.1] using a minimum of algebraic information in place of using Prop.1 one can use instead the fact that X is very strongly (n–2)–pseudoconcave (in the terminology of NADEL–TSUJI [NA–T]) in the sense that X admits a smooth exhaustion function ψ (i.e. $\{\psi < c\} \subset\subset X$ for any real number c) such that $-\psi$ is plurisubharmonic outside some compact subset C of X and such that the Levi form of $-\psi$ has at least 2 positive eigenvalues outside C. Based on this property NADEL–TSUJI [NA–T] gave an analytic proof that X can be compactified to a quasi–projective variety. Their method does not show that there exists a compactification Z such that $S = Z - X$ is of codimension ≥ 2 in Z. It is however still possible to prove that any Kähler metric on X is of finite volume by using slicing arguments.

Let M be a smooth projective–algebraic compactification of X and η be a Kähler form on X. The key point is to prove that η extends as a closed positive (1,1)–current to M. To apply the Bishop Extension Theorem to η one has to show that η is of finite mass with respect to some Kähler metric ν on M. When $n = 2$, X is strongly pseudoconcave (cf. ANDREOTTI–GRAUERT [AG]) so that the infinity of any compactification of X can be blown down to isolated points using GRAUERT [GR], i.e., X can be compactified to a complex–analytic space Z by adjoining a finite number of isolated singularities. In this case one can show that η extends to a closed positive (1,1)–current on M by using the Bishop Extension Theorem for closed positive currents, the projection argument in the proof of [(2.2), Prop. 1.1] and the following extension theorem of Sibony's.

THEOREM 1 (SIBONY [SI])
Let G be a strongly pseudoconvex domain on $\mathbb{C}^n$ and $E \subset G$ be a compact pluripolar set. Let T be an arbitrary closed positive (p,p)–current on $G - E$. Then, for an arbitrary compact subset K of G, Mass(T;K–E) is automatically finite. Consequently, T can be extended trivially as a closed positive (p,p)–current on G.

In arbitrary dimensions one can slice M by algebraic surfaces S to show that for "most" S, $\eta|_{S-D}$ extends as a closed positive (1,1)–current on S. To prove the extension of η to M one uses the following

THEOREM 2 (HARVEY–POLKING [HP])
Suppose T is a closed positive (1,1)–current defined on $\Delta^* \times \Delta^{n-1}$. Assume that for a set $E \subset \Delta^{n-1}$ of positive Lebesgue measure the current $T|\Delta^*\times\{z\}$ is defined and is of finite mass for $z \in E$. Then, T extends trivially as a closed positive (1,1)–current to Δ^n.

The fact that X is very strongly (n–2)–pseudoconcave was proved first in a special case by ANDREOTTI–GRAUERT [AG,1960] and then in general by BOREL [BO3,1970]. The proof is much more elementary than the compactification theorem of SATAKE–BAILY–BOREL. The proof of NADEL–TSUJI [NA–T] that X can be compactified complex–analytically does not use the algebraic structure of X at all.

CHAPTER 9 THE IMMERSION PROBLEM FOR COMPLEX HYPERBOLIC SPACE FORMS

§1 The Equi–dimensional Case

(1.1) The study of holomorphic mappings between Hermitian locally symmetric manifolds is a major motivation for proving metric rigidity theorems. The rigidity theorem for holomorphic mappings [Ch.6, (5.1), Thm.1] was motivated by the Superrigidity Theorem of Margulis (cf. ZIMMER [ZIM]). We recall

THEOREM 1 ([Ch.6, (5.1), Thm.1])
Let $X = \Omega/\Gamma$, $X' = \Omega'/\Gamma'$ be quotients of bounded symmetric domains Ω and Ω'. Suppose Ω is irreducible and of rank ≥ 2 and (X,g) is of finite volume with respect to a canonical metric g. Let f: $X \longrightarrow X'$ be a holomorphic mapping which is not totally degenerate. Then, up to a normalizing constant f is necessarily a totally geodesic isometric immersion.

In [Ch.8, (1.2), Thm.1] we gave a generalization of the above theorem by replacing X' by a Kähler manifold (Y,h) of seminegative holomorphic bisectional curvature. Thm.1 does not address the problem when (X,g) is of rank 1, i.e., when $\Omega \cong B^n$, the complex Euclidean unit ball. We call such manifolds complex hyperbolic space forms. We discuss here the situation when both X and X' are complex hyperbolic space forms and X is of finite volume. Beyond the case when X is a hyperbolic Riemann surface, for which very little can be said, there is in 2 dimensions an example of MOSTOW [MOS2] giving rise to a surjective holomorphic map f: $B^2/\Gamma \longrightarrow B^2/\Gamma'$ with both Γ and Γ' cocompact such that the induced map f_*: $\Gamma \longrightarrow \Gamma'$ has an infinite kernel (cf. [(2.3), Thm.1]). In particular, f is not a holomorphic covering map. Furthermore, there is an example of LIVNE [LIV] of a surjective holomorphic map f: $B^2/\Gamma \longrightarrow \Delta/\Gamma'$ with both Γ and Γ' cocompact. Consequently, to formulate rigidity problems for holomorphic mappings between complex hyperbolic space forms it is necessary to impose additional geometric/algebraic conditions.

In this chapter we address the immersion problem between complex hyperbolic space forms, as follows

PROBLEM

Let $f: X \longrightarrow Y$ be a holomorphic immersion between hyperbolic space forms such that X is of finite volume and of complex dimension ≥ 2. Is f necessarily a totally geodesic isometric immersion?

We will discuss two different results. The first result (MOK [MOK7]) is in the equidimensional case, where we will give an affirmative answer based on a Hermitian metric rigidity theorem. The second result (CAO–MOK [CM]) concerns the case when X is compact and $\dim_{\mathbb{C}}(Y) \leq 2\dim_{\mathbb{C}}(X)$. Our affirmative answer in this case comes from some integral formula concerning characteristic classes and a study of some holomorphic foliation on X. While this result does not follow from the kind of metric rigidity theorems we proved it is nonetheless motivated by such theorems. In our formulation we will assume implicitly that the canonical metric on an n–dimensional complex hyperbolic space form has been chosen such that the Ricci curvature is equal to $-(n+1)$. The first result is given by

THEOREM 2 (MOK [MOK7, 1988])

Let X be an n–dimensional hyperbolic space form of finite volume and Y be an arbitrary complex hyperbolic space form of the same dimension. Then, any holomorphic immersion $f: X \longrightarrow Y$ is necessarily an isometric immersion.

We note first of all that in case X is compact Thm.1 is an immediate consequence of the Ahlfors–Schwarz lemma for volume forms (cf. [Ch.8, (1.1), Thm.4]) of YAU [YAU], which implies the uniqueness of Kähler–Einstein metrics of Ricci curvature $-(n+1)$ on X. To see this suppose g, g' are two such metrics. By the Ahlfors–Schwarz lemma the identity map $\iota: (X,g) \longrightarrow (X,g')$ and its inverse are both volume–decreasing. Thus, ι preserves the volume and is hence an isometry, as a Kähler–Einstein metric is completely determined by its volume form. Suppose now $f: X \longrightarrow Y$ is as in the theorem and g and h are the canonical metrics on X and Y resp. (of Ricci curvature $-(n+1)$). Then, both g and f^*h are Kähler–Einstein metrics on X of Ricci curvature $-(n+1)$, so that $g = f^*h$, i.e., f is an isometric immersion.

In general, the argument using the Ahlfors–Schwarz lemma implies the uniqueness of complete Kähler–Einstein metric of Ricci curvature $-(n+1)$ on an arbitrary complex manifold. Thm.1 would therefore follow if we know that f is proper. If both X and Y are of finite volume, one can show that f is proper by using their minimal compactifications $\overline{X}_{min}$ and $\overline{Y}_{min}$. (In the non–arithmetic case cf. [MOK7] for the existence.) The holomorphic map $f: X \to Y$ extends holomorphically to $\bar{f}: \overline{X}_{min} \to \overline{Y}_{min}$ by [Ch.8, (1.1), Thm.2] of Borel (or by using the Ahlfors–Schwarz lemma [Ch.6, (5.1), Thm.4].). If f were not proper then there exists an isolated $P \in \overline{X}_{min} - X$, open neighborhoods U and W of P and $Q = \bar{f}(P) \in Y$ in $\overline{X}_{min}$ and Y resp. such that $\Phi := f|_{U-\{P\}} : U - \{P\} \to W - \{Q\}$ is a holomorphic covering map. If we choose $W \cong \Delta^n$, such a mapping Φ cannot possibly exist as $W - \{Q\}$ is simply–connected for $n \geq 2$.

The point of Thm.2 is that it applies to the general case when no additional assumptions are imposed on the complex hyperbolic space form Y. We also note that from obvious examples Thm.1 does not hold for $n = 1$. In our proof of Thm.1 we will need compactifications of X even when X is non–arithmetic. We have

THEOREM 3 (SIU–YAU [SY2])
Let (Y,s) be a complete Kähler manifold of finite volume such that for some negative constant $-C$

$$-C \leq \text{Sectional curvatures} \leq -1.$$

Then, Y can be compactified to a non–singular projective–algebraic variety M such that $M - Y$ can be blown down to a finite number of points to obtain a compact complex space Z.

The statement of Thm.3 itself is not enough to justify the integral formulas that we will encounter. However, based on the description on the "cusps" of M given in [SY2] in the case of complex hyperbolic space forms (X,g) it is possible to give the following precise description

THEOREM 4 (cf. [MOK7])
If in Thm.3 $(Y,s) = (X,g)$ is a complex hyperbolic space form of finite volume, then the non–singular compactification M can be obtained by adjoining a finite number of abelian varieties A_i. Furthermore, writing $D = M - X = \amalg A_i$ the canonical metric on K_X defines a Hermitian metric on $K_X \otimes [D]|_X \cong (K_M \otimes$

$[D])|_X$ of logarithmic growth with respect to M (in the sense of [Ch.8, (1.1), Def.1]).

Here the abelian varieties have negative normal bundles in M so that they can be blown down to isolated points.

(1.2) The Hermitian metric rigidity theorems proved so far arise from zeros of holomorphic bisectional curvatures. In the rank–1 situation the Poincaré metric carries strictly negative holomorphic bisectional curvature. In order to apply our methods we work instead with the Hermitian holomorphic vector bundle $V_X :=$ $S^{n+1}T_X \otimes K_X$, where S^k denotes the k–th symmetric power. The point of choosing V_X is that V_X is of seminegative curvature in the sense of Griffiths but is not of strictly negative curvature. Write g' and h' for the canonical metrics on V_X and V_N induced by g and h resp. Given a local biholomorphism $f: X \to Y$ there is an induced bundle map $\Phi: V_X \to V_Y$ over $f: X \to Y$, so that it makes sense to consider f^*h'. It is here that we use the assumption that f is a holomorphic immersion, which allows us to invert $\Lambda^n df: K_X^{-1} \to K_Y^{-1}$. Our approach is to prove that g' is up to a normalizing constant the unique Hermitian metric of seminegative curvature on V_X. By considering $g'+f^*h'$ we can then prove that $f^*h' = \text{Const}.g'$. The deduction from there that f is an isometric immersion is very simple.

We formulate first of all

PROPOSITION 1

Let (X,g) be a complex hyperbolic space form of complex dimension $n \geq 2$. Let g' denote the induced Hermitian metric on $V_X = S^{n+1}T_X \otimes K_X$. Then, (V_X,g') is of seminegative, but not strictly negative curvature, in the sense of Griffiths.

Proof:

We write $X = B^n/\Gamma$ and compute the curvature tensor at the origin $o \in B^n$ using the Euclidean coordinates (z_i). Denote the Kähler form of the Poincaré metric on B^n by $2\pi\omega$, normalized so that $(B^n,2\pi\omega)$ is of constant holomorphic sectional curvature -2. We have $2\pi\omega = \sqrt{-1}\partial\bar{\partial}\log(1-\|z\|^2)$. Denote by R the curvature of (X,g). We have at $o \in B^n$ (cf. [Ch.4, (3.1)])

$$R_{i\bar{j}k\bar{l}} = -(\delta_{ij}\delta_{kl} + \delta_{il}\delta_{kj}),$$

so that for unit vectors α and β of type (1,0) we have

$$R_{\alpha\bar{\alpha}\beta\bar{\beta}} = -(1 + |(\alpha,\beta)|^2), \tag{1}$$

where (.,.) denotes the Hermitian inner product on $T_o(B^n)$. In particular,

$$\begin{aligned} -2 \leq R_{\alpha\bar{\alpha}\beta\bar{\beta}} &\leq -1; \\ R_{i\bar{j}} &= -(n+1)\delta_{ij}. \end{aligned} \tag{2}$$

Consider now $V_X = S^{n+1}T_X \otimes K_X \subset (\otimes^{n+1}T_X) \otimes K_X := W_X$. (V_X,g') is a direct summand of the Hermitian holomorphic vector bundle W_X (with a canonical metric g''). We will do curvature computations by identifying $x \in X$ with $o \in B^n$. Denote by Θ the curvature of (W_X,g''). $\Theta|V_X$ is the curvature of (V_X,g'). For $\beta \in T_x(X)$ define $\beta^k \in S^kT_x(X)$ to be the k–th symmetric power of β. Let $e \in K_x(X)$ be an element of unit length and write $\beta_k = \beta^k \otimes e$. For any two unit vectors $\alpha, \beta \in T_x(X)$ we have by straight–forward calculation

$$\Theta(\beta_k,\beta_k;\alpha,\bar{\alpha}) = k\, R_{\beta\bar{\beta}\alpha\bar{\alpha}} - R_{\alpha\bar{\alpha}}. \tag{3}$$

From (1), (2) and (3) choosing $k = n + 1$ we have

$$\begin{aligned} \Theta(\beta_{n+1},\overline{\beta_{n+1}};\alpha,\bar{\alpha}) &= -(n+1)(-R_{\beta\bar{\beta}\alpha\bar{\alpha}} - 1) \\ &= -(n+1)|(\alpha,\beta)|^2 \leq 0. \end{aligned} \tag{4}$$

In particular for $\alpha \perp \beta$ we have

$$\Theta(\beta_{n+1},\overline{\beta_{n+1}};\alpha,\bar{\alpha}) = 0. \tag{5}$$

To prove Prop.1 it remains to show that (W_X,g'') is of seminegative curvature, i.e., $\Theta(\mu,\bar{\mu};\alpha,\bar{\alpha}) \leq 0$ for any $\mu \in W_x(X)$ and any $\alpha \in T_x(X)$. Write $\{e_i\}$ for an orthonormal basis of $T_x(X)$ such that $e_1 = \alpha$. For an (n+1)–tuple I =

$(i(1),\ldots,i(n+1))$ define $e_I = e_{i(1)} \otimes \ldots \otimes e_{i(n+1)}$. Let $\mu \in W_x(X)$ be given by $\mu = \Sigma \mu_I(e_I \otimes e)$. Then, for μ of unit length, i.e., $\Sigma |\mu_I|^2 = 1$, we have

$$\begin{aligned} \Theta(\mu,\bar{\mu};\alpha,\bar{\alpha}) &= \sum_I \Big[|\mu_I|^2 (-\sum_{1 \le k \le n+1} (|(e_{i(k)},\alpha)|^2 + 1) \Big] + (n+1) \\ &= -\sum_I |\mu_I|^2 . \mathrm{Card}(\{k: i(k) = 1\}) \le 0. \end{aligned} \tag{6}$$

The proof of Prop.1 is thus complete. ▌

In the rank–1 case the irreducible locally homogeneous Hermitian holomorphic vector bundle V_X will now play the same role as that of the tangent bundle T_X in the higher rank case. We define $\mathcal{S} = \mathcal{S}(X) = \{(x,[\beta_{n+1}]) \in \mathbb{P}(V_X): \beta \neq 0\}$. Here $[v]$ denotes the projectivization of a non–zero vector $v \in V_x(X)$. Clearly $\mathcal{S}(X)$ is isomorphic as a holomorphic fiber bundle over X to the $\mathbb{P}^{n-1}$–bundle $\mathbb{P}(T_X)$. Such an isomorphism is given by $(x,[\beta_{n+1}]) \longrightarrow (x,[\beta])$. Let $L \longrightarrow \mathbb{P}(V_X)$ denote the tautological line bundle with the induced Hermitian metric $\hat{g}$. Recall that $2\pi\omega$ denotes the Kähler form of the Poincaré metric on X and π: $\mathbb{P}(V_X) \longrightarrow X$ denote the canonical base projection. We are going to show

PROPOSITION 2

Let μ be any Hermitian metric on V_X. Then, for the induced Hermitian metric $\hat{\mu}$ on L, we have

$$\int_{\mathcal{S}} [-c_1(L,\hat{\mu})] \wedge [-c_1(L,\hat{g})]^n \wedge (\pi^*\omega)^{n-2} = 0.$$

Consequently, if (V,μ) is of seminegative curvature, then on $\mathcal{S}$

$$[-c_1(L,\hat{\mu})] \wedge [-c_1(L,\hat{g})]^n \wedge (\pi^*\omega)^{n-2} \equiv 0.$$

<u>Proof</u>:

We will first establish Prop.2 under the assumption that X is compact. In this case imitating the proof of the analogous integral formula in [Ch.6, (2.1), Prop.1] it suffices to show that given $[\beta_{n+1}] \in \mathcal{S}$, the zero–eigenspace of $c_1(L,\hat{g})$ at $[\beta_{n+1}]$ are tangential to $\mathcal{S}$. As usual we work with the universal covering B^n and define $\mathcal{S}(B^n) \subset \mathbb{P}(V_{B^n})$ similarly. Denote by g_0 the canonical metric on B^n and by g_0'

the induced metric on V_{B^n}. Since the Euclidean coordinates (z_i) serve as complex geodesic coordinates for (B^n, g_0) at the origin o they also give rise to special coordinates for the Hermitian holomorphic vector bundle (V_{B^n}, g_0') at o. (Here and henceforth the suffix o will be used to indicate that we are working with B^n.) It is immediate from the definition of $\mathcal{S}(B^n)$ that with respect to the trivialization $\mathbb{P}(V_{B^n}) \cong \mathbb{P}(V_0) \times B^n$ we have the identification $\mathcal{S}(B^n) \cong \mathcal{S}_0 \times B^n$, i.e., $\mathcal{S}(B^n)$ is parallel in the Euclidean sense. Prop.2 in case X is compact now follows immediately from the argument of [Ch.6, (2.1), Prop.1]. When X is non–compact the justification of the integral formula given in [Ch.8, (2.1)] works equally well in the rank–1 situation because of [(1.1), Thm.3] and because the determinant bundle of V_X is a negative power of the canonical line bundle. ▌

We proceed to complete the proof of [(1.1), Thm.1]

Proof of [(1.1), Thm.1]

Recall that $f: X \longrightarrow Y$ is a local biholomorphism between complex hyperbolic space forms (X,g) and (Y,h). Write g' and h' for the induced metrics on V_X and V_Y and define $\mu = g' + f^*h'$. As explained f^*h' is well–defined because $\Lambda^n(df): K_X \longrightarrow K_Y$ is invertible. For the induced metrics $\hat{g}$ and $\hat{\mu}$ let $\hat{\mu} = e^u\hat{g}$. From the same argument as [Ch.6, §3] we deduce the identity

$$\sqrt{-1}\partial u \wedge \bar{\partial} u \wedge [-c_1(L,\hat{g})]^n \wedge (\pi^*\omega)^{n-2} \equiv 0 \quad \text{on } \mathcal{S}. \tag{1}$$

From the proof in [Ch.6, (3.1)] using Moore's Ergodicity Theorem one deduces that u is constant on $\mathcal{S}$. It remains to show that this implies that $h = \text{Const}.g$ on X. The fact that $\hat{\mu} = \text{Const}.\hat{g}$ on $\mathcal{S}$ is equivalent to saying that in terms of local holomorphic coordinates such that $g_{i\bar{j}}(x) = \delta_{ij}$, for any $x \in X$ and any unit vector $\alpha \in T_x(X)$ we have

$$1 + \frac{h^{n+1}_{\alpha\bar{\alpha}}}{\det(h_{i\bar{j}})} = C, \tag{2}$$

where C is independent of the point $x \in X$. It follows that for any $x \in X$,

$$h_{\alpha\bar{\alpha}}(x) = [(C-1)\det(h_{i\bar{j}}(x))]^{\frac{1}{n+1}}, \tag{3}$$

so that h is proportional to g at every point $x \in X$. From (2) it follows that

$$1 + (c(x))^{n+1} = C, \tag{4}$$

which forces $c(x)$ to be independent of $x \in X$. Thus, $h = cg$ on X and the proof of [(1.1), Thm.1] is completed. ▌

§2 Holomorphic Immersions Between Compact Hyperbolic Space Forms

(2.1) As in §1 we will normalize the canonical metrics on N–dimensional hyperbolic space forms to be of Ricci curvature $-(N+1)$. In this section we give a proof of

THEOREM 1 (CAO–MOK [CM])
Let (X,g) be a compact n–dimensional complex hyperbolic space form. Let (Y,h) be an m–dimensional hyperbolic space form such that $m \le 2n-1$. Let $f: X \to Y$ be a holomorphic immersion. Then, f is necessarily a totally geodesic isometric immersion.

Throughout the discussion we will assume that $f: X \hookrightarrow Y$ is a holomorphic embedding. The proof goes through without any significant modification to the general case of holomorphic immersions. There are two sources of motivation for the formulation and proof of Thm.1. The first motivation comes from the proof of [(1.1), Thm.1] in the equi–dimensional case. We identify X as a submanifold of Y. The arguments in §1 show that up to a normalizing constant the Hermitian metric g' on V_X arising from g is the unique Hermitian metric of seminegative curvature on V_X. The Hermitian metric $h|_X$ gives rise to a Hermitian metric s on $V_X = S^{n+1}T_X \otimes K_X$. Because of the positive factor K_X in general there is no reason to suspect that (V_X,s) is of semipositive curvature. However, in case $X \hookrightarrow Y$ is a hypersurface by integrating over $\mathcal{S}_X \subset \mathbb{P}(V_X(X))$ one can show that

$$0 = \int_{\mathcal{S}(X)} [-c_1(L,s)]^{2n-1} = \int_X \lambda,$$

where λ is pointwise a nonnegative (n,n)–form. This allows one to prove the vanishing theorem $\lambda \equiv 0$ on X, which gives information about the second fundamental form of the holomorphic embedding f. One can in fact proceed along this line to prove that f is a totally geodesic isometric embedding, but the pointwise computation fails as long as the $m - n \geq 2$.

The second motivation is the dual problem on complex projective spaces. In this direction we have

THEOREM 2 (FEDER [FE,1965])

Let $f : \mathbb{P}^n \to \mathbb{P}^m$ be a holomorphic immersion between complex projective spaces such that $m \leq 2n - 1$. Then, f is necessarily projective linear.

In other words f identifies $\mathbb{P}^n$ with a projective subspace of $\mathbb{P}^m$. The dimension assumption is sharp. Given any n, by the Veronese embedding ν (cf. [App.III.3]) one can embed $\mathbb{P}^n$ into some $\mathbb{P}^N$ (actually $N = \frac{1}{2}(n^2 + 3n)$), isometrically with respect to some choice of Fubini Study metrics. By taking a generic point $Q \in \mathbb{P}^N - \nu(\mathbb{P}^n)$ and a generic hyperplane $H \subset \mathbb{P}^N$ one can project $\nu(\mathbb{P}^n)$ onto the hyperplane H so that the resulting map $\mathbb{P}^n \to H \cong \mathbb{P}^{N-1}$ remains a holomorphic immersion. (The projection is defined by assigning to $x \in \nu(\mathbb{P}^n)$ the unique point on H lying on the rational line joining Q to x.) By a standard dimension count as in the Whitney Embedding Theorem one can always immerse $\mathbb{P}^n$ holomorphically into $\mathbb{P}^{2n}$.

<u>Proof of Thm.2</u>:

The proof is obtained by computing Chern classes. Denote by N the holomorphic normal bundle of the holomorphic immersion f, i.e., $N = f^*T_{\mathbb{P}^m}/T_{\mathbb{P}^n}$. Then, as smooth vector bundles we have $f^*T_{\mathbb{P}^m} \cong T_{\mathbb{P}^n} \oplus N$. For a smooth complex vector bundle V of rank r over a complex p–dimensional manifold M denote by $c(V) = 1 + c_1(V) + \dots + c_k(V) \in H^*(M,\mathbb{Z})$ the total Chern class of V, where $c_i(V)$ denote the i–th Chern class $\in H^{2i}(M,\mathbb{Z})$, and $k = \min(p,r)$. Since $f^*T_{\mathbb{P}^m} \cong T_{\mathbb{P}^n} \oplus N$ we have on $\mathbb{P}^n$

$$c(f^*T_{\mathbb{P}^m}) = c(T_{\mathbb{P}^n})\, c(N) \qquad (1)$$

Denote by ξ the generator of $H^2(\mathbb{P}^n,\mathbb{Z}) \cong \mathbb{Z}$ representing the first Chern class of the hyperplane section line bundle $\mathcal{O}(1)$. The graded cohomology ring $H^*(\mathbb{P}^n,\mathbb{Z})$ is generated by ξ. From the Euler sequence $0 \to \mathcal{O} \to \mathcal{O}(1)^{n+1} \to T_{\mathbb{P}^n} \to 0$ on $\mathbb{P}^n$ and the fact $c(\mathcal{O}(1)) = 1 + \xi$ we have the formula

$$c(T_{\mathbb{P}^n}) = (1+\xi)^{n+1}. \qquad (2)$$

Let δ denote the degree of the map f. To prove Thm.1 it is equivalent to show that $\delta = 1$. From (1), (2) and the same formula for $\mathbb{P}^m$ we have

$$c(N) = (1+\delta\xi)^{m+1}(1+\xi)^{-n-1}. \qquad (3)$$

where inversion in $H^*(\mathbb{P}^n,\mathbb{Z})$ is performed formally. Since $m \leq 2n-1$, N is of rank $m - n \leq n-1$, so that $c_n(N) = 0$. We may assume without loss of generality that $m = 2n-1$. Computing using (3) formally we have on the other hand

$$\begin{aligned} c_n(N) &= \sum_{0\leq i\leq n} (-1)^i \begin{bmatrix}2n\\ i\end{bmatrix} \xi^i \begin{bmatrix}2n-i\\ n-i\end{bmatrix} (\delta\xi)^{n-i}, \\ &= \sum_{0\leq i\leq n} (-1)^i \begin{bmatrix}2n\\ i\end{bmatrix} \begin{bmatrix}2n-i\\ n-i\end{bmatrix} \delta^{n-i}\xi^n, \end{aligned} \qquad (4)$$

where $\begin{bmatrix}p\\ q\end{bmatrix}$ denotes the binomial coefficient $\frac{p!}{q!\,(p-q)!}$. Thus,

$$\begin{bmatrix}2n\\ i\end{bmatrix}\begin{bmatrix}2n-i\\ n-i\end{bmatrix} = \frac{(2n)!}{i!\,(2n-i)!}\frac{(2n-i)!}{(n-i)!\,n!} = \frac{(2n)!}{(n!)^2}\frac{n!}{i!\,(n-i)!} = \begin{bmatrix}2n\\ n\end{bmatrix}\begin{bmatrix}n\\ i\end{bmatrix}.$$

It follows from (4) that

$$\begin{aligned} c_n(N) &= \begin{bmatrix}2n\\ n\end{bmatrix}\left[\sum_{0\leq i\leq n}\begin{bmatrix}n\\ i\end{bmatrix}\delta^{n-i}(-1)^i\right]\xi^n \\ &= \begin{bmatrix}2n\\ n\end{bmatrix}(\delta-1)^n\xi^n. \end{aligned} \qquad (5)$$

Finally, since $c_n(N) = 0$ when $m \leq 2n-1$ we obtain immediately $\delta = 1$, proving Thm.2. ∎

(2.2) We proceed to give a proof of [(2.1), Thm.1], which can be regarded as dual to Feder's Theorem [(2.1), Thm.2]. Recall that (X,g) and (Y,h) are complex hyperbolic space forms of dimensions n and m resp. with X compact and

$m \leq 2n-1$, as in Feder's Theorem. As explained before for the sake of presentation we will make the assumption that $f: X \hookrightarrow Y$ is a holomorphic embedding, allowing us to identify X as a submanifold of Y. We are going to show that $(X,h|_X) \hookrightarrow (Y,h)$ is a totally geodesic Kähler submanifold. In particular $(X,h|_X)$ is of constant holomorphic sectional curvature -2 and hence Kähler–Einstein with Ricci curvature $-(n+1)$. The identity map $(X,g) \longrightarrow (X,h|_X)$ is then necessarily an isometry because of the uniqueness of Kähler–Einstein metrics with a fixed negative Einstein constant. To prove that $X \subset Y$ is totally geodesic it is equivalent to prove that the second fundamental form S vanishes identically on X. Denote by $N = N_{X|Y}$ the real normal bundle of the embedding $X \hookrightarrow Y$ of Riemannian manifolds. Here $S: T_x^{1,0}(X) \times T_x^{1,0}(X) \longrightarrow N_x^{1,0}$, the (1,0) component of the complexified normal bundle $N^{\mathbf{C}} = N \otimes_{\mathbf{R}} \mathbf{C}$. Thus in holomorphic coordinates we write $S = (S^k_{ij})$, where i, j and k are unbarred indices. Denote by R' resp. R the curvature tensors of (Y,h) resp. $(X,h|_X)$. At $x \in X$ fix holomorphic coordinates $(w_i)_{1 \leq i \leq n}$ on a neighborhood U of x in X such that $h_{i\bar{j}}(x) = \delta_{ij}$. Extend the coordinate system to $(w_i)_{1 \leq i \leq m}$ on a neighborhood V of x in Y such that $h_{i\bar{j}}(x) = \delta_{ij}$ remains true for $1 \leq i,j \leq m$. Then, for $\alpha, \beta, \mu, \nu \in T_x^{1,0}(X)$ we have by [Ch.4, (3.1)] and the Gauss–Codazzi equation

$$R'_{i\bar{j}k\bar{l}} = -(\delta_{ij}\delta_{kl} + \delta_{il}\delta_{kj}), \tag{1}$$

$$R_{\alpha\bar{\beta}\mu\bar{\nu}} = R'_{\alpha\bar{\beta}\mu\bar{\nu}} - \sum_{n+1 \leq j \leq m} S^j_{\alpha\mu} \overline{S^j_{\beta\nu}}, \tag{2}$$

where $S(\alpha,\mu) = \Sigma_{n+1 \leq j \leq m} S^j_{\alpha\mu} \frac{\partial}{\partial w_j}$. For the Ricci curvatures we have from (1) and (2)

$$R_{\alpha\bar{\beta}}(X) = -(n+1)\,(h_{\alpha\bar{\beta}} + 2\pi\sigma_{\alpha\bar{\beta}}), \text{ where}$$

$$\sigma_{\alpha\bar{\beta}} = \frac{1}{2\pi(n+1)} \sum_{1 \leq i \leq n, n+1 \leq j \leq m} S^j_{\alpha i} \overline{S^j_{\beta i}}. \tag{3}$$

Denote by ω the Kähler form of (Y,h). The Ricci form ρ of $(X,h|_X)$ is given by $\rho = -(n+1)\,(\omega + \sigma)$, where $\sigma = \sqrt{-1}\, \Sigma\, \sigma_{i\bar{j}}\, dw^i \wedge d\overline{w^j}$ at x. σ is a closed nonnegative (1,1)–form on X. To prove $S \equiv 0$ by (3) it suffices to show that $\sigma \equiv 0$. To this end we imitate the proof of Feder's Theorem [(2.1), Thm.2]. As $N \cong (T_Y|_X)/T_X$ as a smooth bundle it can be given a holomorphic structure when T_X

and T_Y are identified with the holomorphic tangent bundles on X and Y in the standard way. Thus, we identify Y with the holomorphic normal bundle of the embedding $X \hookrightarrow Y$. Analogous to the proof of Feder's Theorem we have

$$c(T_Y|_X) = c(T_X)\, c(N), \tag{4}$$

The trouble is that the cohomology ring $H^*(X,\mathbf{R})$ (resp. $H^*(Y,\mathbf{R})$) over $\mathbf{R}$ is not generated by an element in $H^2(X,\mathbf{R})$ (resp. $H^2(Y,\mathbf{R})$). In particular, one cannot define the degree δ analogously. To compute $c_n(N)$ we first represent $c(T_Y)$ using the Kähler form $2\pi\omega$ of (Y,h). The computation is given by the formula in the dual case of the projective space and the Proportionality Principle. We have

$$c(T_Y) = (1-[\omega])^{m+1}, \tag{5}$$

where for a real closed differential form φ, $[\varphi]$ is used to denote the cohomology class over $\mathbf{R}$ that it represents. On the other hand since X is itself a complex hyperbolic space form and $c_1(X) = [(1/2\pi)\rho] = [\omega|_X + \sigma]$ on X, we have again by the Proportionality Principle

$$c(T_X) = (1-[\omega|_X + \sigma])^{n+1}. \tag{6}$$

Thus, not only is $c_1(X)$ expressed in terms of the Kähler metric $h|_X$ but also the higher Chern classes are expressed in terms of c_1 using the Proportionality Principle. For simplicity we write ω for $\omega|_X$. From (4), (5) and (6) we have in $H^*(X,\mathbf{R})$

$$(1-[\omega])^{m+1} = (1-[\omega+\sigma])^{n+1}\, c(N). \tag{7}$$

We can still formally invert the cohomology class $(1-[\omega+\sigma])^{n+1}$, which gives

$$\begin{aligned} c_n(N) &= \sum_{0\le i\le n} (-1)^i \begin{bmatrix} 2n \\ i \end{bmatrix} [\omega]^i \begin{bmatrix} 2n-i \\ n-i \end{bmatrix} [\omega+\sigma]^{n-i} \\ &= \begin{bmatrix} 2n \\ n \end{bmatrix} \sum_{0\le i\le n} (-1)^i \begin{bmatrix} n \\ i \end{bmatrix} [\omega]^i [\omega+\sigma]^{n-i} \\ &= \begin{bmatrix} 2n \\ n \end{bmatrix} ([\omega]-[\omega+\sigma])^n = (-1)^n \begin{bmatrix} 2n \\ n \end{bmatrix} [\sigma]^n. \end{aligned} \tag{8}$$

Since $m \le 2n-1$ we have $c_n(N) = 0$. As σ is a nonnegative (1,1)–form the vanishing of $[\sigma]^n$ means that $\sigma^n \equiv 0$ on X. Thus, computation using Chern classes yields the partial vanishing $\sigma^n \equiv 0$ in the case of complex hyperbolic space forms.

(2.3) In this subsection we use the partial vanishing $\sigma^n \equiv 0$ of (2.2) to deduce $\sigma \equiv 0$, implying the vanishing of the second fundamental form S of X in Y. To this end we study certain foliations arising from the nonnegative closed (1,1)–form σ. At each $x \in X$ there is a complex subspace $\mathcal{N}_x(\sigma) \subset T_x^{1,0}(X)$ defined as the null–space of σ. Suppose σ does not vanish identically, so that $\mathcal{N}_x(\sigma) \neq T_x^{1,0}(X)$ for some x. We hope to show that $x \longrightarrow \mathcal{N}_x$ defines a holomorphic foliation $\mathcal{F}$ on a dense open subset of X and to prove subsequently that such an $\mathcal{F}$ cannot possibly exist by using special properties of $\mathcal{F}$ arising from the geometry of X and $\mathcal{F}$. We note first of all that in general it is possible to construct a nonnegative closed (1,1)–form φ whose associated complex vector spaces $\mathcal{N}_x(\varphi)$ do define a holomorphic foliation. It suffices to take a compact hyperbolic space form X of dimension 2 such that the first Betti number $b_1(X) \neq 0$. (Such an X exists by MUMFORD [MUM]). Let μ be a non–trivial (closed) holomorphic 1–form on X. Then $\varphi := \sqrt{-1}\, \mu \wedge \overline{\mu}$ is a nonnegative closed (1,1)–form. Outside the zero–set of μ the distribution $x \longrightarrow \mathcal{N}_x(\varphi)$ defines a holomorphic foliation $\mathcal{F}$. By using the example of LIVNE [LIV] it is possible even to give an example where $\mathcal{F}$ arises from level sets of a holomorphic mapping of X into a hyperbolic Riemann surface Σ so that in particular all leaves are closed. To make our approach work one must therefore prove that $\mathcal{F}$ has special geometric properties in our case.

Fix $x \in X$ and the coordinates as above we have $\sigma_{\alpha\overline{\alpha}} = 0$ if and only if $S^j_{\alpha i} = 0$ for all i, j, $1 \le i \le n$, $n+1 \le j \le m$. Recall the formula at $x \in X$

$$R_{\alpha\overline{\beta}\mu\overline{\nu}} = R'_{\alpha\overline{\beta}\mu\overline{\nu}} - \sum_{n+1\le j\le m} S^j_{\alpha\mu}\, \overline{S^j_{\beta\nu}}$$

It follows that we have

LEMMA 1

For $\alpha \in T_x^{1,0}(X)$ of unit length $\alpha \in \mathcal{N}_x(\sigma)$ if and only if $R_{\alpha\overline{\alpha}\alpha\overline{\alpha}} = -2$. (Recall that (Y,h) is of constant holomorphic sectional curvature −2.) Furthermore, for $\alpha \in \mathcal{N}_x(\sigma)$, $R_{\alpha\bar\beta\mu\bar\nu} = R'_{\alpha\bar\beta\mu\bar\nu}$ for all $\beta, \mu, \nu \in T_x^{1,0}(X)$. In terms of the Ricci curvature $\alpha \in \mathcal{N}_x(\sigma)$ if and only if $R_{\alpha\bar\alpha} := R_{\alpha\bar\alpha}(X) = -(n+1)$, and, for $\alpha \in \mathcal{N}_x(\sigma)$ we have $R_{\alpha\bar\mu} = 0$ for $\mu \perp \alpha$. In particular, $\mathcal{N}_x(\sigma)$ is an eigenspace of the Ricci form Ric(X) at x.

Fix an open subset U of X such that $\dim_{\mathbb{C}} \mathcal{N}_x(\sigma)$ is maximum and hence constant on U. We assert

PROPOSITION 1

The distribution $x \longrightarrow Re\,\mathcal{N}_x(\sigma)$ is integrable on U. The integral submanifolds are totally geodesic in Y (and hence *a fortiori* in X).

Proof:

Our basic tool is a commutation formula of Berger's as follows. Denote by ∇ covariant differentiation on $(X, h|_X)$ and write R for R(X). We have

$$\Delta R_{\alpha\bar\alpha\alpha\bar\alpha} - \nabla_{\bar\alpha}\nabla_{\alpha} R_{\alpha\bar\alpha} = \sum_{i,k} \left(|R_{\alpha\bar i\alpha\bar k}|^2 + R_{\alpha\bar i i\bar k} R_{k\bar\alpha\alpha\bar\alpha} - 2\,|R_{\alpha\bar\alpha i\bar k}|^2 \right) \quad (1)$$

This formula is obtained from standard commutation formulas and the Bianchi identity. It is particularly useful for the study of compact Kähler–Einstein manifolds of semipositive bisectional curvature, as initiated by BERGER [BER2]. For Kähler–Einstein manifolds the Ricci tensor is parallel, so that $\nabla_{\bar\alpha}\nabla_{\alpha} R_{\alpha\bar\alpha} = 0$. In our case we will instead make use of the formula (1) to show that $\alpha \in \mathcal{N}_x(\sigma)$ implies $\nabla_{\bar\alpha}\nabla_{\alpha} R_{\alpha\bar\alpha} = 0$. Assume that $h_{\alpha\bar\alpha} = 1$ and that furthermore the holomorphic local coordinates $(w_i)_{1 \le i \le m}$ have been chosen such that $\alpha = \partial/\partial w_1$. Write the right hand side of (1) as $F(R)_{\alpha\bar\alpha\alpha\bar\alpha}$, where F(R) is a tensor of type (2,2) quadratic in R. Compare (1) with the same formula (1)′ obtained for an n–dimensional complex hyperbolic space form with curvature R′ and covariant differentiation ∇'. From $\nabla' R' \equiv 0$ we know from (1)′ that $F(R')_{\alpha\bar\alpha\alpha\bar\alpha} = 0$. On the other hand by Lemma 1 and the curvature formula $R'_{i\bar j k\bar l} = -(\delta_{ij}\delta_{kl} + \delta_{il}\delta_{kj})$ on Y it follows that

$$F(R')_{\alpha\bar\alpha\alpha\bar\alpha} - F(R)_{\alpha\bar\alpha\alpha\bar\alpha} = \sum_{i>n \text{ or } k>n} \left(|R'_{\alpha\bar i\alpha\bar k}|^2 + R'_{\alpha\bar i i\bar k} R'_{k\bar\alpha\alpha\bar\alpha} - 2|R'_{\alpha\bar\alpha i\bar k}|^2 \right)$$

$$= R'_{\alpha\bar\alpha\alpha\bar\alpha}\sum_{i>n} R'_{\alpha\bar\alpha i\bar i} - 2\sum_{i>n} |R'_{\alpha\bar\alpha i\bar i}|^2 = \sum_{i>n} R'_{\alpha\bar\alpha i\bar i}(R'_{\alpha\bar\alpha\alpha\bar\alpha} - 2R'_{\alpha\bar\alpha i\bar i}) = 0 . \quad (2)$$

Hence,

$$\Delta R_{\alpha\bar\alpha\alpha\bar\alpha} - \nabla_{\bar\alpha}\nabla_{\alpha}R_{\alpha\bar\alpha} = 0 \quad (3)$$

To prove $\nabla_{\bar\alpha}\nabla_{\alpha}R_{\alpha\bar\alpha} = 0$ it suffices therefore to show that $\Delta R_{\alpha\bar\alpha\alpha\bar\alpha} = 0$. To this end consider a geodesic $\gamma = \{\gamma(t): -\epsilon < t < \epsilon\}$ on $(X,h|_X)$ with $\gamma(o) = x$ and $\partial_t\gamma(o) = \eta \in T^R_x(X)$. Denote by $\alpha(t) \in T_{\gamma(t)}(X)$ the unit vector obtained by parallel transport of α along γ. Define $f(t) = R(\alpha(t),\overline{\alpha(t)};\alpha(t),\overline{\alpha(t)})$. It suffices to show that for any choice of η, $\nabla^2_\eta f(t) = 0$. To see this write $\alpha(t) = \alpha_t + t\beta_t$, where $\alpha_t \in \mathcal{N}_{\gamma(t)}(\sigma)$ and $\beta_t \perp \alpha_t$. Expanding $f(t)$ we have by Lemma 1

$$\begin{aligned} f(t) = {} & R(\alpha_t,\overline{\alpha_t};\alpha_t,\overline{\alpha_t}) + 4t\,Re R(\alpha_t,\overline{\alpha_t};\alpha_t,\overline{\beta_t}) + t^2[4R(\alpha_t,\overline{\alpha_t};\beta_t,\overline{\beta_t}) + \\ & 2Re R(\alpha_t,\overline{\beta_t};\alpha_t,\overline{\beta_t})] + O(t^3). \end{aligned} \quad (4)$$

By Lemma 1 all terms on the right, except the $O(t^3)$ term, are the same as those given by the curvature tensor R' of the ambient manifold. It follows readily that $f(t) = -2 + O(t^3)$, yielding immediately $\partial_t^2 f(o) = 0$, as desired. We have established thus far $\nabla_\alpha\nabla_{\bar\alpha}R_{\alpha\bar\alpha} = 0$ for $\alpha \in \mathcal{N}_x(\sigma)$. The same argument also shows that $\nabla_\alpha\nabla_{\bar\alpha}R_{\beta\bar\beta} = 0$ for $\alpha, \beta \in \mathcal{N}_x(\sigma)$. Over U write $Re\mathcal{N}(\sigma)$ for the smooth vector bundle $\cup_{x\in U} Re\mathcal{N}_x(\sigma)$. To prove Prop.1 we need to show that for any $Re\mathcal{N}(\sigma)$–valued vector fields A and B extending $Re\alpha$, $Re\beta \in Re\mathcal{N}_x(\sigma)$ ($\alpha, \beta \in \mathcal{N}_x(\sigma)$) to a neigborhood U on X, we have $\nabla_A B(x) \in Re\mathcal{N}_x(\sigma)$. Equivalently this means that $Re\mathcal{N}(\sigma)$ is invariant under parallel transport along its integral curves. Let $\lambda = \{\gamma(t): -\epsilon < t < \epsilon\}$ be a curve on $(U,h|_U)$ with $\gamma(o) = x$ and $\partial_t\gamma(o) = Re\alpha$. Let $\beta(t)$ be the parallel transport of $\beta = \beta_o = \beta(o)$ along γ. Then, $0 = \nabla_{\bar\alpha}\nabla_\alpha R_{\beta\bar\beta}(x) = \partial_t^2 Ric(\beta(t),\overline{\beta(t)})(o)$. Write $\beta(t) = \beta_t + t\mu_t$ with $\beta_t \in Re\mathcal{N}_{\gamma(t)}(\sigma)$ and $\beta_t \perp \mu_t$. By Lemma 1 and our choice of U all eigenvalues of Ric(X) on U other than $-(n+1)$ must be $\leq -(n+1+c)$ for some positive number c independent of the choice of $x \in U$, shrinking U if necessary. We have again by Lemma 1

$$Ric(\beta(t);\overline{\beta(t)}) = -(n+1)\|\beta_t\|^2 - t^2 Ric(\mu_t;\overline{\mu_t})$$

$$\leq -(n+1)(\|\beta_t\|^2 + t^2\|\mu_t\|^2) - ct^2\|\mu_t\|^2$$
$$\leq -(n+1) - ct^2\|\mu_t\|^2. \tag{5}$$

so that

$$0 = \nabla_{\bar{\alpha}}\nabla_{\alpha}R_{\beta\bar{\beta}}(x) = \partial_t^2\mathrm{Ric}(\beta(t),\overline{\beta(t)})(o) \leq -c\|\mu_o\|^2,$$
$$\text{i.e.,}\quad \mu_o = 0. \tag{6}$$

From (6) we deduce that $\mathcal{N}(\sigma)$ is infinitesimally invariant under parallel transport along integral curves of $Re\,\mathcal{N}(\sigma)$. It follows that the integral submanifolds S are complex–analytic (since $Re\,\mathcal{N}(\sigma)$ is invariant under the J–operator of Y) and totally geodesic in X. Given $x \in S$ and $\alpha \in T_x^{1,0}$ it follows from definition that $R_{\alpha\bar{\alpha}\alpha\bar{\alpha}}(S) = R_{\alpha\bar{\alpha}\alpha\bar{\alpha}} = R'_{\alpha\bar{\alpha}\alpha\bar{\alpha}}$. It follows from the Gauss–Codazzi equation that the second fundamental form of S in Y vanishes, i.e., that S is also totally geodesic in Y, as asserted in Prop.1. ▌

Proof of [(2.1), Thm.1]:

Write $\mathcal{F}$ for the foliation on $U \subset X$ obtained from the integrable distribution $Re\,\mathcal{N}(\sigma)$. We proved that the integral submanifolds S are totally geodesic Kähler submanifolds of Y. We are going first of all to deduce that $\mathcal{F}$ is a holomorphic foliation. For this we have to use the geometry of the complex hyperbolic space form Y. Write $\pi: B^m \to Y$ for the universal covering map. Shrinking U if necessary we may assume that on each connected component V of $\pi^{-1}(U)$, $\pi|_V$ is a biholomorphism. Denote also by $\mathcal{F}$ the corresponding holomorphic foliation on V. We also write σ for the (1,1)–form $\pi^*(\sigma)$ on $V \cap \pi^{-1}(X)$. In terms of the Euclidean coordinates on B^m, the totally geodesic Kähler submanifolds passing through the origin o are given by $B^m \cap P$, where P is a complex vector subspace. Since $\mathrm{Aut}(B^m)$ are given by projective linear transformations on $\mathbb{P}^m \supset \mathbb{C}^m$, it follows that the totally geodesic Kähler submanifolds are precisely the intersection of B^m with affine complex subspaces of $\mathbb{C}^m$. One deduces readily from the definition of $\mathcal{N}(\sigma)$ that the leaf S_x of $\mathcal{F}$ passing through $x \in V \subset B^m$ is in fact $V \cap A_x$, where A_x is the maximal affine complex subspace passing through x such that $A_x \cap W \subset V \cap W$ for some open neighorhood W of x in B^m. To show that $\mathcal{F}$ is holomorphic it suffices to show that the variation $x \to$

A_x is holomorphic. Shrinking V if necessary we may assume that V is the common zero set of a finite number of holomorphic functions f_i. A vector ξ of type (1,0) at x is tangent to A_x if and only if $\partial_\xi^k(f_i)(x) = 0$ for all nonnegative integers k. Identifying ξ at any x with a vector in $\mathbf{C}^m$ in the usual way the common zeros of the system of equations $\{\partial_\xi^k(f_i)(x) = 0\}_{i,k}$ on $(\xi,x) \in \mathbf{C}^m \times V$ define the complex–analytic subvariety $\{(\xi,x) \in \mathbf{C}^m \times V : \xi \in \mathcal{N}_x(\sigma)\} \subset \mathbf{C}^m \times V$. It follows that the distribution $x \longrightarrow \mathcal{N}_x(\sigma)$ and hence the foliation $\mathcal{F}$ on V is holomorphic.

We proceed to use the complex–analyticity of $\mathcal{F}$ to show that $\sigma \equiv 0$. We may assume that $o \in V$. Choose a system of local holomorphic coordinates $(z_i)_{1 \leq i \leq n}$ on V such that for $z' = (z_1,\dots,z_{n-1})$ and $z_n = w$ the point $(z';w) \in \Delta^{n-1} \times \Delta(\epsilon)$ corresponds to the point $H(z';w) = (f_1(z'),\dots,f_m(z')) + w(g_1(z'),\dots,g_m(z')) \in V \subset B^m$ in terms of Euclidean coordinates on B^m. Write $\sigma = \sqrt{-1}\,\Sigma\, \sigma_{i\bar{j}}(z)\, dz^i \wedge d\bar{z}^j$. By the choice of coordinates $\sigma_{n\bar{n}} \equiv 0$. Since σ is positive semi–definite it follows that $\sigma_{n\bar{j}} \equiv \sigma_{j\bar{n}} \equiv 0$ for $1 \leq j \leq n$. Furthermore from $d\sigma \equiv 0$ it follows that $\partial_n \sigma_{i\bar{j}} \equiv \partial_i \sigma_{n\bar{j}} \equiv 0$ for $1 \leq i, j \leq n-1$. (Similarly $\partial_{\bar{n}} \sigma_{i\bar{j}} \equiv 0$. Consequently, we have

$$\sigma = \sqrt{-1} \sum_{1 \leq i,j \leq n-1} \sigma_{i\bar{j}}(z')\, dz^i \wedge d\bar{z}^j. \tag{1}$$

The mapping H can be defined on $\Delta^{n-1} \times \mathbf{C}$. H is not necessarily a holomorphic immersion on $\Delta^{n-1} \times \mathbf{C}$. However, since X is (complex–)analytic we have $H(\Delta^{n-1} \times \mathbf{C}) \cap B^m \subset \pi^{-1}(X)$. To prove $\sigma \equiv 0$ our idea is to evaluate $\|\sigma\|$ on the boundary $H(\Delta^{n-1} \times \mathbf{C}) \cap \partial B^m$. The set $E \subset \Delta^{n-1} \times \mathbf{C}$ where H fails to be a holomorphic immersion is a complex–analytic subset. Since H is a holomorphic immersion at the origin it follows that for any $a \in \Delta^{n-1}$, $E \cap (\{a\} \times \mathbf{C})$ is a discrete set. Since $H^{-1}(\partial B^m) \cap (\{a\} \times \mathbf{C})$ is a circle C_a one can always choose a generic point $(a;b)$ on C_a such that H is a holomorphic immersion at $(a;b)$. We are going to evaluate $\|\sigma\|$ at $(a;b)$, so to speak. Let $\{P_\nu\}$ be a sequence of points on $\pi^{-1}(X) \subset B^m$ such that $P_\nu = H(a;b_\nu)$ with $b_\nu \longrightarrow b$. Denote by ds_E^2

the Euclidean metric on $\mathbb{C}^n$. Since H is a holomorphic immersion at (a;b) from the explicit expression of the Poincaré metric on B^m it follows that $H^*(h) \geq c_\nu(ds_E^2)$ with $c_\nu \to \infty$ as $\nu \to \infty$. As $\sigma_{i\bar{j}}(a;b_\nu) = \sigma_{i\bar{j}}(a)$ is independent of ν for for $1 \leq i, j \leq n-1$, it follows by (1) that $\|\sigma(P_\nu)\| \to 0$ as $\nu \to \infty$. Fix a sufficiently small positive number δ, and let B_ν denote the geodesic ball on $(\pi^{-1}(X),h|_{\pi^{-1}(X)})$ centered at P_ν of radius δ. We may assume that $B_\nu \subset H(\Delta^{n-1} \times \mathbb{C})$. From $H^*(h) \geq c_\nu(ds_E^2)$ with $c_\nu \to \infty$ it follows that B_ν shrinks to a point on ∂B^m as $\nu \to \infty$. As H is a holomorphic immersion on a neighborhood of (a;b) it follows that there exist positive numbers $\epsilon_\nu \to 0$ such that $\|\sigma(x_\nu)\| \leq \epsilon_\nu$ for $x_\nu \in B_\nu$. After passing to a subsequence we may assume that the sequence of points $\pi(P_\nu) = Q_\nu$ converges to some point Q on X. It follows that on the geodesic ball $B(Q;\delta)$ on $(X,h|_X)$, $\|\sigma\|^2$ vanishes identically. By the identity theorem on real-analytic functions we have $\|\sigma\|^2 \equiv 0$ on X. This proves $\sigma \equiv 0$ and consequently that the second fundamental form S of X in Y vanishes identically. The proof of [(2.1), Thm.1] is completed under the assumption that f: $X \to Y$ is an embedding. The proof works without any significant modification for the general case of holomorphic immersions f. ∎

(2.4) We give some remarks concerning holomorphic mappings between complex hyperbolic space forms. We start with the example of MOSTOW [MOS2].

THEOREM 1 (MOSTOW [MOS2])
There exist cocompact discrete subgroups Γ_0 and Γ_0' of $\mathbb{P}SU(2,1)$ and a surjective homomorphism $\Phi: \Gamma_0 \to \Gamma_0'$ such that $\mathrm{Ker}(\Phi)$ is infinite. Furthermore, the abelian groups $\Gamma_0/[\Gamma_0,\Gamma_0]$ and $\Gamma_0'/[\Gamma_0',\Gamma_0']$ ([.,.] denoting the commutator subgroup) are finite.

The subgroups $\Gamma_0, \Gamma_0' \subset \mathbb{P}SU(2,1)$ are generated by complex reflections. We argue that Thm.1 leads to a surjective holomorpic mapping f: $B^2/\Gamma \to B^2/\Gamma'$

between complex hyperbolic space forms such that for the induced map f_* on fundamental groups $\mathrm{Ker}(f_*)$ is infinite. We use the Strong Rigidity Theorem for Kähler manifolds of SIU [SIU2]. One can always find a torsion–free subgroup $\Gamma' \subset \Gamma'_0$ of finite index (cf. SATAKE [SA2, Ch.IV, Lemma(7.2), p.196]). Define $\Gamma \subset \Gamma_0$ by $\Gamma = \Phi^{-1}(\Gamma')$. Then, Γ is torsion–free and $\Phi|_\Gamma\colon \Gamma \to \Gamma'$ has an infinite kernel. One can always find a smooth mapping $f_0\colon X := B^2/\Gamma \to B^2/\Gamma' := Y$ such that $(f_0)_* = \Phi|_\Gamma$. By the existence theorem for harmonic maps into compact Riemannian manifolds of negative Riemannian sectional curvature (EELLS–SAMPSON [ES]) there always exists a harmonic map $f\colon B^2/\Gamma \to B^2/\Gamma'$ homotopic to f_0. We claim that f is holomorphic or anti–holomorphic. By the Strong Rigidity Theorem of SIU [SIU2] f is either holomorphic or anti–holomorphic if $\mathrm{rank}_{\mathbf{R}}(df) \geq 3$ at some point. On the other hand, if $\mathrm{rank}_{\mathbf{R}}(df) \leq 1$ everywhere then f must map into a geodesic, contradicting the fact that $\Phi|_\Gamma\colon \Gamma \to \Gamma'$ is surjective. It remains to consider the case when $\mathrm{rank}_{\mathbf{R}}(df) = 2$ on a dense open set. By the $\partial\bar{\partial}$–Bochner–Kodaira formula of SIU [SIU2] if follows that the level sets of f give rise to a holomorphic foliation $\mathcal{F}$ on X – finite point set such that all leaves are closed (cf. SIU [SIU4]). Consequently the map $f\colon X \to Y$ factors through the leaf space $\mathfrak{R}$ of $\mathcal{F}$, which is a compact Riemann surface. Since f_* is surjective it follows that $\mathfrak{R}$ must be a hyperbolic Riemann surface. On the other hand such a Riemann surface $\mathfrak{R}$ supports non–trivial holomorphic 1–forms ν. By lifting ν to X it follows that $b_1(X) \neq 0$. However from Thm.1 we deduce $\mathrm{Card}(\Gamma/[\Gamma,\Gamma]) < \infty$, so that $b_1(X) = 0$, a contradiction. It follows that in fact f is either holomorphic or anti–holomorphic. By replacing Γ' by $\bar{\Gamma}'$ (in $\mathbf{P}SU(2,1) \subset \mathbf{P}GL(3,\mathbf{C})$) and replacing f by its conjugate if necessary we obtain the example $f\colon B^2/\Gamma \to B^2/\Gamma'$ with the desired properties.

It is not clear if [(2.1), Thm.1] is in any sense optimal. While the dual theorem of Feder [(2.2), Thm.2] is optimal we do not have an example of a holomorphic immersion $f\colon X \to Y$ between complex hyperbolic space forms with X compact and of dimension ≥ 2 such that f is not a totally geodesic isometric immersion. One can easily modify the example of MOSTOW [MOS2] to give an example f: $X \longrightarrow Y$ with X compact and of dimension 2 and Y of dimension 3 such that f is a branched covering onto its image f(X), which is a totally geodesic Kähler submanifold of Y. So far there is no example of a holomorphic mapping f: $X \longrightarrow Y$ between complex hyperbolic space forms with X compact and of complex dimension ≥ 2 such that the image f(X) is not a totally geodesic Kähler submanifold.

It is plausible that [(2.1), Thm.1] of Cao–Mok remains valid for X of finite volume. In this case the major difficulty is the justification of the decomposition of the Chern character in case X is non–compact and of finite volume.

In another direction CORLETTE [COR] proved a rigidity theorem for compact complex hyperbolic space forms $X = B^n/\Gamma$ in the context of rigidity of flat bundles over such manifolds. His rigidity theorem can be formulated in terms of homomorphisms $\rho: \Gamma \longrightarrow \mathbb{P}SU(n,1)$. Let $2\pi\omega$ denote the Kähler form of the Poincaré metric on B^n (of constant Ricci curvature $-(n+1)$), which gives rise to the Eilenberg–MacLane cohomology class $[\omega] \in H^2(\mathbb{P}SU(n,1),\mathbf{R})$. Its m–th exterior power is given by $[\omega]^m \in H^{2m}(\mathbb{P}SU(n,1),\mathbf{R})$. On the other hand there is the fundamental class $[\Gamma]$ of $H^2(\Gamma,\mathbf{R}) \cong H^2(X,\mathbf{R})$. Let $\rho: \Gamma \longrightarrow \mathbb{P}SU(n,1)$ be a homomorphism. In [Cor] the homological volume $vol(\rho)$ is defined to be the evaluation of $\frac{1}{m!}\rho^*([\omega]^m)$ at $[\Gamma]$. In terms of $vol(\rho)$ one has

THEOREM 2 (CORLETTE [COR])

Let $X = B^m/\Gamma$ be a compact complex hyperbolic space form. Denote by g the canonical metric on X of constant Ricci curvature $-(n+1)$. Suppose $\rho: \Gamma \longrightarrow \mathbb{P}SU(n,1)$ is a homomorphism such that $vol(\rho) = Vol(X,g)$. Then, there is a totally geodesic holomorphic embedding of B^m into B^n which is equivariant with respect to ρ.

Mostow's example [MOS2] shows that the Superrigidity Theorem of Margulis (cf. ZIMMER [ZIM]) is no longer valid for simple Lie groups associated to rank–1 symmetric manifolds of non–compact type beyond $\mathbb{P}SU(1,1)$. In the case of $\mathbb{P}SU(n,1)$ the following problem remains open.

PROBLEM

Let $\Gamma \subset \mathbb{P}SU(m,1)$ be a lattice with $m \geq 1$. Suppose $\rho\colon \Gamma \longrightarrow \mathbb{P}SU(n,1)$ is an injective homomorphism such that $\rho(\Gamma) \subset \Gamma'$ for some lattice $\Gamma' \subset \mathbb{P}SU(n,1)$. Does ρ extend to a homomorphism $\mathbb{P}SU(m,1) \longrightarrow \mathbb{P}SU(n,1)$?

When Γ is cocompact and $\rho(\Gamma) \subset \Gamma'$ for some cocompact discrete subgroup of $\mathbb{P}SU(n,1)$, one can use the existence theory for harmonic mappings (EELLS–SAMPSON [ES]) and the Strong Rigidity Theorem of SIU [SIU2] to show that ρ is represented by a holomorphic or anti–holomorphic map $f\colon = B^m/\Gamma \longrightarrow B^n/\Gamma'$. So far we do not know of any geometric property of f that can be deduced from the fact that the induced map $f_* = \rho$ on fundamental groups is injective.

CHAPTER 10 THE HERMITIAN METRIC RIGIDITY THEOREM ON LOCALLY HOMOGENEOUS HOLOMORPHIC VECTOR BUNDLES

§1 Homogeneous Hermitian Vector Bundles on Bounded Symmetric Domains

(1.1) For the notations and the background for this chapter cf. Chap.5 (on bounded symmetric domains) and App.I (on semisimple Lie algebras). Let Ω be an irreducible bounded symmetric domain and write $\Omega = G_0/K$ as usual with $G_0 = \mathrm{Aut}_0(\Omega)$. We recall that K is not semisimple. In fact $\mathfrak{k} = \mathfrak{k}_s + \mathfrak{z}$ for a one-dimensional center $\mathfrak{z}$ (cf. [Ch.3, (2.1), Prop.2]) and for $\mathfrak{k}_s := [\mathfrak{k},\mathfrak{k}]$ the semisimple part of $\mathfrak{k}$. Write $K_s \subset K$ for the subgroup corresponding to the semisimple Lie subalgebra $\mathfrak{k}_s$. In general the compact semisimple Lie group K_s has a finite fundamental group. If $Z \subset K$ denotes the subgroup corresponding to the center $\mathfrak{z}$ then Z is the center of K and we have $K \cong (K_s \times Z)/\Phi$ for a finite group Φ.

V_0 will always denote a finite-dimensional complex vector space. $GL(V_0)$ will always mean $GL(V_0,\mathbb{C})$, etc. Let $\rho\colon K \to GL(V_0)$ be an irreducible (finite-dimensional) complex representation of K on a complex vector space V_0. We will henceforth equip V_0 with a $\rho(K)$-invariant Euclidean metric s_0 so that ρ is a unitary representation on V_0. Write π for $d\rho\otimes_{\mathbb{R}}\mathbb{C}\colon \mathfrak{k}^{\mathbb{C}} \to \mathfrak{gl}(V_0)$. We define a smooth complex vector bundle V over $\Omega = G_0/K$ as follows. Introduce an equivalence relation $\mathfrak{R}$ on $G_0 \times V_0$ by specifying that $(\gamma,v)\,\mathfrak{R}\,(\gamma',v')$ if and only if $\gamma' = \gamma k$ for some $k \in K$ and $v' = \rho(k)^{-1}(v)$. The fibered product $G_0\times_K V_0$ is defined to be the set of equivalence classes $(G_0\times V_0)/\mathfrak{R}$. There is a projection $\eta\colon G_0\times_K V_0 \to G_0/K \cong \Omega$ defined by $\eta(\gamma,v) = \gamma K \in G_0/K$. The fibers of η are complex vector spaces isomorphic to V_0 and it is clear that $\eta\colon V := G_0\times_K V_0 \to \Omega$ defines the structure of a smooth complex vector bundle on V. We sometimes identify V_0 with the fiber of η over $o = eK$. Furthermore V inherits a Hermitian metric s from the $\rho(K)$-invariant Euclidean metric s_0 on V_0. $\eta\colon (V,s) \to \Omega$ is thus a Hermitian vector bundle homogeneous under the action of G_0.

Denote by X_c the compact dual of Ω and let $\Omega \subset\subset \mathbb{C}^N \subset X_c$ be the Harish-Chandra and the Borel embeddings of Ω. We use the notation $G^{\mathbb{C}}$ to denote the centerless connected simple Lie group with Lie algebra $\mathfrak{g}^{\mathbf{C}}$, i.e. $G^{\mathbf{C}}$ is the inner automorphism group $\mathrm{Int}(\mathfrak{g}^{\mathbf{C}})$, so that $G_0 \subset G^{\mathbf{C}}$ is the subgroup corresponding to the non–compact real form $\mathfrak{g}_0 \subset \mathfrak{g}^{\mathbf{C}}$. Write $P \subset G^{\mathbf{C}}$ for the (complex) parabolic subgroup corresponding to $\mathfrak{p} = \mathfrak{m}^- + \mathfrak{k}^{\mathbf{C}}$. From $[\mathfrak{m}^-,\mathfrak{m}^-] = 0$ it is immediate that one can extend $\pi\colon \mathfrak{k}^{\mathbf{C}} \longrightarrow GL(V_0)$ to a complex representation of $\mathfrak{p}$ by defining $\pi|_{\mathfrak{m}^-} \equiv 0$. Since $K \hookrightarrow P = K^{\mathbf{C}}M^-$ is a homotopy equivalence (as can be easily deduced from [App.I, §5, Thm.3], applied to the semisimple part $K_s \cong (K\times Z)/\Phi$) the extended π lifts to a representation $P \longrightarrow GL(V_0)$. Consider the fibered product $V^h = G^{\mathbf{C}}\times_P \mathbf{C}^n$ defined similar to the above. It is clear that there is a canonical projection $\eta^h\colon V^h \longrightarrow G^{\mathbf{C}}/P \cong X_c$ making V^h into a holomorphic vector bundle over X_c. Moreover, it follows readily from the Borel Embedding Theorem that the natural map $V = G_0\times_K V_0 \longrightarrow G^{\mathbf{C}}\times_P \mathbf{C}^n = V^h$ is an open embedding of V into V^h identifying V as the part of V^h lying above Ω. For $(\gamma,v) \in G^{\mathbf{C}}\times V_0$ denote by $\{\gamma,v\}$ the element in $G^{\mathbf{C}}\times_P V_0 = V^h$ it defines. Then the G_0–action on V extends to a $G^{\mathbf{C}}$–action on V^h defined by $\gamma\{\gamma',v\} = \{\gamma\gamma',v\}$. $G^{\mathbf{C}}$ acts as a group of holomorphic bundle transformations on V^h. Thus, V inherits the structure of a holomorphic vector bundle from V^h such that G_0 acts on V as holomorphic bundle transformations. For $\gamma \in G^{\mathbf{C}}$ the orbit of $v \in V^h$ under γ will simply be denoted by $\gamma(v)$. From now on we will always understand that (V,s) is equipped with such a holomorphic structure. To indicate the relationship between V and π we sometimes write $\pi = \pi_V$ (or $\pi = \pi(V)$) and $V = V_\pi$. Henceforth by a Hermitian vector bundle (E,h) we will always mean a Hermitian *holomorphic* vector bundle.

(1.2) Using the Harish–Chandra Embedding $\Omega \hookrightarrow \mathbf{C}^N$ we are going to introduce fiber coordinates on homogeneous Hermitian vector bundles (V,s). We will use such coordinates to compute the curvature of (V,s).

In terms of Harish–Chandra coordinates the subgroup $M^+ = \exp(\mathfrak{m}^+)$ acts on $\mathbb{C}^N \subset X_c$ by Euclidean translations. We identify $V^h|\mathbb{C}^N$ with $\mathbb{C}^N \times V_o$ via the map $\lambda\{\exp(m^+),v\} = (m^+,v) \in m^+ \times V_o \cong \mathbb{C}^N \times V_o$. M^+ acts on V^h by $\exp(m^+)(\xi,v) = (\xi+m^+,v)$. Let k be the rank of V. Introduce Euclidean coordinates on V_o such that at o the Hermitian metric is given by $s_{\alpha\bar{\beta}}(o) = \delta_{\alpha\beta}$, $1 \leq \alpha,\beta \leq k$. The system of holomorphic fiber coordinates $\{v^\alpha\}_{1\leq\alpha\leq k}$ thus obtained will also be referred to as Harish–Chandra (fiber) coordinates on (V,s). We assert that they form a special holomorphic coordinate system adapted to s at o. We observe first of all that the symmetry σ_o at o acts on V^h by $\sigma_o\{\gamma,v\} = \{\sigma_o\gamma,v\} = \{\sigma_o\gamma\sigma_o^{-1},\rho(\sigma_o)v\}$, i.e., $\sigma_o(z,v) = (-z,\rho(\sigma_o)v)$ in terms of Harish–Chandra base and fiber coordinates on $V^h|\mathbb{C}^N$. Write $\{\epsilon_\alpha\}_{1\leq\alpha\leq k}$ for the holomorphic basis of $V^h|\mathbb{C}^N$ using Harish–Chandra coordinates; and write $\{\epsilon^\alpha\}_{1\leq\alpha\leq k}$ for the dual basis. Since $\rho(\sigma_o) = \pm id$ it follows that for the Hermitian metric $s(z) = \Sigma s_{\alpha\bar{\beta}}(z)\epsilon^\alpha\otimes\bar{\epsilon}^\beta$ on $\Omega \subset\subset \mathbb{C}^N$, $s_{\alpha\bar{\beta}}(z)$ are even functions, showing in particular $ds_{\alpha\bar{\beta}}(o) = 0$, as asserted.

We proceed to compute the curvature of the homogeneous vector bundle (V,s) over Ω. Normalize the canonical metric g_o on Ω so that at the origin o = eK, g_o is represented by the identity matrix in terms of the Harish–Chandra coordinates. Identify the fibre $\eta^{-1}(o)$ at o = eK with V_o and denote by (.,.) the (K–invariant) Hermitian inner product s_o at V_o. We have

PROPOSITION 1

Let $\eta: (V,s) \longrightarrow \Omega$ be an irreducible Hermitian homogeneous vector bundle over an irreducible bounded symmetric domain (Ω,g_o). Write $\pi = \pi_V: \mathfrak{k}^{\mathbb{C}} \longrightarrow \mathfrak{gl}(V_o)$ for the associated irreducible representation of $\mathfrak{k}^{\mathbb{C}}$. Then, the curvature tensor Θ of (V,s) is given at o by

$$\Theta_{v\bar{v}'\xi\bar{\xi}'} = -(\pi[\xi,\bar{\xi}]v,v'),$$

for $\xi, \xi' \in m^+ \cong T_o(\Omega)$ and $v, v' \in V_o$.

Proof:

We will give a proof using Harish–Chandra coordinates. For a derivation of Prop.1

in a general context using Maurer–Cartan forms cf. GRIFFITHS–SCHMID [GS].

In terms of the Harish–Chandra base and fiber coordinates $\{z^i\}_{1\leq i\leq N}$, $\{v^\alpha\}_{1\leq\alpha\leq k}$ resp. we have

$$s_{\alpha\overline{\beta}}(z) = \delta_{\alpha\beta} - \sum_{1\leq i,j\leq N} \Theta_{\alpha\overline{\beta}i\overline{j}} z^i \overline{z}^j + O(|z|^4), \qquad (1)$$

(where $|.|$ denotes the Euclidean norm on $\mathbf{C}^N$) since $s_{\alpha\overline{\beta}}(z)$ is an even function on Ω. The point $<\xi> := (\xi_1,...,\xi_N) \in \mathbf{C}^N$ is translated to the origin o by $\exp(-\xi) \in M^+ \in G^{\mathbf{C}}$. Fixing ξ we want to expand $s_{\alpha\overline{\beta}}(<t\xi>)$ as a function of t. We have $\xi + \overline{\xi} = 2Re\xi \in Re(\mathfrak{m}^+) = \mathfrak{m}$. Define $\tau_t := \exp(-2tRe\xi) = \exp(-t\xi - t\overline{\xi})$. For t real sufficiently small τ_t gives a transvection along the geodesic l on (Ω,g_0) joining $<\xi> \in \Omega$ to o. In terms of Harish–Chandra coordinates l is a line segment on $\Omega \subset \mathbf{C}^N$. From the proof of the Harish–Chandra Embedding Theorem for ξ sufficiently small we have a decomposition $\tau_t = \kappa_t \exp(-t\overline{\xi}) \exp(-t\xi)$ where $\kappa_t \in K^{\mathbf{C}}$ (cf. [Ch.5]). $\xi_t := \exp(-t\xi)$ acts on V^h by translation $\xi_t(z,v) = (z-t\xi, v)$ in Harish–Chandra coordinates, moving the base point $<t\xi>$ to the base point o, $\exp(-t\overline{\xi}) \in M^-$ fixes the origin and acts trivially on V_0; and $\kappa_t \in K^{\mathbf{C}}$ acts on V_0 via the complexification of $\rho: K \longrightarrow GL(V_0)$. Thus $s_{\alpha\overline{\beta}}(<\xi>) = (\kappa_t(e_\alpha),\kappa_t(e_\beta))$. We have

$$\kappa_t = \exp(-2tRe\xi)\exp(t\xi)\exp(t\overline{\xi}). \qquad (2)$$

Observe that $Re\xi = \xi + \overline{\xi}$. From the Campbell–Hausdorff formula (which for matrices A, B gives the expansion $e^{-t(A+B)}e^{tA}e^{tB} = 1 + \frac{1}{2}t^2(AB-BA) + ...$, cf. e.g. VARADARAJAN [VA]) we deduce

$$\partial_t^2\kappa_t = [\xi,\overline{\xi}] \in \sqrt{-1}\mathfrak{k} \subset \mathfrak{k}^{\mathbf{C}}. \qquad (3)$$

It follows that

$$s_{\alpha\overline{\beta}}(<t\xi>) = (\kappa_t(e_\alpha),\kappa_t(e_\beta))$$

$$= (e_\alpha + \tfrac{1}{2}t^2(\pi[\xi,\bar\xi]e_\alpha) \,;\, e_\beta + \tfrac{1}{2}t^2(\pi[\xi,\bar\xi]e_\beta) + O(t^4).$$

$$= \delta_{\alpha\beta} + \tfrac{1}{2}t^2\{(\pi[\xi,\bar\xi]e_\alpha, e_\beta) + (e_\alpha, \pi[\xi,\bar\xi]e_\beta)\} + O(t^4). \qquad (4)$$

Since $[\xi,\bar\xi] \in \sqrt{-1}\mathfrak{k}$ and $\pi(\mathfrak{k})$ consists of skew Hermitian matrices, $\pi[\xi,\bar\xi]$ is Hermitian symmetric, thus by (4) and in Harish–Chandra coordinates

$$s_{\alpha\bar\beta}(t\xi_1,\ldots,t\xi_N) = \delta_{\alpha\beta} + t^2(\pi[\xi,\bar\xi]e_\alpha, e_\beta) + O(|z|^4). \qquad (5)$$

By polarization we obtain

$$s_{\alpha\bar\beta}(z^1,\ldots,z^N) = \delta_{\alpha\beta} + (\pi[\partial/\partial z_i, \partial/\partial \bar z_j]e_\alpha, e_\beta)z^i\bar z^j + O(|z|^4). \qquad (6)$$

Comparing with (1) this yields immediately the curvature formula in the proposition. ▌

Let (E,h) be a Hermitian vector bundle. (E,h) is said to be indecomposable if and only if it cannot be written as a direct sum of two proper Hermitian vector subbundles. We have

PROPOSITION 2

An irreducible Hermitian homogeneous vector bundle (V,s) over Ω is indecomposable (as a Hermitian vector bundle).

To prove Prop.2 we need

LEMMA 1

$[\mathfrak{m}^+,\mathfrak{m}^-] = \mathfrak{k}^{\mathbb{C}}$.

Proof:

From the Cartan decomposition $\mathfrak{g}_0 = \mathfrak{k} + \mathfrak{m}$ we have $[\mathfrak{k},\mathfrak{m}] \subset \mathfrak{m}$ and $[\mathfrak{m},\mathfrak{m}] \subset \mathfrak{k}$. We assert that in fact $[\mathfrak{m},\mathfrak{m}] = \mathfrak{k}$. To see this consider $\mathfrak{u} = \mathfrak{m} + [\mathfrak{m},\mathfrak{m}]$. From the Jacobi identity it is clear that $\mathfrak{u}$ is invariant under $ad(\mathfrak{k})$. It follows that $[\mathfrak{g}_0,\mathfrak{u}] \subset ad(\mathfrak{k})(\mathfrak{u}) + [\mathfrak{m},\mathfrak{m}] + [\mathfrak{m},[\mathfrak{m},\mathfrak{m}]] \subset \mathfrak{u} + [\mathfrak{m},\mathfrak{m}] + [\mathfrak{m},\mathfrak{k}] \subset \mathfrak{u}$. Consequently, $\mathfrak{u}$ is a non-trivial ideal in the simple real Lie algebra $\mathfrak{g}_0$, showing $\mathfrak{u} = \mathfrak{g}_0$ and so $[\mathfrak{m},\mathfrak{m}]$

$= \mathfrak{k}$, as asserted. From $[\mathfrak{m}^+,\mathfrak{m}^+] = [\mathfrak{m}^-,\mathfrak{m}^-]$ it follows that $\mathfrak{k}^{\mathbf{C}} = [\mathfrak{m}^{\mathbf{C}},\mathfrak{m}^{\mathbf{C}}] = [\mathfrak{m}^+,\mathfrak{m}^-]$, proving the lemma. ▌

We now give the proof of Prop.1 using the curvature formula in Prop.1

Proof of Prop.2:

Write π for π_V. Suppose $V \cong W \oplus W'$ is a direct sum decomposition of V as a Hermitian vector bundle. At $o = eK$ we have $\Theta_{w\bar{w}',\xi\bar{\xi}'} = 0$ for any $\xi, \xi' \in \mathfrak{m}^+$ and any $w \in W_o$, $w' \in W_o'$. From the curvature formula this means $(\pi[\xi,\bar{\xi}']w,w') = 0$. By Lemma 1 it follows that $\pi(\mathfrak{k}^{\mathbf{C}})W_o$ is orthogonal to W_o' and hence contained in W_o. As $\mathfrak{k}^{\mathbf{C}}$ acts irreducibly on V_o we see that $W_o = 0$ or V_o, showing that (V,s) is indecomposable as a Hermitian vector bundle, as desired. ▌

(1.3) To describe representations of K, recall the decomposition $\mathfrak{k} = \mathfrak{k}_s + \mathfrak{z}$ into the semisimple part $\mathfrak{k}_s$ and the center $\mathfrak{z}$, and the corresponding decomposition $K = (K_s \times Z)/\Phi$ for some finite subgroup $\Phi \subset K_s \times Z$. Every complex representation $\rho: K \longrightarrow \mathfrak{gl}(n)$ restricts to a complex representation $\rho|K_s$. By Schur's Lemma for ρ irreducible $\rho(Z)$ must act as scalar multiplications, so that ρ is irreducible if and only if $\rho|K_s$ is. We prove

LEMMA 1

Let $\rho_s: K_s \longrightarrow GL(n,\mathbf{C})$ be a (an irreducible) complex representation on the semisimple compact Lie group K_s. Then, there exists a (an irreducible) complex representation $\rho: K \longrightarrow GL(n,\mathbf{C})$ such that $\rho_s \equiv \rho|K_s$.

Proof:

Fix an isomorphism $Z \cong S^1 := \{e^{i\theta}: \theta \in \mathbf{R}\}$ of the center Z of K and write z_θ for the element in Z corresponding to $e^{i\theta}$. For p an integer define a homomorphism $\rho': K_s \times Z \longrightarrow GL(V_o)$ by $\rho'(k_s z_\theta) = e^{pi\theta}\rho_s(k_s)$ for $k_s \in K_s$. Writing $K = (K_s \times Z)/\Phi$ to define $\rho: K \longrightarrow GL(n,\mathbf{C})$ it suffices to find an integer p such that ρ' contains Φ in its kernel. $\Phi \subset K_s \times Z$ consists of all elements

(k_s, z_θ) such that $k_s z_\theta = 1$ in K, so that $\Phi = \{(z_{-\theta}, z_\theta): z_{-\theta} \in K_s\} \cong K_s \cap Z$, which is a finite cyclic group of order q, say. By Schur's Lemma $\rho_s(K_s \cap Z) \subset S^1.id$ so that $\rho_s(z_\varphi) = e^{ki\varphi}$ for every $z_\varphi \in K_s \cap Z$. It suffices to take $p = k + tq$ for any integer t to guarantee that $\Phi \subset \mathrm{Ker}(\rho')$. ∎

It may actually happen that K_s has a non–trivial fundamental group so that some complex representation $\pi_s: \mathfrak{k}_s \longrightarrow \mathfrak{gl}(n,C)$ cannot be lifted to a representation of K_s. For instance, when $\Omega = D_n^{II}$ or D_n^{III}, $K = U(n)/\{\pm 1\}$ and $K_s = SU(n)/\{\pm 1\}$. In this case any faithful representation of $SU(n)$ (e.g., the standard representation) cannot be lifted to K_s. Since $\pi_1(K_s)$ is finite it is clear that some tensor power of any π_s can be lifted to K_s.

From the proof of Lemma 1 it follows that the group of 1–dimensional representations of K is infinite cyclic with some generator χ. Write (L,s) for the associated homogeneous Hermitian line bundle. On the other hand the canonical line bundle (K_Ω, θ) of $\Omega \cong G_0/K$ with a canonical metric θ is homogeneous and non–trivial. Choose L so that $\pi(K_\Omega) = p\pi(L)$ for a positive integer p. We write $L^q = K_\Omega^{q/p}$ and call (L^q, s^q) a fractional power of (K_Ω, θ).

We proceed to describe the weight space decomposition of irreducible representations of $\mathfrak{k}$. Let $\mathfrak{h}_s \subset \mathfrak{k}_s$ be a (real) Cartan subalgebra. Then $\mathfrak{h} := \mathfrak{h}_s + \mathfrak{z} \subset \mathfrak{k}$ is a Cartan subalgebra of $\mathfrak{k}$. Since $\mathfrak{z}$ is the center of $\mathfrak{k}$ any $\mathfrak{k}_s$–root $\gamma \in \sqrt{-1}\mathfrak{h}_s^*$ defines a $\mathfrak{k}$–root in $\sqrt{-1}\mathfrak{h}^*$ vanishing on $\mathfrak{z}$. We identify γ with its extension. Fix a fundamental system of positive $\mathfrak{k}_s$–roots Σ_s. Let $\rho: K \longrightarrow GL(n,C)$ be an irreducible complex representation on $V_0 \cong C^N$, $\pi = d\rho \otimes_{\mathbf{R}} C$ and α be the highest $\mathfrak{k}_s$–weight of $\pi|\mathfrak{k}_s^C$ and v^α be a corresponding non–zero weight vector. Thus, for any $h_s \in \mathfrak{h}_s$, $\pi(h_s)v^\alpha = \alpha(h_s)v^\alpha$ and any weight η of $\pi|\mathfrak{k}_s^C$ is of the form $\alpha - \gamma$ for some positive $\mathfrak{k}_s$–root γ. On the other hand as $\rho|Z$ is given by $\rho(z_\theta) = e^{ki\theta}.id$ for any fixed $z \in \mathfrak{z}$ and any $v \in V_0$ we have $\pi(z)v = \sqrt{-1}cv$ for some real constant c. Identify $\eta \in \sqrt{-1}\mathfrak{h}_s^*$ as an element of $\sqrt{-1}\mathfrak{h}^*$ by defining $\eta|\mathfrak{z} \equiv 0$ we know therefore that there exists an element $\delta \in \sqrt{-1}\mathfrak{h}^*$, $\delta|\sqrt{-1}\mathfrak{h}_s \equiv 0$, such that η is a $\mathfrak{k}_s$–weight of $\pi|\mathfrak{k}_s^C$ if and only if $\eta + \delta$ is a

$\mathfrak{k}_s$–weight of π. We will henceforth call $\omega = \alpha + \delta$ the highest weight of π in the sense that any $\mathfrak{k}$–weight η' of π is of the form $\eta' = \eta + \delta = \omega - \gamma$ for some positive $\mathfrak{k}_s$–root γ. The notion of a lowest weight of π is similarly defined.

A Cartan subalgebra $\mathfrak{h}^{\mathbf{C}}$ of $\mathfrak{k}^{\mathbf{C}}$ is at the same time a Cartan subalgebra of $\mathfrak{g}^{\mathbf{C}}$, which is semisimple. Extend Σ_s to a fundamental system Σ of positive $\mathfrak{g}^{\mathbf{C}}$–roots by adding a non–compact root (cf. [Ch.5, (1.1)]). Then, the highest weight ω of π belongs to the weight lattice Λ of $\mathfrak{g}^{\mathbf{C}}$. We note that, in contrast to the semisimple situation, ω may or may not be in the highest Weyl chamber of $\mathfrak{h}^*_{\mathbf{R}}$ with respect to Σ.

Remarks

For a non–singular representation π, i.e., one for which $<\omega,\varphi> \neq 0$ for any $\mathfrak{k}$–root φ, the number of Weyl reflections on $\sqrt{-1}\mathfrak{h}^*$ needed to move ω into the highest Weyl chamber determined by Σ is of great importance in the study of vanishing theorems on cohomology groups for the associated locally homogeneous Hermitian vector bundles (V_X,s) on quotients $X = \Omega/\Gamma$ of finite volume, cf. Griffiths–Schmid [GS].

§2 An Extension of the Hermitian Metric Rigidity Theorem and Applications

(2.1) Let Ω be an irreducible bounded symmetric domain, $\Gamma \subset Aut_o(\Omega) = G_0$ be a torsion–free discrete group of automorphisms such that $X = \Omega/\Gamma$ is of finite volume with respect to the canonical metric. Write $K \subset G_0$ for the isotropy subgroup at $o \in \Omega$. Let $\rho: K \longrightarrow GL(V_o)$ be an irreducible complex representation of K and (V,s) be the associated G_0–homogeneous Hermitian vector bundle on Ω. Since Γ acts on the left by isometries on (V,s) the latter descends to a Hermitian holomorphic vector bundle (V_X,s) on X, where we use the same symbol s for the canonical quotient metric. We say that (V_X,s) is locally homogeneous. We are going to derive a metric rigidity theorem for certain locally homogeneous Hermitian vector bundles V on X. The rigidity theorem will also apply to the case when X is of rank 1, encompassing the example used in [Ch.9, §1] in the study of the immersion problem for hyperbolic space forms. In §1 we wrote down the curvature formula of (V,s) in terms of the representation $\pi = \pi_V$

on the complexified Lie algebra $\mathfrak{k}^{\mathbf{C}}$. Here as in the above we adopt throughout the normalization on the Hermitian metrics h and hence on the curvature as given in §1. Regarding the curvatures of (V,s) we prove

PROPOSITION 1

Let Θ denote the curvature tensor of (V,s). Define

$$M_V = \max\{\Theta_{v\bar{v}\xi\bar{\xi}}: v \in V_o,\ \xi \in \mathfrak{m}^+,\ \|v\| = \|\xi\| = 1\}$$

$$m_V = \min\{\Theta_{v\bar{v}\xi\bar{\xi}}: v \in V_o,\ \xi \in \mathfrak{m}^+,\ \|v\| = \|\xi\| = 1\}.$$

Then, M_V is attained if $\xi = \alpha$ is a characteristic vector with respect to a Cartan subalgebra $\mathfrak{h} \subset \mathfrak{k}$ and a choice of the positive Weyl chamber in $\mathfrak{h}^*_{\mathbf{R}}$, and $v = v^\lambda$ is a corresponding lowest weight vector of the representation $\pi: \mathfrak{k}^{\mathbf{C}} \to \mathfrak{gl}(V_o)$. On the other hand, m_V is attained if ξ is the same as above and $v = v^\omega$ is a corresponding highest weight vector.

Proof:

Fix a Cartan subalgebra $\mathfrak{h} \subset \mathfrak{k}$ and a choice of the positive Weyl chamber C in $\mathfrak{h}^*_{\mathbf{R}}$. Let Δ be the space of $\mathfrak{g}^{\mathbf{C}}$–roots and $\Psi \subset \Delta^+_M$ be a maximal strongly orthogonal set of non–compact positive roots obtained by descending from the highest root μ with respect to $\mathfrak{h}^{\mathbf{C}}$ and C. Write $\Psi = \{\psi_1 = \mu, \psi_2, \ldots, \psi_r\}$, $e_i = e_{\psi_i}$ for associated unit root vectors and α for e_1. We may assume without loss of generality that $\xi = \Sigma \xi_i e_i$. Recall that in terms of the Hermitian inner products (.,.) on (V,s), we have

$$\Theta_{v\bar{v}\xi\bar{\xi}} = -(\pi[\xi,\bar{\xi}]v,v) = -\sum_{1\le i,j\le r} \xi_i\bar{\xi}_j\,(\pi[e_i,\bar{e}_j]v,v). \tag{1}$$

As Ψ is strongly orthogonal for $i \neq j$, $\psi_i - \psi_j$ is not a root so that $[e_i,\bar{e}_j] = 0$. For $i = j$ we have $[e_i,\bar{e}_i] = H_i \in \mathfrak{h}^*_{\mathbf{R}}$. Hence,

$$\Theta_{v\bar{v}\xi\bar{\xi}} = -\sum_{1\le i\le r} |\xi_i|^2(\pi(H_i)v,v). \tag{2}$$

Denote by Φ the set of weights of π. Write $v = \Sigma_{\eta\in\Phi} v^\eta$ for a decomposition of v into root vectors. Then, $\pi(H_i)v = \Sigma_{\eta\in\Phi}\pi(H_i)v^\eta = \Sigma_{\eta\in\Phi}\eta(H_i)v^\eta$, so that

$$\Theta_{v\bar{v}\xi\bar{\xi}} = -\sum_{1\le i\le r} |\xi_i|^2(\pi(H_i)v,v) = -\sum_{1\le i\le r,\eta\in\Phi} |\xi_i|^2\eta(H_i)\|v^\eta\|^2. \tag{3}$$

We examine the largest and smallest possible values of $\eta(H_i) = \langle\eta,\psi_i\rangle$. Each $\eta \in \Phi$ is of the form $\eta = \omega - \gamma$ for some positive root γ. If $\psi_i = \psi_1 = \mu$, we have $\langle\eta,\mu\rangle = \langle\omega-\gamma,\mu\rangle = \langle\omega,\mu\rangle - \langle\gamma,\mu\rangle$. Since μ is a highest weight vector of the adjoint representation of $\mathfrak{g}^{\mathbf{C}}$ on $\mathfrak{g}$, we have $\langle\gamma,\mu\rangle \ge 0$ for every positive root γ. Consequently, $\langle\eta,\mu\rangle \le \langle\omega,\mu\rangle$ for any $\eta \in \Phi$. For a general $\psi \in \Psi$ we know by [App.(III.1), Cor.1] that the Weyl group of Δ permutes the roots in Ψ, so that each root $\psi \in \Psi$ is the highest root relative to the fixed Cartan subalgebra $\mathfrak{h}^{\mathbf{C}}$ and some choice of the positive Weyl chamber $C(\psi)$. Write $\omega(\psi)$ for the highest weight of π with respect to $\mathfrak{h}^{\mathbf{C}}$ and $C(\psi)$. It follows that for any $\psi \in \Psi$ and $\eta \in \Phi$ we have $\langle\eta,\psi\rangle \ge \langle\omega(\psi),\psi\rangle = \langle\omega,\mu\rangle$. The same argument shows that $\langle\eta,\psi\rangle \le \langle\lambda,\mu\rangle$ for the lowest weight of π with respect to $\mathfrak{h}^{\mathbf{C}}$ and C. Putting this into the curvature formula (3) it is then immediate that for $\xi = \Sigma\xi_ie_i$ and $v \in V_o$ of unit length the minimum (resp. maximum) of $\Theta_{v\bar{v}\xi\bar{\xi}}$ is attained when $\xi = e_1$ and $v = v^\omega$ (resp. v^λ) is a highest (resp. lowest) weight vector with respect to $\mathfrak{h}^{\mathbf{C}}$ and C. The proof of Prop.1 is completed. ▌

From now on we fix a Cartan subalgebra $\mathfrak{h} \subset \mathfrak{k}$ and a positive Weyl chamber C unless otherwise specified. We define the notion of homogeneous vector bundles of properly seminegative curvature.

DEFINITION 1

Let (V,s) be a Hermitian homogeneous vector bundle on Ω and Θ be the curvature tensor of (V,s). We say that (V,s) is of properly seminegative curvature (or simply that (V,s) is properly seminegative) if and only if (V,s) is of seminegative curvature in the sense of Griffiths and there exist non–zero vectors $v \in V_o$, $\xi \in \mathfrak{m}^+$ such that $\Theta_{v\bar{v}\xi\bar{\xi}} = 0$.

By Prop.1 for (V,s) irreducible we have

PROPOSITION 2

Let (V,s) be a homogeneous Hermitian vector bundle. Denote by $\mu \in \Delta$ the highest root of $\mathfrak{g}^{\mathbf{C}}$ and by λ the lowest root of $\pi = \pi_V$. Then, (V,s) is of properly seminegative curvature if and only if $<\lambda,\mu> = 0$.

We give some computational examples illustrating Props. 1 & 2.

(1) Suppose Ω is irreducible and of rank ≥ 2, then the tangent bundle (T_Ω, g_0), equipped with the canonical metric, is properly seminegative. Consider the case when $\Omega = D^I_{p,q}$ is of type I. Then, $\Omega \cong SU(p,q)/S(U(p)\times U(q))$. Choose the Cartan subalgebra $\mathfrak{h}$ of $\mathfrak{s}(\mathfrak{u}(p)\times\mathfrak{u}(q))$ to consist of purely imaginary diagonal matrices of trace 0. Write E_{ij} for the $(p+q)$ by $(p+q)$ matrix with zero entries excepting an entry of unity at the (i,j)–th entry. Write λ_i for the linear form on $\mathfrak{h}^{\mathbf{C}}$ (the set of all complex diagonal matrices of trace 0) which takes the value 1 at E_{ii} and 0 at E_{jj} for $j \neq i$, $1 \leq j \leq p+q$. (Thus, $\lambda_1 + ... + \lambda_{p+q}$ is zero on $\mathfrak{h}^{\mathbf{C}}$.) $\mathfrak{h}^*_{\mathbf{R}} = \{\Sigma a_i\lambda_i: a_i \text{ real}, \Sigma a_i = 0\}$. The inner product $<.,.>$ on $\mathfrak{h}^*_{\mathbf{R}}$ induced by the Killing form and duality is then simply $<\Sigma a_i\lambda_i, \Sigma b_j\lambda_j> = \frac{2}{p+q}\Sigma a_i b_j$. One can choose the positive Weyl chamber so that we have in the set Δ of $\mathfrak{g}^{\mathbf{C}}$–roots the subsets of

positive roots:	$\{\lambda_i - \lambda_j: i < j, 1 \leq i,j \leq p+q\}$;
simple roots:	$\{\sigma_i: 1 \leq i \leq p+q-1\}$, $\sigma_i = \lambda_i - \lambda_{i+1}$;
positive non–compact roots:	$\{\lambda_i - \lambda_j: 1 \leq i \leq p, p+1 \leq j \leq q\}$.

We have $\lambda_i - \lambda_j = \sigma_i + ... \sigma_{j-1}$. In particular, the highest (positive) non–compact root is given by $\mu = \lambda_1 - \lambda_{p+q} = \sigma_1 + ... + \sigma_{p+q-1}$, the lowest positive non–compact root by $\nu = \lambda_p - \lambda_{p+1} = \sigma_p$. For ρ: $S(U(p)\times U(q)) \longrightarrow U(pq)$ the isotropy representation on $\mathfrak{m}^+ \cong M(p,q;\mathbf{C})$ we have thus the highest weight $\omega = \mu$ and the lowest weight $\lambda = \nu$. We have $<\lambda,\mu> = <\lambda_p - \lambda_{p+1}, \lambda_1 - \lambda_{p+q}> = 0$ if $p, q \geq 2$; i.e., (T_Ω, g_0) is properly seminegative for $\Omega = D^I_{p,q}$ with $\min(p,q) \geq 2$.

(2) Consider the rank–1 case $\Omega = D^I_{n,1} \cong B^n$ Write $T = T_{B^n}$, K for the

canonical line bundle and consider now the irreducible homogeneous vector bundle $V = S^{n+1}T\otimes K$. By the formulas given in [App(I.4), Prop.3] the lowest weight vector $\lambda(\pi)$ of the corresponding irreducible represenation $\pi = \pi_V$ is then given by $\lambda(\pi) = (n+1)\nu + \kappa$, where κ is the unique weight of K given by

$$\kappa = -(\lambda_1-\lambda_{n+1}) - \dots - (\lambda_n-\lambda_{n+1}) = -(\lambda_1+\dots+\lambda_n) + n\lambda_{n+1}, \text{ so that}$$

$$\lambda(\pi) = (n+1)(\lambda_n-\lambda_{n+1}) - (\lambda_1+\dots+\lambda_n) + n\lambda_{n+1} = -\lambda_1-\dots-\lambda_{n-1}+n\lambda_n-\lambda_{n+1};$$

$$<\lambda,\mu> = <-\lambda_1-\dots-\lambda_{n-1}+n\lambda_n-\lambda_{n+1}\, ;\, \lambda_1-\lambda_{n+1}> = 0,$$

showing that the (Hermitian) homogeneous vector bundle $V = S^{n+1}T\otimes K$ is properly seminegative. This is the bundle V used in [Ch.9, §1] in studying the immersion problem for hyperbolic space forms.

(2.2) We are now ready to state a generalization of the Hermitian metric rigidity theorem of Chapters 6 and 9 to the case of properly seminegative locally homogeneous Hermitian vector bundles. For simplicity we will only formulate it for the case when $X = \Omega/\Gamma$ is locally irreducible. The generalization to the locally reducible case is quite obvious.

THEOREM 1

Let Ω be an irreducible bounded symmetric domain and $X = \Omega/\Gamma$ be a quotient of Ω by a torsion–free discrete group of automorphisms Γ such that X is of finite volume in the canonical metric. Let (V_X,s) be an irreducible Hermitian locally homogeneous vector bundle on X. Suppose (V_X,s) is of properly seminegative curvature. Then, for any Hermitian metric h of seminegative curvature on V_X we have $h = cs$ for some constant c on X.

Write $\Omega = G_0/K$ as usual with $G_0 = \mathrm{Aut}_o(\Omega)$ and let $\Omega \subset\subset \mathbb{C}^N \subset X_c$ be the Harish–Chandra and the Borel embeddings of Ω. Denote by (V,s) the corresponding Hermitian vector bundle on Ω. Write ω for the highest weight of $\pi = \pi_V$ and v^ω for a corresponding non–zero weight vector at $o = eK$. As explained in (1.1) there is a corresponding homogeneous vector bundle V^h on G_c

in such a way that V can be identified in a canonical way as the part of V^h lying over Ω. We will make this identification from now on. The complex Lie group $G^{\mathbf{C}}$ acts on V^h and hence on $\mathbf{P}(V^h)$ holomorphically. For $v \in V^h$ denote by $[v]$ the projectivized element in $\mathbf{P}(V^h)$ defined by v. Write $\mathcal{M}(V) = G_0[v^\omega]$. The quotient $\mathcal{M}(V)/\Gamma \subset \mathbf{P}(V_X)$ is then denoted by $\mathcal{M}(V_X)$. We call $\mathcal{M}(V_X)$ the characteristic variety of V_X over X, etc. From [App.(I.5), Thm.1] and the proof of [Ch.6, (1.1) Props. 1 & 2] we have immediately

PROPOSITION 1

The characteristic bundle $\mathcal{M}(V) \to \Omega$ of V over Ω is holomorphic and is equal to the part of $G^{\mathbf{C}}[v^\omega]$ lying over Ω. Moreover, in terms of the Harish–Chandra embedding $\Omega \hookrightarrow \mathbf{C}^N$, $\mathcal{M}(V)$ is parallel on Ω, i.e., identifying $\mathbf{P}(V)$ with $V \times \mathbf{P}(\mathfrak{m}^+)$ using Euclidean coordinates we have $\mathcal{M}(V) \cong \mathcal{M}_o \times \mathbf{C}^N$, where $\mathcal{M}_o$ is the fiber of $\mathcal{M}(V)$ over $o = eK$.

Proof of Thm.1:

To prove Thm.1 we use an integral formula on the characteristic variety $\mathcal{M}(V)$ of V which is the obvious analogue of that appearing in [Ch.6, (2.1), Prop.1]. Using Prop.1 and the proof of [Ch.6, (2.1), Thm.1] we obtain immediately that $\nabla_\zeta h_{v\overline{w}} = 0$ for any $x \in X$, $[v] \in \mathcal{M}_x$, $\zeta \in T_x(X)$ such that $\Theta_{v\overline{v}\zeta\overline{\zeta}} = 0$ and for any $w \in T_x(X)$. To finish the proof of Thm.1 it suffices to provide a polarization argument in place of [Ch.6, (2.1), Prop.3]. There we use the property that (Ω, g_o) is of seminegative curvature in the dual sense of Nakano. Here in the general case we are going to deduce $\nabla h \equiv 0$ directly from representation–theoretic considerations.

Write ρ: $K \to U(V_o)$ (U denoting the unitary group) for the unitary representation giving rise to V. By the proof of [(2.1), Prop.1] $\Theta(v_o,\overline{v_o};\zeta_o,\overline{\zeta_o}) = 0$ for $\zeta_o = e_\nu \in \mathfrak{m}^+$ corresponding to the lowest positive non–compact root ν. Equivalently, $-\nu$ is the highest weight of the conjugate isotropy representation $\overline{\tau}$: $K \to U(\mathfrak{m}^-)$ on $\mathfrak{m}^-$ and $\overline{\zeta_o}$ is a corresponding highest weight vector (cf. [App.(I.3), Prop.2]). Conjugating the Cartan sub–algebra $\mathfrak{h}^{\mathbf{C}} \subset \mathfrak{k}^{\mathbf{C}}$ by $\kappa \in K$ the highest weight vector v_o is conjugated to a highest weight vector of $\rho(k)v_o$ with respect to the conjugate Cartan subalgebra $ad(\kappa)\mathfrak{h}^{\mathbb{C}}$ and some choice of positive Weyl chamber C_κ in $ad(\kappa)\mathfrak{h}^{\mathbb{C}}$. Similarly ζ_0 is transformed to a highest weight vector $\overline{\tau}(\kappa)\overline{\zeta_0}$ with respect to C_κ. Consider the set $\Sigma_0 := \{[v\otimes\overline{\zeta}] \subset \mathbb{P}(V_0\otimes\mathfrak{m}^-) : [v] \in \mathcal{M}_0, \zeta \in \mathfrak{m}^+ \text{ and } \Theta_{v\overline{v}\zeta\overline{\zeta}} = 0.\}$ From the preceding discussion one sees that $\Sigma := \{[\rho(\kappa)v_0\otimes\overline{\tau}(\kappa)\zeta_0]: \kappa \in K\}$ is contained in Σ_0. By definition Σ is the orbit of $[v_0\otimes\overline{\zeta_0}]$ in $\mathbb{P}(V_0\otimes\mathfrak{m}^-)$ for the representation $\rho\otimes\overline{\sigma} : K \to GL(V_0\otimes\mathfrak{m}^-)$. Let $W \subset V_0\otimes\mathfrak{m}^-$ be the K–subspace generated by $v_0\otimes\overline{\zeta_0}$. Then, W is irreducible, has highest weight $\eta = \omega - \nu$ with respect to $\mathfrak{h}^{\mathbb{C}}$ and C, and $v_0\otimes\overline{\zeta_0}$ is a highest weight vector w^η (cf. [App.(I.4), Prop.3]). It follows that Σ is complex–analytic by the easy part of the Borel–Weil Theorem (cf. [App.(I.5), Thm.1]). In other words, near v_0 and ζ_0 we can vary v holomorphically and ζ anti–holomorphically while maintaining the curvature property $\Theta_{v\overline{v}\zeta\overline{\zeta}} = 0$. This suffices to replace the polarization argument in [Ch.6, (2.1), Prop.3] to complete the proof of Thm.1. in case X is compact.

In the non–compact case one has to justify the integral formula used in [Ch.8, (1.2)]. To generalize the argument there one uses the determinant line bundle $\det(V_X)$ in place of the canonical line bundle. Recall that every homogeneous Hermitian line bundle on Ω is a fractional power of the canonical line bundle. Thus, $\det(V_X) = K_X^p$ for some rational number $p > 0$ when (V_X,s) is properly seminegative and non–trivial. As already remarked in [Ch.9, (1.1), Thm.3] the metric description of K_X in terms of toroidal compactifications M as in [Ch.8, (1.1), Thm.4] and the description of $K_X \otimes [D]$, $D = M - X$, in terms of Satake–Baily–Borel compactifications as in [Ch.8, (1.1), Thm.2] extends to the non–arithmetic (rank–1) case. The proof of Thm.1 is completed. ∎

(2.3) We give here an application of the Hermitian metric rigidity theorem [(2.2), Thm.1] to study homomorphisms of irreducible locally homogeneous Hermitian vector bundles (V_X,s) over $X \cong \Omega/\Gamma$. Consider first of all the case when (V_X,s) is properly seminegative and non–trivial. Then, [(2.1), Thm.1] implies immediately that $\Gamma(X,V_X^*) = 0$. In fact, if $\sigma \in \Gamma(X,V_X^*)$, then $h = s +$

$Re(\sigma\otimes\bar{\sigma})$ defines a Hermitian metric of seminegative curvature on V_X, so that $Re(\sigma\otimes\bar{\sigma})$ = Const. s over X by the metric rigidity theorem, which is a plain contradiction since V_X is not the trivial line bundle. As a special case, when Ω is irreducible and of rank ≥ 2, then the holomorphic tangent bundle (T_X,g) is properly seminegative, so that $\Gamma(X,T_X^*) = 0$, i.e., there are non non–trivial holomorphic 1–forms on X. In case X is compact, this is a special case of a vanishing theorem of MATSUSHIMA [MA1] on Betti numbers. (In the non–compact case classical methods only yield the non–existence of closed L^2 holomorphic 1–forms.) We remark that in the rank–1 case the first Betti number of compact quotients $X = B^n/\Gamma$ can be zero (cf. MUMFORD [MUM2] & MOSTOW [MOS2]) or non–zero (cf. LIVNE [LIV]).

For any irreducible Hermitian vector bundle (V,s), $V = V_\pi$, we defined in [(2.1), Prop.1] M_V and m_V as the maximum and minimum of curvatures on unit vectors (under the normalization of §1). Let μ be the highest (positive) non–compact root. Denote by $\omega(\pi)$ resp. $\lambda(\pi)$ the highest (resp. lowest) weight of π. By [(2.1), Prop.1] we have $M_V = -<\lambda(\pi),\mu>$ (resp. $m_V = -<\omega(\pi),\mu>$). We make the same assumptions on $X = \Omega/\Gamma$ as in [(2.2), Thm.1]. For the sake of simplifying notations we will from now on drop the subscript X when referring to homogeneous vector bundles on X. We are going to prove

THEOREM 1

Let (V,s) and (V′,s′) be irreducible Hermitian locally homogeneous vector bundles on X. Write $\Omega = G_0/K$ as usual and write $\pi\colon \mathfrak{k}^{\mathbb{C}} \to \mathfrak{gl}(V_0)$, $\pi'\colon \mathfrak{k}^{\mathbb{C}} \to \mathfrak{gl}(V_0')$ for the irreducible representations of $\mathfrak{k}^{\mathbb{C}}$ associated to V and V′ resp. Suppose either (i) $<\lambda(\pi),\mu> \leq <\lambda(\pi'),\mu>$ or (ii) $<\omega(\pi),\mu> \leq <\omega(\pi'),\mu>$. Then, either $\Gamma(X,Hom(V,W)) = 0$ or $V \cong W$, in which case we have $\Gamma(X,Hom(V,W)) \cong \Gamma(X,End(V)) \cong \mathbb{C}$.

We remark first of all that the conditions on the highest and lowest weights are necessary, as given by the examples of compact hyperbolic space forms $X = B^n/\Gamma$ with non–vanishing first Betti numbers. A holomorphic 1–form gives a section $s \in \Gamma(X,T^*) = \Gamma(X,Hom(1,T^*)) \cong \Gamma(X,Hom(T,1))$ for the trivial bundle 1 and the holomorphic tangent bundle T. Denoting by $\tau\colon \mathfrak{k}^{\mathbb{C}} \to \mathfrak{gl}(\mathfrak{m}^+)$ the complexified isotropy representation and by 0 the trivial representation we have $\pi(1) = 0$, $\pi(T) = \tau$. We have in the notations of (2.1) $\lambda(0) = \omega(0) = 0$, $\mu = \omega(\tau) = \lambda_1 - \lambda_{n+1}$ and $\lambda(\tau) = \lambda_n - \lambda_{n+1}$. Thus,

$$<\lambda(\tau),\mu> = <\lambda_n-\lambda_{n+1}, \lambda_1-\lambda_{n+1}> = \frac{2}{n+1} > 0 = <\lambda(0),\mu> \quad \text{while}$$
$$\frac{4}{n+1} = <\lambda_1-\lambda_{n+1}; \lambda_1-\lambda_{n+1}> = <\omega(\tau),\mu> > <\omega(0),\mu> = 0.$$

In the proof of Thm.1 we need:

LEMMA 1

Let (E,h) be a Hermitian vector bundle over a complex manifold M, $S \subset E$ a holomorphic vector subbundle such that the second fundamental form σ of the inclusion $(S,h|_S) \hookrightarrow (E,h)$ vanishes identically. Then, the orthogonal complement $S^\perp$ of S is a holomorphic vector subbundle of E.

Proof:

It suffices to show that for each local $S^\perp$–valued smooth section η'', $\bar\partial\eta''$ is a $S^\perp$–valued (0,1) form. Let η' be an arbitrary local S–valued smooth section. Thus, for (.,.) denoting the Hermitian inner product and ∇ denoting the Hermitian connection on (E,h), we have $(\eta',\eta'') = 0$ where defined, so that for ξ a tangent vector of type (1,0) at $x \in M$, $0 = \nabla_\xi(\eta',\eta'') = (\nabla_\xi\eta',\eta'') + (\eta',\bar\partial_\xi\eta'')$ where defined, and the condition that $S^\perp$ is holomorphic is equivalent to the condition that $\nabla_\xi\eta' \in S$ for arbitrary η' and ξ as defined. On the other hand the second fundamental form $\sigma(\xi,\eta')(x) = pr_{S^\perp}(\nabla_\xi\eta')$. The vanishing of σ therefore implies that $S^\perp \subset E$ is holomorphic, as desired. ∎

Proof of Thm.1:

Observe first of all that $Hom(V,V') \cong V^* \otimes V' \equiv V' \otimes V^* \cong Hom(V'^*,V^*)$ and that $\omega(\pi) = -\lambda(\pi^*)$ for any irreducible representation $\pi\colon \mathfrak{k}^{\mathbb{C}} \to \mathfrak{gl}(\mathfrak{m})$, so that the condition $<\omega(\pi),\mu> \le <\omega(\pi'),\mu>$ is equivalent to the condition $<\lambda(\pi^*),\mu> \ge <\lambda(\pi'^*),\mu>$. Replacing π'^* by π and π^* by π' it suffices therefore to establish Thm.1 under the assumption (1) $<\lambda(\pi),\mu> \le <\lambda(\pi'),\mu>$.

Consider first of all the special case when $<\lambda(\pi),\mu> = 0$ and $<\lambda(\pi'),\mu> \geq 0$. By [(2.1), Prop.1] (V,s) is of properly seminegative curvature and (V′,s′) is of seminegative curvature. For any $\tau \in \Gamma(X,\mathrm{Hom}(V,V'))$ we obtain a Hermitian metric h of seminegative curvature on V by defining $h = s + \tau^*s'$. By [(2.2), Thm.1] we conclude that there is a constant c on X such that $h = cs$. Consequently $\tau^*s' = (c-1)s$, showing that either τ is trivial or it is an isometry up to normalizing constants. In the latter case we claim that

(*) V and V′ are isomorphic as holomorphic homogeneous vector bundles.

Replacing φ by a constant multiple we assume that φ is an isometry and identify V as a Hermitian vector subbundle of V′. By the same polarization argument as in the proof of [Ch.6, (1.1), Prop.3] one can also show that the second fundamental form of V in V′ vanishes identically. This implies by Lemma 1 that the orthogonal complement $V^\perp$ of V in V′ is a holomorphic vector bundle and that $V' \cong V \oplus V^\perp$ isometrically. By [(1.2), Prop.1] V′ is however indecomposable as a Hermitian vector bundle, proving that in fact $V' = V$, hence (*). To complete the proof in the special case it suffices to observe that identifying V′ with V any $\varphi \in \Gamma(X,\mathrm{End}(V))$ must be a multiple of the identity *id*. In fact, fixing any $x \in X$ there exists a constant c such that $\varphi - c.id$ is not injective at some point. As $\varphi - c.id$ cannot then be an isometry it must be identically zero, proving that $\varphi \equiv c.id$ on X, as desired.

Let K_X denote the canonical line bundle and denote by χ the infinitesimal character (unique weight) of $\pi(K_X) = \kappa$. To prove Thm.1 we assume for the time being that for some integer p we have $<\lambda(\pi),\mu> + p<\chi,\mu> = 0$. Under this assumption we have $\pi(V \otimes K_X^p) = \pi \otimes \chi^p$ with $<\lambda(\pi \otimes \chi^p),\mu> = 0$. Since $\mathrm{Hom}(V,V') \cong V^* \otimes V' \cong (V \otimes K^p)^* \otimes (V' \otimes K^p) \cong \mathrm{Hom}(V \otimes K^p, V' \otimes K^p)$ by tensoring with K^p we return to the special situation where $0 = <\lambda(\pi),\mu> \leq <\lambda(\pi'),\mu>$.

In the general case p can only be taken to be a fraction. There are two ways to complete the proof. One way is to replace V (resp. V′) by the irreducible component W (resp. W′) of highest weight of the symmetric tensor power S^qV (resp. S^qV') for some positive integer q. We have $\lambda(W) = q\lambda(V)$. For an appropriate choice of q we can always make sure that $<q\lambda(\pi),\mu> + p<\chi,\mu> = 0$ for an integer p. One can then conclude that $\Gamma(X,Hom(W,W')) = 0$ or any $\varphi \in \Gamma(X,Hom(W,W'))$ is an isometry (onto) up to a normalizing constant. Suppose now $\varphi \in \Gamma(X,Hom(V,V'))$. φ gives rise to a section $\varphi_q \in \Gamma(X,Hom(W,W'))$ by taking tensor powers and it is immediate to see that φ_q is an isometry (up to a normalizing constant) if and only if it is true for φ. This completes the proof of Thm.1. Another way to complete the proof of Thm.1 is to modify the formulation of the Hermitian metric rigidity theorem [(2.2), Thm.1]. Denote by θ the canonical metric on K_X and by $(L,\hat{s}) \to \mathbb{P}(V)$ the Hermitian tautological line bundle on $\mathbb{P}(V)$ associated to (V,s). Write η: $\mathbb{P}(V) \to X$ for the base projection. It suffices to show that for a Hermitian locally homogeneous holomorphic vector bundle (V,s) such that $\beta(s) = c_1(L,\hat{s}) - p\eta^*c_1(K_X,\theta) \geq 0$ and such that β is not strictly positive definite, then any Hermitian metric h satisfying $\beta(h) \geq 0$ must be s up to a normalizing constant. The modifications needed in the proof of the latter statement are obvious. ▌

We retain the assumption that Ω is irreducible and of complex dimension $\geq$ 2. By replacing V by the trivial line bundle and V′ by V we have

COROLLARY 1

Suppose (V,s) is an irreducible non–trivial locally homogeneous Hermitian vector bundle on $X = \Omega/\Gamma$ and $\Gamma(X,V) \neq 0$. Then, (V,s) is of strictly positive curvature in the sense of Griffiths.

REMARKS

In case X is compact the specal case in Thm.1 that V is simple, i.e., $\Gamma(X,End(V)) \cong \mathbb{C}$ for irreducible locally homogeneous Hermitian vector bundles follow from the fact that V is stable with respect to the canonical polarization since (V,h) is Hermitian–Einstein with respect to the canonical metric on X and and since (V,h) is indecomposable as a Hermitian vector bundle (cf. SIU [SIU, (1.5) & (1,6), p.18ff.]).

Corollary 1 can be generalized to

COROLLARY 1′

Let Ω be a bounded symmetric domain of complex dimension ≥ 2 and $X = \Omega/\Gamma$ be an irreducible quotient of finite volume of Ω by a torsion–free discrete group Γ of automorphisms. Suppose (V,s) is an irreducible non–trivial locally homogeneous Hermitian vector bundle on $X = \Omega/\Gamma$ and F is a locally flat Hermitian vector bundle on X such that $\Gamma(X,V\otimes F) \neq 0$. Then, (V,s) is of strictly positive curvature in the sense of Griffiths.

Proof:

In case F is trivial Cor.1′ follows from an obvious generalization of Thm.1 to the irreducible, locally reducible case using Moore's Ergodicity Theorem on some subvariety $\mathscr{M}$ of $\mathbb{P}(V)$, as is done in Chapter 6. To prove Cor.1′ it suffices to show that when (V^*,s^*) is non–trivial and of properly seminegative curvature, $\Gamma(X,V) = 0$. Cor.1′ will then follow as in the proof of Thm.1. Let h be a flat Hermitian metric on F. Any $\sigma \in \Gamma(X,V\otimes F)$ can be interpreted as a holomorphic bundle homomorphism φ: $V^* \rightarrow F$. As (F,h) is locally flat $s^* + \varphi^*h$ is a Hermitian metric of seminegative curvature. It follows from the metric rigidity theorem [(2.2), Thm.1] that φ is an isometry. By the same polarization argument as in the proof of [Ch.6, (1.1), Prop.3] one shows that in fact the second fundamental form of the isometry φ vanishes identically, so that (V^*,s^*) is locally flat, which is a plain contradiction. The proof of Cor.1′ is completed. ▌

REMARKS

Cor.1′ can be interpreted as a vanishing theorem on certain classes of automorphic forms on Ω. In this regard it is related to a vanishing theorem of MATSUSHIMA–SHIMURA [MA–S] when $\Omega \cong \Delta^n$ and when $\Gamma \subset \mathrm{Aut}(\Delta^n)$ is an irreducible cocompact lattice. For higher cohomology groups it should be noted that in [MA–S] the vanishing of certain harmonic forms can be obtained by transforming such forms to certain (holomorphic) automorphic forms using the Bochner–Kodaira formula and proving the non–existence of such automorphic forms. It is easy to obtain along this line a generalization to the finite volume case as long as one works with L^2–harmonic forms (cf. LAI–MOK [LM]). For zero–dimensional cohomology groups Cor.1′ yields a stronger result since no condition on square–integrability is imposed on the holomorphic sections.

CHAPTER 11

A RIGIDITY THEOREM FOR HOLOMORPHIC MAPPINGS BETWEEN IRREDUCIBLE HERMITIAN SYMMETRIC MANIFOLDS OF COMPACT TYPE

§1 Formulation of the Problem

(1.1) The rigidity theorem for holomorphic mappings between Hermitian locally symmetric manifolds of non–compact type ([Ch.6, (5.1), Thm.1]) naturally leads one to ask the dual question on Hermitian symmetric manifolds of compact type. Since there are many canonical (Kähler–Einstein) metrics on an irreducible Hermitian symmetric manifold of compact type, one should formulate the rigidity problem differently, as follows.

PROBLEM

Let X_c and Y_c be (the underlying complex manifolds of) Hermitian symmetric manifolds of compact type with X_c irreducible. Let $f: X_c \longrightarrow Y_c$ be a holomorphic mapping. Find sufficient conditions on f to guarantee that f is a totally geodesic isometric embedding with respect to some choice of canonical metrics on X_c and Y_c.

In the rank–1 situation $X_c \cong \mathbb{P}^n$ it is known that for any surjective holomorphic mapping $f: \mathbb{P}^n \longrightarrow Y$ onto a compact projective–algebraic manifold Y is necessarily biholomorphic to $\mathbb{P}^n$, although in general f is only a branched covering. This is proved by LAZARSFELD [LA] using Mori's characterization of the projective space ([MO]) in terms of rational curves. On the other hand, it is also known that a holomorphic immersion $f: \mathbb{P}^n \longrightarrow \mathbb{P}^m$ is necessarily projective–linear whenever $m \leq 2n - 1$, by [Ch.9, (2.1), Thm.2] of Feder.

Since the rigidity theorem for holomorphic mappings on Hermitian locally symmetric manifolds of non–compact type was proved using a Hermitian metric rigidity theorem on such manifolds [Ch.6, (1.1), Thm.1], one could hope that the dual Kähler metric rigidity theorem for Hermitian symmetric manifolds of compact type can lead to a rigidity theorem on such manifolds. We first note however that the exact analogue (with respect to some choices of canonical metrics) of [Ch.6, (5.1), Thm.1] is false. To see this let $Q^n \subset \mathbb{P}^{n+1}$ be the hyperquadric defined by

$\Sigma z_i^2 = 0$ in terms of the homogeneous coordinates $[z_0,\ldots,z_{n+1}]$ on $\mathbb{P}^{n+1}$. One can embed $\mathbb{P}^n$ into Q^{2n+1} simply by the projective–linear map $\tau\colon \mathbb{P}^n \to Q^{2n} \subset \mathbb{P}^{2n+1}$ defined by $\tau([z_0,\ldots,z_n]) = [z_0,\ldots,z_n,i\,z_0,\ldots,i\,z_n]$. Let $\nu\colon \mathbb{P}^n \to \mathbb{P}^n$ be a holomorphic mapping of degree ≥ 2, which exists by using the Veronese embedding (cf. [App.III.3]) and taking projections. Then, in general the composition $\sigma = \tau\circ\nu$ gives a holomorphic mapping $\sigma|Q^n\colon Q^n \to Q^{2n}$ which is not even a holomorphic immersion. On any irreducible Hermitian symmetric manifold X_c one can construct similarly a holomorphic mapping $f\colon X_c \to Y_c$ into some irreducible Hermitian symmetric manifold Y_c such that f is not a holomorphic immersion. In such constructions the dimensions of Y_c are in general large in comparison to X_c. One might conjecture that for X_c of rank ≥ 2 and Y_c irreducible any holomorphic mapping $f\colon X_c \to Y_c$ is necessarily a totally geodesic isometric embedding (noting that X_c is simply–connected) provided that the dimension of Y_c is small enough in comparison to X_c. The dimension restrictions would possibly depend on the ranks and the dimensions of nullity of the manifolds involved. This general conjecture appears to be rather difficult. In this chapter we deal only with the equi–dimensional case. We present a theorem of TSAI [TSA] which affirms the conjecture in this case.

THEOREM 1

Let X_c and Y_c be two equi–dimensional irreducible Hermitian symmetric manifolds of compact type with $\mathrm{rank}(Y_c) \geq 2$. Let $f\colon X_c \to Y_c$ be a holomorphic map which is not totally degenerate. Then, f is a biholomorphism.

The proof of [(1.1), Thm.1] is achieved by a differential–geometric study of the minimal rational curves on Y_c. By using the fact that $H_2(X_c,\mathbb{Z}) \cong \mathbb{Z}$ for any irreducible Hermitian symmetric manifold of compact type it is easy to see that as long as $f(X_c)$ is not a single point, it must be a finite map (i.e., fibers of f are finite). Otherwise some fiber of f would be positive–dimensional so that some algebraic curve S would be mapped to a point, implying that $f_*\colon H_2(X_c,\mathbb{Z}) \cong \mathbb{Z} \to H_2(Y_c,\mathbb{Z}) \cong \mathbb{Z}$ is zero, which in turn implies that every algebraic curve on X_c is mapped to a point, since any algebraic curve on a Kähler manifold represents a non–trivial homology class. Consequently, either f is a finite map or f is totally degenerate, as asserted. In the former case f is a branched covering onto Y. The key to the proof of Tsai's Theorem is to show that

PROPOSITION 1

Let $x \in X_c$ be a point where f is a local biholomorphism. Fix an open neighborhood V of x in X_c such that $f|_V$ is a biholomorphism and write $U = f(V)$. Let C be a minimal rational curve on Y_c passing through $y = f(x) \in Y_c$. Then, for any choice of canonical metric g_c on X_c, C is totally geodesic in V with respect to f_*g_c.

To simplify notations we will from now on drop the suffix c. Fix a canonical metric h on Y and denote by ∇ its connection. In the proof of the Kähler metric rigidity theorem on Y ([Ch.7, (2.1), Thm.1]) we developed a method for showing that a minimal rational curve C on Y is totally geodesic with respect to a given Kähler metric s of semipositive bisectional curvature. The proof is partly global and partly local. The global part consists of an integral formula on the dual characteristic bundle $\mathcal{S}^*$. The local computation consists of using the partial vanishing of ∇s^*, s^* denoting the dual metric on the cotangent bundle, and the Kähler condition on s. In the present case we are comparing the Kähler metric f_*g to a background metric h on Y_c. However f_*g is only defined on the open set U. The key difficulty to proving Prop.1 is therefore to establish the partial vanishing of $\nabla(f_*g)^*$ by a semilocal argument. We will write f_*g as g by identifying V with its image $U = f(V)$. As usual we will identify $T^{1,0}(Y)$ with the holomorphic tangent bundle T_Y. Recall that for any $\mu \in T_y(Y)$, $\mathcal{N}_\mu$ denotes the null space of the positive semidefinite Hermitian form $H_\mu(\xi,\eta) = R_{\mu\bar{\mu}\xi\bar{\eta}}(Y)$ and $\Phi: T_Y \to T_Y^*$ is the fiber–by–fiber conjugate–linear map defined by contraction with the metric tensor h and conjugation. To prove Prop.1 it suffices to establish

PROPOSITION 1.1

At any point $y \in V$ and any minimal rational curve C on Y passing through y, we have, in terms of the connection ∇ on (Y,h),

$$\nabla_\alpha g^{\zeta^*\bar{\eta}} = 0$$

for $\alpha \in T_y(C)$, $\zeta^* \in (\mathcal{N}_\alpha)$, and for any $\eta \in T_y^*(Y)$.

For the proof of Prop.1 and subsequently Thm.1 we need to study the space of minimal rational curves on Y, to which we now turn.

§2 Minimal Rational Curves on Hermitian Symmetric Manifolds of Compact Type

(2.1) Let (M,s) denote an irreducible Hermitian symmetric manifold of compact type. Denote by $\mathcal{S} = \mathcal{S}(M)$ the characteristic bundle over $\mathcal{S}$, as defined in [Ch.6, (1.2), Def.1]. From the proof of the Polysphere Theorem [Ch.5, (1.1), Thm.1] we know that for each $x \in M$, $[\alpha] \in \mathcal{S}$, there exists a unique totally geodesic rational curve C passing through x such that $T_x(C) = \mathbb{C}\alpha$. In terms of the first canonical embedding $\nu: M \hookrightarrow \mathbb{P}^N$, C is a rational line in $\mathbb{P}^N$. C is a minimal rational curve in the sense that $C.K_M^{-1}$ is minimum among all (rational) curves. Furthermore any minimal rational curves arise this way (cf. [Ch.7, (3.1), Prop.1]. We study first of all the splitting of the holomorphic tangent bundle over the minimal rational curves C. We use

THEOREM 1 (GROTHENDIECK [GRO])

Let V be a holomorphic vector bundle of rank r over the Riemann sphere $\mathbb{P}^1$. Then, V is isomorphic to a direct sum of line bundles, i.e.,

$$V \cong \mathcal{O}(a_1) \oplus \ldots \oplus \mathcal{O}(a_r)$$

for some integers $a_1, \ldots, a_r$.

Here $\mathcal{O}(a)$ denotes the unique holomorphic line bundle over $\mathbb{P}^1$ of degree a. We call the splitting in the theorem the Grothendieck decomposition of V over $\mathbb{P}^1$. For a proof of Thm.1 cf. also GRAUERT–REMMERT [GR, Ch.VII, p.232ff].

By [App.III.1, Prop.1] over M a unit characteristic vector is equivalently a unit vector of type (1,0) realizing the maximum of holomorphic sectional curvatures. In this chapter we will always normalize the canonical metric so that the maximum of holomorphic sectional curvature is 2. Recall that for the Hermitian bilinear form $H_\alpha(\xi,\eta) = R_{\alpha\bar{\alpha}\xi\bar{\eta}}(M)$ associated to α, we have a decomposition $T_x(M) \cong \mathbb{C}\alpha \oplus \mathcal{H}_\alpha \oplus \mathcal{N}_\alpha$ of eigenspaces of H_α corresponding to the eigenvalues 2, 1 and 0 resp. We called $n(M) := \dim_{\mathbb{C}} \mathcal{N}_\alpha$ the dimension of nullity of M. We are going to relate this eigenspace decomposition with the Grothendieck decomposition over minimal rational curves. We prove using Thm.1

PROPOSITION 1

Let (M,s) be an n–dimensional irreducible Hermitian symmetric manifold of compact type. Let C be a minimal rational curve over M. Denote by q the dimension of nullity $n(M)$ and define $p = n - 1 - q$. Then, for the holomorphic tangent bundle T_M over M we have the Grothendieck splitting

$$T_M|_C \cong \mathcal{O}(2) \oplus \mathcal{O}^p(1) \oplus \mathcal{O}^q.$$

Proof:

Since M is homogeneous as a complex manifold T_M is generated by global holomorphic sections. In particular, $T_M|_C$ is generated by global holomorphic sections, so that all components in the Grothendieck decomposition $T_M|_C \cong \mathcal{O}(a_1) \oplus \dots \oplus \mathcal{O}(a_r)$ are necessarily of degree ≥ 0. Since $T_C \cong \mathcal{O}(2)$ and the inclustion $C \subset M$ gives rise to an injective bundle homomorphism $T_C \hookrightarrow T_M|_C$ we have $0 \neq \mathrm{Hom}(T_C, T_M|_C) \cong \Gamma(C, \oplus\, \mathcal{O}(a_i-2))$, so that for some i, $1 \leq i \leq r$, $a_i \geq 2$. Prop.1 will clearly follow from

(i) $K_M^{-1}|_C \cong \mathcal{O}(2+p)$;

(ii) there are at most q trivial components in the Grothendieck decomposition $T_M|_C \cong \mathcal{O}(a_1) \oplus \dots \oplus \mathcal{O}(a_n)$.

To prove (i) we note first of all that by our normalization (M,s) is of constant Ricci curvature $2 + p$. Denote by θ the canonical metric on $K_M^{-1}|_C$ induced by s. It follows that

$$\deg(K_M^{-1}|_C) = \int_C c_1(K_M^{-1}|_C, \theta) = \frac{1}{2\pi}\,(2 + p)\,\mathrm{Area}(C, h|_C) \qquad (1)$$

On the other hand since $(C, s|_C)$, being totally geodesic in (M,s), is of constant holomorphic sectional curvature 2, we deduce from a similar formula that $2 = \deg(T_C) = (1/2\pi)(2\mathrm{Area}(C, h|_C)$, so that $\mathrm{Area}(C, h|_C) = 2\pi$. Consequently it follows from (1) that $\deg(K_M^{-1}|_C) = 2 + p$, proving (i).

To prove (ii) we consider the Grothendieck decomposition $T_M^*|_C \cong \mathcal{O}(-a_1) \oplus \dots \oplus \mathcal{O}(-a_n)$ of the dual bundle $T_M^*|_C$ over C. By using the Hermitian metric s^* on $T_M^*|_C$ induced by s, $(T_M^*|_C, s^*|_C)$ is of seminegative curvature in the sense of Griffiths. Denote its curvature form by Θ. For any trivial component $L \cong \mathcal{O}$ in $T_M^*|_C$ by the curvature–decreasing property of Hermitian holomorphic vector subbundles we know that $(L, s^*|_L)$ is of seminegative curvature. From the triviality of L it follows that $(L,s^*|_L)$ is flat. Consequently, for $\alpha \in T_x(C)$, $\Theta_{\eta^*\bar{\eta}^*\alpha\bar{\alpha}} = 0$ for any $\eta^* \in L_x$. If $\eta \in T_x(M)$ is obtained from η^* by lifting using the Kähler metric s, we have $R_{\eta\bar{\eta}\alpha\bar{\alpha}} = -\Theta_{\eta^*\bar{\eta}^*\alpha\bar{\alpha}} = 0$. In other words $n(M) = \dim_{\mathbb{C}} \mathcal{N}_\alpha$ must be at least equal to the number r of trivial summands in the Grothendieck decomposition of $T_M|_C$, i.e., $q = n(M) \geq r$, proving (ii). The proof of Prop.1 is completed. ∎

Regarding the characteristic bundle $\mathcal{S} = \mathcal{S}(M)$ over M, we will need to use another structure of $\mathcal{S}$ as a holomorphic fiber bundle. Let $\mathscr{C}$ denote the Chow space of minimal rational curves on M. $\mathscr{C}$ is a normal complex–analytic space. Let G_c denote the identity component of Aut(M,s). Then, $\mathscr{C}$ is the G_c–orbit of a single minimal rational curve C, so that $\mathscr{C}$ is a homogeneous compact complex manifold. By [Ch.7, (3.1), Prop.1], for each $x \in X$ and $[\alpha] \in \mathcal{S}_x$, there exists a unique minimal rational curve C passing through x such that $T_x(C) = \mathbb{C}\alpha$. We define a map $\tau: \mathcal{S} \longrightarrow \mathscr{C}$ by associating to each $[\alpha] \in \mathcal{S}_x$ this unique minimal rational curve $[C] \in \mathscr{C}$. (We use the notation [C] when we think of C as an element of the Chow space $\mathscr{C}$.) We assert

Proposition 2

$\tau: \mathcal{S} \longrightarrow \mathscr{C}$ is holomorphic and realizes $\mathcal{S}$ as a $\mathbb{P}^1$–bundle over $\mathscr{C}$.

Proof:

Denote by $\hat{C}$ the lifting of C to $\mathcal{S}$ defined by $\hat{C} = \{[\alpha] \in \mathcal{S}_x : x \in C \text{ and } T_x(C) = \mathbb{C}\alpha\}$. Then, obviously for $C \neq C'$, their liftings $\hat{C}$ and $\hat{C}'$ are disjoint. Define $\mathscr{E} \subset \mathscr{C} \times M$ by $\mathscr{E} = \{([C],x) \in \mathscr{C} \times M : x \in C\}$. $\mathscr{E}$ is a complex–analytic subspace. The compact Lie group G_c acts on $\mathscr{E}$ in a natural way. It follows from the proof of the Polysphere Theorem [Ch.5, (1.1), Thm.1] that for each $[\alpha] \in \mathcal{S}$ and the associated minimal rational curve C there is a subgroup $G_\alpha \subset G_c$ with G_α isomorphic to $\mathbb{P}SU(2)$ such that G_α preserves C and acts transitively on it. As G_c acts transitively on $\mathscr{C}$ it follows readily that G_c acts transitively on $\mathscr{E}$ as a group of biholomorphisms. In particular, $\mathscr{E} \subset \mathscr{C} \times M$ is a complex submanifold. The lifting $C \longrightarrow \hat{C}$ defines clearly a bijective holomorphic map Φ from $\mathscr{E}$ to $\mathcal{S}$, hence a biholomorphism, and the mapping $\tau: \mathcal{S} \longrightarrow \mathscr{C}$ is simply $pr_1 \circ \Phi^{-1}$, where $pr_1: \mathscr{C} \times M \longrightarrow \mathscr{C}$ is the canonical projection onto the first factor.

As $\tau: \mathcal{S} \longrightarrow \mathcal{C}$ is holomorphic map between complex manifolds and all fibers are isomorphic (to $\mathbb{P}^1$) as complex manifolds it follows from Fischer–Grauert [FG] that $\tau: \mathcal{S} \longrightarrow \mathcal{C}$ realizes $\mathcal{S}$ as a holomorphic $\mathbb{P}^1$–bundle over $\mathcal{C}$.

§3 Proof of the Rigidity Theorem for Holomorphic Mappings

(3.1) We return now to the proof of [(1.1), Prop.1]. As in [Ch.7, (2.1), Thm.1], [(1.1), Prop.1] would follow from curvature identities associated to the Hermitian metrics h^*, g^* and $h^* + g^*$. (Note that contrary to the notations there, the background metric is now h instead of g.) Denote by $\Theta(h^*)$ the curvature of (T^*_Y, h^*), etc. It suffices to prove that for all $y \in U$, C a minimal rational curve passing through y, $\alpha \in T_y(C)$ and $\zeta^* \in \Phi(\mathcal{N}_\alpha)$, we have

$$(*)\qquad \begin{cases} \Theta_{\zeta^*\overline{\zeta^*}\alpha\overline{\alpha}}(h^*) = 0; \\ \Theta_{\zeta^*\overline{\zeta^*}\alpha\overline{\alpha}}(g^*) = 0; \\ \Theta_{\zeta^*\overline{\zeta^*}\alpha\overline{\alpha}}(h^*+g^*) = 0. \end{cases}$$

We proceed now to give a

Proof of [(1.1), Prop.1]:

It suffices to prove the curvature identities (*) above. Recall that the holomorphic mapping $f: X \longrightarrow Y$ between the irreducible n–dimensional Hermitian symmetric manifolds X and Y is a biholomorphism on an open set $V \subset X$ and that we write $U = f(V)$. Let $y = f(x) \in U$ be an arbitrary point, $x \in V$ and C be an arbitrary minimal rational curve on Y passing through y. We have $T_y(C) = \mathbb{C}\alpha$ with $[\alpha] \in \mathcal{S}_y(Y)$ Recall from [(2.1), Prop.1] that we have the Grothendieck decomposition $T^*_M|_C \cong \mathcal{O}(-2) \oplus \mathcal{O}^p(-1) \oplus \mathcal{O}^q$ for $q = n(Y)$ and $p = n - 1 - q$. Denote by $E \subset T^*_M|_C$ the subbundle corresponding to the trivial summands. Equivalently E is the subbundle generated by global holomorphic sections of $T^*_M|_C$, so that E is well–defined independent of the choice of isomorphism in the Grothendieck decomposition. Let $\zeta^* \in E_x$. From the proof of [(2.1), Prop.1] we know that $\Theta_{\zeta^*\overline{\zeta^*}\alpha\overline{\alpha}}(h^*) = 0$. Since $E \longrightarrow C$ is a trivial vector bundle there is a unique holomorphic section σ of $E \subset T^*_M|_C$ over C such that $\sigma(y) = \zeta^*$. The proof there showed in fact that $\|\sigma\|_h{}^* = $ Const. in terms of the metric $h^*|_C$.

Recall that we identify V with U and write g for f_*g, etc. We proceed to prove that $\Theta_{\zeta^*\overline{\zeta}^*\alpha\overline{\alpha}}(g^*) = 0$. In [Ch.7, (2.2), Prop.3] this was obtained by a global integral formula over X. Here we have to supply a semiglobal argument by restriction to the curve C. Let D be the connected component of $f^{-1}(C)$ passing through x. Assume for the time being that D is non-singular. $f^*\sigma = \rho$, $\rho(x) \neq 0$, is a holomorophic section of T_X^* over D, possibly with zeros *a priori*. There is a unique holomorphic line subbundle $F \subset T_X^*|_C$ such that for the associated coherent subsheaf $\mathcal{O}(F)$ of $\mathcal{O}(T_X^*)$, we have $\mathcal{O}_D\rho \subset \mathcal{O}(F)$, $\mathcal{O}_D$ denoting the structure sheaf on D. Since (T_X^*,g^*) is of seminegative curvature, $(F,g^*|_F)$ is also of seminegative curvature, so that $\deg(F) \leq 0$. On the other hand since $\rho \in \Gamma(D,F)$ is non-trivial, we have $\deg(F) \geq 0$, proving therefore that $\deg(F) = 0$ and that in fact ρ is a nowhere-vanishing holomorphic section of constant length with respect to g^*. We have therefore proved that $\|\sigma\|_{g^*}$ = Const. over $C \cap V$. Let L denote the holomorphic line subbundle in $T_Y^*|_C$ generated by the nowhere-vanishing section σ and $\Theta^L(g^*)$ denote the curvature of $(L,g^*|_C)$. From $\|\sigma\|_{g^*}$ = Const. we conclude that $\Theta^L_{\zeta^*\overline{\zeta}^*\alpha\overline{\alpha}}(g^*) = 0$. Since $\Theta^L_{\zeta^*\overline{\zeta}^*\alpha\overline{\alpha}}(g^*) \leq \Theta_{\zeta^*\overline{\zeta}^*\alpha\overline{\alpha}}(g^*) \leq 0$ we have in fact $\Theta_{\zeta^*\overline{\zeta}^*\alpha\overline{\alpha}}(g^*) = 0$, as desired.

Finally, $\Theta_{\zeta^*\overline{\zeta}^*\alpha\overline{\alpha}}(g^*+h^*) = 0$ is obtained in the same way using the fact that $\|\sigma\|^2_{g^*+h^*} = \|\sigma\|^2_{g^*} + \|\sigma\|^2_{h^*}$ = Const. We have thus established (*) under the additional assumption that D is nonsingular. In general let $\nu: \hat{D} \to D$ be a normalization of D and consider the holomorphic vector bundle $W = \nu^*(T_X^*|_D)$ over $\hat{D}$. $(W,\nu^*(g^*))$ is of seminegative curvature. The holomorphic section $\sigma \in \Gamma(C,T_Y^*|_C)$ gives rise to a holomorphic section $\rho = \nu^*f^*\sigma$ of W over $\hat{D}$. The rest of the argument can be carried through without modification. ▌

As a corollary to our proof of [(1.1), Prop.1] we obtain

Corollary 1
$n(X) \geq n(Y)$. Consequently, $\dim_{\mathbb{C}}\mathcal{S}_o(X) \leq \dim_{\mathbb{C}}\mathcal{S}_o(Y)$.

Proof:
We have $\dim_{\mathbb{C}}\Gamma(\hat{D},W) \geq \dim_{\mathbb{C}}\Gamma(C,T_X^*|_C) = n(Y)$, which implies by the argu–

ments given above that for $x \in D \cap V$ and $T_x(D) = \mathbb{C}\xi$ we have $\dim_{\mathbb{C}} \mathcal{N}_\xi \geq n(Y)$. From [App.III.1, Prop.1] we conclude that $n(X) \geq \dim_{\mathbb{C}} \mathcal{N}_\xi \geq n(Y)$. The inequality $\dim_{\mathbb{C}} \mathcal{S}_0(X) \leq \dim_{\mathbb{C}} \mathcal{S}_0(Y)$ follows readily from $\dim_{\mathbb{C}} \mathcal{S}_0(X) = n - 1 - n(X)$, $\dim_{\mathbb{C}} \mathcal{S}_0(Y) = n - 1 - n(Y)$ ([Ch.6, (1.2), Prop.4]).

(3.2) We proceed now to deduce the rigidity theorem for holomorphic mappings [(1.1), Thm.1] of TSAI [TSA] from [(1.1), Prop.1]. First of all we prove

PROPOSITION 1

For any minimal rational curve C on Y, any irreducible component D of $f^{-1}(C)$ is a minimal rational curve on X.

Proof:

Write $R \subset X$ for the branching locus of $f\colon X \longrightarrow Y$. Our proof of [(1.1), Prop.1] shows that for every $y \in Y - f(R)$ and every minimal rational curve C passing through y, every irreducible component D of $f^{-1}(C)$ is totally geodesic in (X,g) over $X - R$. It follows readily that except possibly for self–intersections D is smooth and totally geodesic in (X,g). By abuse of language we will still say that D is totally geodesic in X. As the space of such $[C]$ is dense in $\mathcal{C}$ and $f\colon X \longrightarrow Y$ is a finite map, we conclude that in fact what is said is in fact true for any minimal rational curve C on Y.

To prove that D is a minimal rational curve on X we use the fact that D is totally geodesic in (X,g) for any choice of canonical (Kähler–Einstein) metric g on X. Fix any such canonical metric g. Fix D and pick a smooth point $x \in X$. Write $T_x(D) = \mathbb{C}\alpha$. Let r be the rank of X as a Hermitian symmetric manifold. By the Polysphere Theorem [Ch.5, (1.1), Thm.1] there exists a polysphere $P \cong (\mathbb{P}^1)^r$ passing through x such that P is totally geodesic in (X,g) and such that $\alpha \in T_x(P)$. Since both D and P are totally geodesic in (X,g) it follows that in fact D lies in P. Furthermore from the proof of the Polysphere Theorem there exists a compact Lie subgroup $G_P \subset G_c$ with $G_P \cong (\mathbb{P}SU(2))^r$ stabilizing P (as a set) and acting on $P \cong (\mathbb{P}^1)^r$ in the standard way as isometries. Write $G_P^{\mathbb{C}} \cong$

$(\mathbb{P}SL(2,\mathbb{C}))^r$ for the associated complex Lie group. $G_P^{\mathbb{C}}$ stabilizes $P \cong (\mathbb{P}^1)^r$, acting on it in the standard way as a transitive group of (holomorphic) automorphisms. The curve $D \subset P$ is totally geodesic in (X,Φ^*g) with respect to any automorphism Φ of the complex manifold X. Since P is $G_P^{\mathbb{C}}$–invariant it follows that P is totally geodesic with respect to Φ^*g for any $\Phi \in G_P^{\mathbb{C}}$, so that D is totally geodesic in (P,Φ^*g) for any such Φ. Equivalently, it means that $\Phi(D) \subset P$ is totally geodesic with respect to $(P,g|_P)$ for any such Φ. We identify x with the origin o of P in the identification $o \in \mathbb{C}^r \subset (\mathbb{P}^1)^r \cong P$. Changing the orders of the $\mathbb{P}^1$–factors if necessary we may assume without loss of generality that $T_x(D) = \mathbb{C}\alpha$ with $\alpha = (1,\alpha_2,...,\alpha_r)$ with respect to the Euclidean coordinates on $\mathbb{C}^r$. To prove that D is a minimal rational curve on X it is equivalent to show that $\alpha_2 = ... = \alpha_r = 0$. From the uniqueness of geodesics it follows that for any $\Phi \in G_P^{\mathbb{C}}$ fixing o (i.e., x) such that $\Phi_*(\alpha)$ is proportional to α we must have $\Phi(D) = D$. In particular, D is Φ–invariant for any $\Phi \in G_P^{\mathbb{C}}$ such that $\Phi(o) = o$ and $d\Phi(o) = \mathrm{id}$. The set of such maps Φ constitutes a non–trivial subgroup given by $\Phi(z_1,...,z_r) = \left[\frac{z_1}{c_1z_1+1}, ..., \frac{z_r}{c_rz_r+1}\right]$, with $c_1,..., c_r \in \mathbb{C}$ arbitrary. By picking Φ with $c_1 = 0$ and c_i arbitrary otherwise it follows that D cannot be invariant under $G_P^{\mathbb{C}}$ unless $\alpha_2 = ... = \alpha_r$. The proof of Prop.1 is completed. ▌

From Prop.1 we are going to deduce

PROPOSITION 1

There exists some minimal rational curve D in X such that $f|_D$ is unramified and such that $f(D) = C$ is a minimal rational curve on Y.

<u>Deduction of [(1.1), Thm.1] from Props. 1 & 2</u>:

We deduce first of all [(1.1), Thm.1] from Props. 1 & 2. Since there is no covering map $\mathbb{P}^1 \to \mathbb{P}^1$ except for biholomorphism it would follow that f maps D biholomorphically onto C, and, as D (resp. C) are generators of $H^2(X,\mathbb{Z})$ (resp. $H^2(Y,\mathbb{Z})$) it follows that f induces an isomorphism $H^2(X,\mathbb{Z}) \cong H^2(Y,\mathbb{Z})$. Consequently, any minimal rational curve D on X must be mapped biholomorphically onto a minimal rational curve C on X. Conversely, by [(3.2), Prop.1], for every minimal rational curve C on X, $f^{-1}(C)$ is a finite union of minimal rational curves D_i, and, as $f|D_i$ is a biholomorphism onto C for each i there can only be one irreducible component. In other words, f establishes a one–to–one correspondence between the spaces $\mathscr{C}_X$ and $\mathscr{C}_Y$ of minimal rational curves on X and Y resp. Take any $y \in Y$ and $x \in X$ such that $f(x) = y$. Then, for any $[\alpha] \in \mathcal{S}_y$ with the associated $[C] \in \mathscr{C}_Y$, $f^{-1}(C)$ is a minimal rational curve D on X, and $df(x)(T_x(D)) = T_y(C)$. Consequently, $df(x)(T_x(X))$ contains the $\mathbb{C}$–linear span M_y of characteristic vectors α at x. Since Y is irreducible and M_y is invariant under the isotropy subgroup $K_y \subset G_c = \mathrm{Aut}_0(Y,h)$ of (Y,h) at y, it follows that $M_y = T_y(Y)$, so that $df(x)\colon T_x(X) \to T_y(Y)$ is an isomorphism. Consequently, f is a holomorphic covering map. As any Hermitian symmetric manifold of compact type is simply connected, f is in fact a biholomorphism, as asserted. It remains now to establish Prop.2.

Proof of Prop. 2:
Recall that $R \subset X$ is the branching locus of f. For $x \in X - R$, $y = f(x)$, f induces an isomorphism $\mu = [df(x)]\colon \mathbb{P}T_x(X) \to \mathbb{P}T_y(Y)$. By [(1.1), Prop.1] $\mu^{-1}(\mathcal{S}_y(Y)) \subset \mathcal{S}_x(X)$. On the other hand by [(3.1), Cor.1] $\dim_{\mathbb{C}}(\mathcal{S}_x(X)) \leq \dim_{\mathbb{C}}\mathcal{S}_y(Y)$. Since $\mathcal{S}_x(X)$ is connected it follows that in fact $\mu^{-1}(\mathcal{S}_y(Y)) = \mathcal{S}_x(X)$. Consequently it follows from [(3.2), Prop.1] that given any minimal rational curve D on X passing through x, $f(D)$ is also a minimal rational curve. Since $X - R$ is dense in X by a continuity argument the same is true for any $x \in X$. To prove Prop.2 it suffices therefore to prove that for some minimal rational curve D on X, $f|_D$ is unramified. We are going to prove this by contradiction. Suppose Prop.2 is false. We claim that for every x on the branching locus $R \subset X$, $f|_D$ is ramified at x for every $[D] \in \mathscr{C}_X$. This is obtained by counting dimensions. Define a subset $B_o \subset X \times \mathscr{C}_X$ as

$$B_o = \{(x,[D]) : x \in R \text{ and } f|_D \text{ is ramified at } x\}.$$

$B_o \subset X \times \mathscr{C}_X$ is a complex–analytic subvariety. Write pr_1 and pr_2 for the canonical projections of $X \times \mathscr{C}_X$ onto X and $\mathscr{C}_X$ resp. If Prop.2 is not valid then $pr_2\colon B_o \to \mathscr{C}_X$ is surjective. Denote by B an irreducible component of B_o which projects onto $\mathscr{C}_X$. Since f is a finite map, when restricted to each minimal rational curve D, $f|_D$ is ramified only at a finite number of points, so that $pr_2\colon B$

$\longrightarrow \mathscr{C}_X$ is a finite map. Consequently, $\dim_{\mathbb{C}} B = \dim_{\mathbb{C}} \mathscr{C}_X = \dim_{\mathbb{C}} \mathcal{S}(X) - 1$, by [(2.1), Prop.2] and [Ch.6, (1.2), Prop.4]

$$\dim_{\mathbb{C}} B = \dim_{\mathbb{C}} \mathscr{C}_X = \dim_{\mathbb{C}} \mathcal{S}(X) - 1 = 2n - 2 - n(X). \qquad (1)$$

Consider now the projection map $pr_1: B \longrightarrow R \subset X$. We have $\dim_{\mathbb{C}} R = n - 1$. Let k be the dimension of the generic fiber. We have by (1)

$$\begin{aligned} \dim_{\mathbb{C}} \mathcal{S}_0(X) \geq k \geq \dim_{\mathbb{C}} B - (n-1) &= 2n - 2 - n(X) - (n-1) \\ &= n - 1 - n(X) = \dim_{\mathbb{C}} \mathcal{S}_0(X). \end{aligned} \qquad (2)$$

It follows from (2) that in fact $k = \dim_{\mathbb{C}} \mathcal{S}_0(X)$ and that $pr_1: B \longrightarrow R$ is surjective. We have therefore shown

(*) If Prop.2 fails, then for every x on the branching locus R of $f: X \longrightarrow Y$ and for every minimal rational curve D on X passing through x, $f|_D$ is ramified at x.

We are going to derive a contradiction from (*). To do this we are going to find a minimal rational curve D such that D does not lie on R and f (not $f|_D$) is ramified at a every point on D. This contradicts with the definition of the branching locus R and establishes Prop.2. First, by (*) for any point x on R, $df(x)(\alpha) = 0$ for any characteristic vector α at x. Since such α span $T_x(X)$ as a vector space over $\mathbb{C}$, we conclude that $df(x) = 0$. Now pick a smooth point x on R. The set of characteristic vectors α at x tangent to some minimal rational curve $D_0 \subset R$ can span at most $T_x(R) \neq T_x(X)$. It follows that there exists some minimal rational curve D through x such that $D \cap R$ is isolated. We claim that f (not $f|_D$) is ramified everywhere on D, as desired. Pick any $z \in D$. f is unramified at z, $f(z) = w$, if and only if $f_z^*: T_w^*(Y) \longrightarrow T_z^*(X)$ is an isomorphism. Write $f(D) = C$, $[C] \in \mathscr{C}_Y$. The set of $\sigma \in \Gamma(C, T_Y^*|_C)$ generate a vector subbundle E of $T_Y^*|_C$ over C of rank $n(Y) \geq 1$. As $(T_X^*|_D, g^*|_D)$ is of semi-negative curvature $f^*\sigma \in \Gamma(D, T_X^*|_D)$ is either zero or nowhere vanishing. However as $df(x) = 0$ for $x \in D \cap R \neq \phi$ we must have $f^*\sigma(x) = 0$, showing that in fact $f^*\sigma(x) \equiv 0$ on D. It follows that for the map on vector spaces $f_z^*: T_w^*(Y) \longrightarrow T_z^*(X)$, $f_z^*(v) = 0$ for every $v \in E_{f(z)}$. Consequently, f is ramified at $z \in D$, as claimed. We have established Prop.2 by contradiction. The proof of the rigidity theorem on holomorphic mappings [(1.1), Thm. 1] of TSAI [TSA] is completed. ▌

APPENDIX

I. SEMISIMPLE LIE ALGEBRAS AND THEIR REPRESENTATIONS

I.1 Semisimple Lie Algebras – General Theorems

The background materials we collect here and in the next section are essentially taken from SERRE [SER2]. For the proofs we refer the reader to JACOBSON [JA]. We fix an underlying field k which is either $\mathbf{R}$ or $\mathbf{C}$. All vector spaces and Lie algebras/groups are understood to be finite dimensional over k. Let $\mathfrak{g}$ be a Lie algebra over k. There is a largest solvable ideal $\mathfrak{r}(\mathfrak{g})$ in $\mathfrak{g}$. $\mathfrak{r}(\mathfrak{g})$ is called the radical of $\mathfrak{g}$.

DEFINITION 1

We say that $\mathfrak{g}$ is semisimple if and only if its radical $\mathfrak{r}(\mathfrak{g})$ is 0. A Lie group G is said to be semisimple if and only if G is connected and its Lie algebra $\mathfrak{g}$ is semisimple.

For a vector space V over k we denote by $GL(V,k)$ the general linear group of V as a k–vector space. The vector space of k–linear endomorphisms on V will be denoted by $\mathfrak{gl}(V,k)$, which can be identified with the Lie algebra of $GL(V,k)$. We write $End(\mathfrak{g}) \subset \mathfrak{gl}(\mathfrak{g},k)$ for the subalgebra of Lie algebra endomorphisms of $\mathfrak{g}$ and $Aut(\mathfrak{g}) \subset GL(\mathfrak{g},k)$ for the linear group of Lie algebra automorphisms of $\mathfrak{g}$. By the inner automorphism group $Int(\mathfrak{g})$ of $\mathfrak{g}$ we mean the Lie subgroup of $Aut(\mathfrak{g})$ generated by $\exp(\mathbf{ad}(x)) \in GL(\mathfrak{g},k)$, $x \in \mathfrak{g}$. Elements of $Int(\mathfrak{g})$ are called inner automorphisms of $\mathfrak{g}$. $Int(\mathfrak{g})$ is also called the adjoint group of $\mathfrak{g}$.

A symmetric bilinear form $B: \mathfrak{g} \oplus \mathfrak{g} \longrightarrow k$ is said to be invariant if and only if $B([x,y],z) + B(y,[x,z]) = 0$. The Killing form $B(x,y) := Tr_k(\mathbf{ad}(x) \circ \mathbf{ad}(y))$ is an invariant symmetric form. Here $\mathbf{ad}(u)(v)$ means $[u,v]$ for $u, v \in \mathfrak{g}$, so that $\mathbf{ad}(u) \in End(\mathfrak{g}) \subset \mathfrak{gl}(\mathfrak{g},k)$. Traces are taken in the matrix algebra $\mathfrak{gl}(\mathfrak{g},k)$. We have

THEOREM 1

$\mathfrak{g}$ is semisimple if and only if the Killing form $B(.,.)$ over k is non–degenerate.

A Lie algebra $\mathfrak{s}$ is said to be simple if it is not abelian and its only ideals are 0 and $\mathfrak{s}$. We have

THEOREM 2

Let $\mathfrak{g}$ be a semisimple Lie algebra, and $(\mathfrak{a}_i)$ be its minimal non–zero ideals. The ideals $\mathfrak{a}_i$ are simple Lie algebras. Moreover, $\mathfrak{g} = \Sigma\mathfrak{a}_i$ is a decomposition of $\mathfrak{g}$ as a direct sum of simple Lie algebras. Any decomposition of $\mathfrak{g}$ as a direct sum of simple Lie algebra is isomorphic to the decomposition $\mathfrak{g} = \Sigma\mathfrak{a}_i$.

For a simple complex Lie algebra $\mathfrak{g}$ the underlying real Lie algebra $\mathfrak{g}_{\mathbf{R}}$ is simple over $\mathbf{R}$. A Lie algebra $\mathfrak{g}_0$ over $\mathbf{R}$ is semisimple if and only if its complexification $\mathfrak{g} := \mathfrak{g}_0 \otimes_{\mathbf{R}} \mathbf{C}$ is semisimple over $\mathbf{C}$. A Lie algebra $\mathfrak{g}_0$ over $\mathbf{R}$ is simple if and only if its complexification $\mathfrak{g}$ is simple over $\mathbf{C}$ or $\mathfrak{g}_0$ is of the form $\mathfrak{s} \oplus \bar{\mathfrak{s}}$ for some simple complex Lie subalgebra $\mathfrak{s} \subset \mathfrak{g}$. The latter possibility occurs precisely when $\mathfrak{g}_0 = \mathfrak{h}_{\mathbf{R}}$ for some complex simple Lie algebra $\mathfrak{h}$. Given a simple Lie algebra $\mathfrak{g}$ over $\mathbf{C}$ there are always a finite number of non–isomorphic real simple Lie algebra $\mathfrak{g}_0$ such that $\mathfrak{g}$ is isomorphic over $\mathbf{C}$ to the complexification of $\mathfrak{g}_0$. Any such real Lie algebra $\mathfrak{g}_0$ is called a real form of $\mathfrak{g}$. For example, for $n \geq 3$ odd the real forms of the simple complex Lie algebra $\mathfrak{sl}(n,\mathbf{C})$ are precisely $\mathfrak{sl}(n,\mathbf{R})$ and $\mathfrak{su}(p,q)$ (cf. [Ch.4, (2.2)] for definition) for nonnegative integers p, q such that $p + q = n$. For the classification of real forms of simple complex Lie algebras cf. HELGASON [HEL, Ch.X, p.517ff.].

REMARKS

The notations here differ from those used in the text, where we deal with symmetric spaces. There the primary objects are the real semisimple Lie groups/algebras; everything is extended to $\mathbf{C}$ by complexification, and we use superscripts as in $\mathfrak{g}^{\mathbf{C}}$ to indicate this. There the subscripts as in $\mathfrak{g}_0$ are reserved for non–compact real Lie groups/algebras, their compact analogues being denoted by $\mathfrak{g}_c$, etc.

I.2 Cartan Subalgebras

Let $\mathfrak{g}$ be an n–dimensional Lie algebra over k and $\mathfrak{a} \subset \mathfrak{g}$ be a subalgebra. By the normalizer $\mathfrak{n}(\mathfrak{a})$ of $\mathfrak{a}$ we mean $\{x \in \mathfrak{g}: [x,\mathfrak{a}] \subset \mathfrak{a}\}$, i.e., the largest Lie subalgebra of $\mathfrak{g}$ containing $\mathfrak{a}$ as an ideal. We have

DEFINITION 1

A subalgebra $\mathfrak{h}$ of $\mathfrak{g}$ is called a Cartan subalgebra if and only if $\mathfrak{h}$ is nilpotent and $\mathfrak{n}(\mathfrak{h}) = \mathfrak{h}$.

For $x \in \mathfrak{g}$ we denote by $P_x(T)$ the characteristic polynomial in T of the endomorphism $ad(x)\colon \mathfrak{g} \longrightarrow \mathfrak{g}$. Write $P_x(T) = \Sigma_{0 \leq i \leq n}\, a_i(x)\, T^i$. We define the rank of $\mathfrak{g}$ as the least integer r such that $a_r(x) \neq 0$ for some $x \in \mathfrak{g}$. An element $x \in \mathfrak{g}$ is said to be regular if and only if $a_r(x) \neq 0$ for $r = \mathrm{rank}_k(\mathfrak{g}) \leq n$. We have $r = n$ if and only if $\mathfrak{g}$ is nilpotent.

Suppose now $k = \mathbf{C}$. Let $\lambda \in \mathbf{C}$ and $x \in \mathfrak{g}$. We define $\mathfrak{g}_x^\lambda$ to be the nilspace of $ad(x) - \lambda \in \mathfrak{gl}(\mathfrak{g},\mathbf{C})$, i.e., $\mathfrak{g}_x^\lambda := \{y \in \mathfrak{g}\colon (ad(x) - \lambda)^k y = 0$ for some integer $k > 0\}$. We have

PROPOSITION 1

Let $\mathfrak{g}$ be a complex Lie algebra; $x \in \mathfrak{g}$ and $\lambda, \mu \in \mathbf{C}$. Then,

(1) $\mathfrak{g}$ is the vector space direct sum of nilspaces $\mathfrak{g}_x^\lambda$.

(2) $[\mathfrak{g}_x^\lambda, \mathfrak{g}_x^\mu] \subset \mathfrak{g}_x^{\lambda+\mu}$.

(3) $\mathfrak{g}_x^0$ is a Lie subalgebra of $\mathfrak{g}$.

PROPOSITION 2

Suppose again that $k = \mathbf{C}$. If $x \in \mathfrak{g}$ is regular, then $\mathfrak{g}_x^0$ is of dimension $r = \mathrm{rank}_{\mathbf{C}}(\mathfrak{g})$ and is a Cartan subalgebra of $\mathfrak{g}$. Conversely, any Cartan subalgebra $\mathfrak{h}$ of $\mathfrak{g}$ is of the form $\mathfrak{g}_x^0$ for some regular element x of $\mathfrak{g}$. Furthermore, all Cartan subalgebras are conjugate under inner automorphisms of $\mathfrak{g}$.

Let $\mathfrak{g}_o$ be a real Lie algebra and $\mathfrak{g}$ be its complexification. We have $\mathrm{rank}_{\mathbf{R}}(\mathfrak{g}_o) = \mathrm{rank}_{\mathbf{C}}(\mathfrak{g})$. $\mathfrak{h}_o \subset \mathfrak{g}_o$ is a Cartan subalgebra if and only if its complexification $\mathfrak{h} \subset \mathfrak{g}$ is a Cartan subalgebra. For $k = \mathbf{R}$ there are in general a finite number of conjugacy classes of Cartan ($\mathbf{R}$–)subalgebras (cf. KOSTANT [KO] for the classification). The key difference is that the set $\mathrm{Reg}(\mathfrak{g})$ of regular elements is always connected ($\mathrm{Reg}(\mathfrak{g})$ being the complement of an affine algebraic variety over $\mathbf{C}$) while the analogous statement fails for $k = \mathbf{R}$.

I.3 Semisimple Lie Algebras – Structure Theory

For the material in this section we refer the reader to SERRE [SER2] and HELGASON [HEL] for proofs. When necessary we give precise references where it is most convenient to look up a statement in a specific form.

For $k = \mathbf{R}$ or $\mathbf{C}$ we say that $x \in \mathfrak{g}$ is semisimple if and only if $ad(x) \in \mathfrak{gl}(\mathfrak{g},\mathbf{C})$ is semisimple, i.e., $ad(x)$ is diagonalizable as a matrix (after extending the ground field to $\mathbf{C}$ in case $k = \mathbf{R}$). For a Lie subalgebra $\mathfrak{h} \subset \mathfrak{g}$ we define its centralizer $\mathfrak{z}(\mathfrak{h}) = \{x \in \mathfrak{g}\colon ad(x)h = 0$ for any $h \in \mathfrak{h}\} \subset \mathfrak{n}(\mathfrak{h})$. Regarding semisimple Lie algebras we have

PROPOSITION 1

Let $\mathfrak{g}$ be a semisimple Lie algebra over $k = \mathbf{R}$ or $\mathbf{C}$. Then, the Cartan subalgebras $\mathfrak{h} \subset \mathfrak{g}$ are abelian subalgebras. In this case $\mathfrak{h}$ is its own centralizer and consists of semisimple elements. Furthermore, the restriction of the Killing form $B = B_{\mathfrak{g}}$ to $\mathfrak{h}$ is non–degenerate.

REMARKS

The Cartan subalgebras of $\mathfrak{g}$ are precisely the maximal abelian subalgebras which consist of semisimple elements.

In case of $k = \mathbf{C}$ because of Prop.1 we can simultaneously diagonalize $\mathfrak{g}$ with respect to $\{ad(h)\colon h \in \mathfrak{h}\}$. For $\rho \in \mathfrak{h}^* := \mathrm{Hom}_{\mathbf{C}}(\mathbf{h},\mathbf{C})$ write $\mathfrak{g}^{\varphi} = \{x \in \mathfrak{g} :$ $[h,x] = \rho(h)x$ for any $h \in \mathfrak{h}.\}$ Thus $\mathfrak{g}^0$ is the centralizer of $\mathfrak{h}$, so that $\mathfrak{g}^0 = \mathfrak{h}$. By a $\mathfrak{g}$–root φ of $\mathfrak{g}$ with respect to $\mathfrak{h}$ we mean a non–zero element of $\mathfrak{h}^*$ for which $\mathfrak{g}^{\varphi} \neq 0$. $\mathfrak{g}^{\varphi}$ is then called a root space and $x \in \mathfrak{g}^{\varphi}$ a root vector of weight φ. Based on Prop.1 and [(I.2), Prop.2] we obtain a root space decomposition of $\mathfrak{g}$. More precisely, we have

PROPOSITION 2 (cf. HELGASON [HEL, Ch.III, Thms.(4.2), p.166])

Let $\mathfrak{g}$ be a complex semisimple Lie algebra. Let $\mathfrak{h} \subset \mathfrak{g}$ be a Cartan subalgebra. Denote by Δ the set of all $\mathfrak{g}$–roots with respect to the complex Cartan subalgebra $\mathfrak{h}$. Then, $\mathfrak{h} = \mathfrak{g}^0$ and $\varphi \in \Delta$ implies $-\varphi \in \Delta$. All root spaces $\mathfrak{g}^{\varphi}$ are one–dimensional over $\mathbf{C}$. We have the (direct sum) root space decomposition $\mathfrak{g} = \mathfrak{h} + \Sigma_{\varphi\in\Delta}\mathfrak{g}^{\varphi}$ as a complex vector space. $\{\mathfrak{h} = \mathfrak{g}^0;\ \mathfrak{g}^{\varphi} + \mathfrak{g}^{-\varphi} : \varphi \in \Delta\}$ are orthogonal with respect to the Killing form B of $\mathfrak{g}$. Furthermore, B is positive definite on the real vector space $\mathfrak{h}^*_{\mathbf{R}} \subset \mathfrak{h}^*$ generated by the roots Δ.

Since $B|_{\mathfrak{h}}$ is non–degenerate we can identify $\mathfrak{h}$ and $\mathfrak{h}^*$ by duality. Thus, we associate to each $\alpha \in \mathfrak{h}^*$ a dual element $H_\alpha \in \mathfrak{h}$ satisfying $\alpha(\eta) = B(H_\alpha,\eta)$ for any $\eta \in \mathfrak{h}$. The image of $\mathfrak{h}^*_{\mathbf{R}}$ under this identification is denoted by $\mathfrak{h}_{\mathbf{R}}$. As B is positive definite on $\mathfrak{h}^*_{\mathbf{R}}$ we can define a positive definite inner product $<.,.>$ on the real vector space $\mathfrak{h}^*_{\mathbf{R}}$ by $<\rho,\rho'> = B(H_\rho,H_{\rho'})$. For $\varphi, \psi \in \Delta$, a φ–string of roots attached to ψ means a set $\{\psi + k\varphi\colon -p \leq k \leq q\}$, where $p, q \geq 0$ are integers. The φ–string is said to be maximal if p and q are maximal. We have

PROPOSITION 3 (cf. HELGASON [HEL, Ch.III, Thm.(4.3), p.168; Thm.(5.5),p.176])
Let φ, ψ be roots in Δ. We have

(1) Let $\{\psi + k\varphi, -p \leq k \leq q\}$ be the maximal φ–string of roots attached to ψ. Then, $q - p = -2\dfrac{<\psi,\varphi>}{<\varphi,\varphi>}$.

(2) $[\mathfrak{g}^\varphi,\mathfrak{g}^{-\varphi}] = \mathbf{C}H_\varphi \subset \mathfrak{h}$; and $[\mathfrak{g}^\varphi,\mathfrak{g}^\psi] = \mathfrak{g}^{\varphi+\psi}$ for $\varphi + \psi \neq 0$. In particular, $[\mathfrak{g}^\varphi,\mathfrak{g}^\psi] = 0$ if $\varphi + \psi \notin \Delta$.

(3) Choose a basis of root vectors $\{e_\varphi\colon \varphi \in \Delta\}$ such that $B(e_\varphi,e_{-\varphi}) = 1$. Then, $[e_\varphi,e_{-\varphi}] = H_\varphi$.

(4) Writing $[e_\varphi,e_\psi] = N_{\varphi,\psi}e_{\varphi+\psi}$ (by (2)) for $\varphi, \psi, \varphi+\psi \in \Delta$ we may further assume that $N_{\varphi,\psi}$ are real and satisfy $N_{\varphi,\psi} = -N_{-\varphi,-\psi}$.

(5) In terms of root vectors $\{e_\varphi\}_{\varphi\in\Delta}$ satisfying (3) and (4) and for p, q defined as in (1) we have $N^2_{\varphi,\psi} = \dfrac{q(1+p)}{2} <\varphi,\varphi>$.

The systems of $\mathfrak{g}$–roots Δ have the following properties.

PROPOSITION 4 (cf. HELGASON [HEL, Ch.III, Thm.(4.3), p.168])

(1) $\Delta \subset V := \mathfrak{h}^*_{\mathbf{R}}$ generates the real vector space V;

(2) $\Delta \equiv -\Delta$;

(3) For each $\varphi \in \Delta$, Δ is invariant under the Weyl reflections σ_φ defined by
$\sigma_\varphi(\rho) = \rho - 2\frac{<\rho,\varphi>}{<\varphi,\varphi>}\varphi$.

(4) For any $\varphi \in \Delta$ the only roots proportional to φ are $\pm\varphi$.

Suppose now $\mathfrak{g}$ is the complexification of a real semisimple Lie algebra $\mathfrak{g}_0$. Since Cartan subalgebras $\mathfrak{h}$ of $\mathfrak{g}$ are conjugate under $\mathrm{Int}(\mathfrak{g})$ we may assume that $\mathfrak{h}$ is the complexification of an R–Cartan subalgebra $\mathfrak{h}_0$ of $\mathfrak{g}_0$. Write $\mathfrak{h}_0^*$ for $\mathrm{Hom}_{\mathbf{R}}(\mathfrak{h}_0,\mathbf{R})$ and τ for the conjugation on $\mathfrak{g}$ induced by $\mathfrak{g}_0$. From the fact that $-\Delta \equiv \Delta$ one can easily deduce that $\mathfrak{h}_{\mathbf{R}}^* = \sqrt{-1}\mathfrak{h}_0^* \subset \mathfrak{h}^*$ (cf. HELGASON [HEL, Ch.VI, Lemma 3.1, p.257] for $\mathfrak{g}_0$ compact; the non–compact case is obtained by duality) and that for $\varphi \in \Delta$, $\mathfrak{g}^{-\varphi} = \tau(\mathfrak{g}^{\varphi})$.

In general for a real vector space V a finite subset $\Delta \subset V$ is said to be a root system if and only if it satisfies (1), (2) and (3) of Prop.3. Δ is said to be a reduced root system if it satisfies in addition (4). In our case $\Delta \subset \sqrt{-1}\mathfrak{h}_0^*$ is a reduced root system by Prop.2. For the theory of root systems cf. SERRE [SER2, Ch.V]. Let $\Phi \subset V$ be a root system in a real vector space V and write $V^* := \mathrm{Hom}_{\mathbf{R}}(V,\mathbf{R})$. We have

DEFINITION 1

A fundamental system of roots of Φ is a subset S of Φ such that

(1) S is a basis of V as a real vector space.

(2) Every root $\varphi \in \Phi$ can be written in the form $\varphi = \Sigma_{\alpha\in S} m_\alpha \alpha$ for $\{m_\alpha\}$ integers such that either all $m_\alpha \geq 0$ or all $m_\alpha \leq 0$. In the two cases we say that φ is a positive resp. negative root with respect to S.

An element α of S is called a simple (positive) root.

REMARKS

S is called a base in SERRE ([SER2]).

DEFINITION 2

For each $\varphi \in \Phi$ let Π_φ denote the hyperplane fixed by the Weyl reflection σ_φ.

A Weyl chamber is a connected component of $V - \cup_{\varphi\in\Phi}\Pi_\varphi$. The Weyl group $W(\Phi)$ is the subgroup of $\mathrm{Aut}_{\mathbf{R}}(V)$ generated by the Weyl reflections σ_φ.

PROPOSITION 5

Let $t \in V^*$ be such that $t(\varphi) \neq 0$ for any $\varphi \in \Phi$. Write $R_t^+ = \{\alpha \in S\colon t(\alpha) > 0\}$. Then, $\Phi = R_t^+ \cup (-R_t^+)$. Define S_t to be the subset of R_t^+ consisting of indecomposable elements α, i.e., α cannot be written in the form $\beta + \gamma$ with β, $\gamma \in R_t^+$. Then, S_t is a fundamental system of roots of Φ; R_t^+ and $-R_t^+$ are resp. the space of positive and negative roots with respect to S_t. Any fundamental system of roots S is of the form $S = S_t$ for some such t. S_t is uniquely determined by the Weyl chamber C_t containing t. The Weyl group $W(\Phi)$ acts simply transitively on the Weyl chambers C and hence on the set of fundamental systems of roots S.

We return now to root systems of a semisimple complex Lie algebra $\mathfrak{g}$.

PROPOSITION 6 (cf. SERRE [SER2, Ch.VI, Remarks, p.47])

Let $\mathfrak{g}$ be a complex semisimple Lie algebra, $\mathfrak{h} \subset \mathfrak{g}$ be a Cartan subalgebra, and $\Delta \subset \mathfrak{h}_{\mathbf{R}}^*$ be the set of $\mathfrak{g}$–roots with respect to $\mathfrak{h}$. Identify the Weyl group $W(\Delta) \subset \mathrm{Aut}_{\mathbf{R}}(\mathfrak{h}_{\mathbf{R}}^*)$ as a subgroup of $\mathrm{Aut}_{\mathbf{R}}(\mathfrak{h}_{\mathbf{R}}) \subset \mathrm{Aut}_{\mathbf{C}}(\mathfrak{h})$ via the identification $\mathfrak{h}_{\mathbf{R}}^* \cong \mathfrak{h}_{\mathbf{R}}$ using the Killing form B. Let Γ be the subgroup of $\mathrm{Int}(\mathfrak{g})$ consisting of elements fixing $\mathfrak{h}$. Then, $W(\Delta)$ is precisely the restriction of Γ to $\mathfrak{h}$.

I.4 Representations of Semisimple Lie Algebras

The basic reference for this section is SERRE [SER2, Chs. VII & VIII]. As above all Lie groups/algebras and vector spaces are understood to be finite–dimensional. Let G_0 be a real Lie group. By a (complex) representation of G_0 on a complex vector space V we mean a homomorphism $\rho\colon G_0 \to GL(V,\mathbf{C})$ of real Lie groups. ρ is said to be faithful if it is injective. For a real Lie algebra $\mathfrak{g}_0$ and a Lie algebra homomorphism $\sigma\colon \mathfrak{g}_0 \to \mathfrak{gl}(V,\mathbf{C})$, σ is called a complex representation of $\mathfrak{g}_0$ on V. Given a real Lie group G_0 with Lie algebra $\mathfrak{g}_0$ and a complex representation $\rho: G_0 \to GL(V,C)$ the differential $d\rho: \mathfrak{g}_o \to \mathfrak{gl}(V,C)$ is a complex representation of $\mathfrak{g}_o$. Conversely, given a representation $\sigma: \mathfrak{g}_o \to \mathfrak{gl}(V,C)$ on a (finite–dimensional) complex vector space V one obtains by Lie's Theorem a unique representation $\rho: G_0 \to GL(V,C)$ for a connected, simply–connected Lie group G_0 with Lie algebra $\mathfrak{g}_o$ such that $d\rho = \sigma$. A complex representation $\rho: G_0 \to GL(V,C)$ is said to be irreducible if and only if for any complex vector subspace $V' \subset V$ invariant under $\rho(G_0)$ we have $V' = 0$ or V. On the other hand $\rho: G_0 \to GL(V,C)$ is said to be completely reducible if there is a decomposition of V into a direct sum of complex vector subspaces V_i such that V_i is invariant under $\rho(G_0)$ and the representations $\rho_i: G_0 \to GL(V_i,C)$ obtained by restriction are irreducible. Similar terminology applies to complex representations of the Lie algebra $\mathfrak{g}_o$. For a complex vector subspace V' of V invariant under $\rho(G_0)$ we will simply call V' a G_0–invariant space.

When $G_0 = G_c$ is a compact Lie group averaging a Euclidean metric on V with respect to some Haar measure on the compact group $\rho(G_c)$ we obtain a Euclidean metric s on V invariant with respect to $\rho(G_c)$, i.e., $\rho(G_c)$ acts as a group of unitary transformations on (V,s). For every G_c–invariant subspace $V' \subset V$ define $V'' \subset V$ to be the s–orthogonal complement of V'. As $\rho(G_c)$ acts as isometries it is clear that V'' is a G–invariant subspace. By induction on the dimension of V we have therefore

PROPOSITION 1

Let G_c be a compact real Lie group. Then, every representation $\rho: G_c \to GL(V,C)$ on a complex vector space V is completely reducible.

We have similarly the notion of (irreducible) representations of complex Lie groups and Lie algebras, where the homomorphisms are understood to be complex Lie group/Lie algebra homomorphisms. Let now G_c be a connected, simply-connected compact real semisimple Lie group with Lie algebra $\mathfrak{g}_c$, $\mathfrak{g}$ be the complexification of $\mathfrak{g}_c$; and V be a (finite–dimensional) vector space. Given a complex representation $\rho: G_c \to GL(V,C)$ we obtain the complex representation $d\rho: \mathfrak{g}_c \to \mathfrak{gl}(V,C)$ and hence by complexification $(d\rho)_C: \mathfrak{g} \to \mathfrak{gl}(V,C)$. To classify finite–dimensional complex representations of G_c it is therefore equivalent to classify irreducible representations $\sigma: \mathfrak{g} \to \mathfrak{gl}(V,C)$ of the complex semisimple Lie algebra $\mathfrak{g}$. Fix from now on a (complex) Cartan subalgebra $\mathfrak{h} \subset \mathfrak{g}$ and write $\mathfrak{g} =$

$\mathfrak{h} + \Sigma_{\varphi \in \Delta} \mathfrak{g}^{\varphi}$ for the corresponding root space decomposition of $\mathfrak{g}$. Given $\eta \in \mathfrak{h}^*$ we define $V^{\eta} := \{v \in V: [h,v] = \eta(h)v\}$ for any $h \in \mathfrak{h}\}$. By a weight of $\sigma: \mathfrak{g} \longrightarrow \mathfrak{gl}(V,C)$ (or just of σ) we mean an element $\eta \in \mathfrak{h}^*$ for which $V^{\eta} \neq 0$. An element $v \in V^{\eta}$ is then called a weight vector with weight η. For a weight η the dimension of V^{η} is called the multiplicity of η. We also fix from now on a Weyl chamber C on $\mathfrak{h}^*_{\mathbf{R}}$ and a corresponding fundamental system of simple (positive) roots S. We define

DEFINITION 1

A weight $\omega \in \mathfrak{h}^*_{\mathbf{R}}$ is said to be a highest (dominant) weight of the representation $\sigma: \mathfrak{g} \longrightarrow \mathfrak{gl}(V,C)$ if and only if any weight η of σ is of the form $\eta = \omega - \gamma$ for some positive root γ.

We have

THEOREM 1 (cf. SERRE [SER2])

Let $\sigma: \mathfrak{g} \longrightarrow \mathfrak{gl}(V,C)$ be an irreducible finite–dimensional representation of the complex semisimple Lie algebra $\mathfrak{g}$. Then, there always exists a unique highest weight ω. The weight space V^{ω} is 1–dimensional. There is a direct sum decomposition $V = \Sigma_{\eta \in \Theta} V^{\eta}$ of V in terms of the set of weights Θ of σ. Moreover, any weight $\eta \in \Theta$ satisfies the integrality condition $2 \dfrac{<\eta,\alpha>}{<\alpha,\alpha>} \in \mathbf{Z}$ for any $\alpha \in \Delta$. Two irreducible representations $\sigma, \sigma' : \mathfrak{g} \longrightarrow \mathfrak{gl}(V,C)$ are isomorphic if and only if their associated highest weights ω and ω' are identical.

For example, when $\mathfrak{g}$ is simple then the adjoint representation $\boldsymbol{ad}: \mathfrak{g} \longrightarrow \mathfrak{gl}(V,C)$ of $\mathfrak{g}$ given by $\boldsymbol{ad}(x)(y) = [x,y]$ is irreducible. The highest weight λ of $\boldsymbol{ad}$ is then called a highest root of $\mathfrak{g}$. The Cartan subalgebra $\mathfrak{h}$ is equal to $\mathfrak{g}^{0}$, so that o is a weight of $\boldsymbol{ad}$ of multiplicity $r = \mathrm{rank}(\mathfrak{g})$. Let G be the corresponding complex Lie group. Suppose G is realized as a linear algebraic group, i.e., a complex Lie subgroup of some $GL(V,C)$ and $\mathfrak{g}$ is identified with the corresponding Lie subalgebra of $\mathfrak{gl}(V,C)$, then $\boldsymbol{ad}$ is derived from the adjoint representation $ad: G \longrightarrow GL(\mathfrak{g},C)$ defined by $ad(\gamma)(g) = \gamma g \gamma^{-1}$ for $\gamma \in G$ and g

$\in \mathfrak{g}$. ad: $G \longrightarrow GL(\mathfrak{g},C)$ is in general not faithful. $G/Ker(ad)$ is the group $Int(\mathfrak{g})$ of inner automorphisms of $\mathfrak{g}$.

REMARKS

(1) The set of all possible $\eta \in \mathfrak{h}^*_{\mathbf{R}}$ satisfying the integrality condition $2\frac{<\eta,\alpha>}{<\alpha,\alpha>} \in \mathbf{Z}$ for all $\alpha \in \Delta$ forms a lattice Λ in $\mathfrak{h}^*_{\mathbf{R}}$. Λ is referred to as the weight lattice of $\mathfrak{g}$ with respect to $\mathfrak{h}$.

(2) In Thm.1 there is also a lowest weight λ in the sense that every weight η of σ is of the form $\eta = \lambda + \beta$ for some positive root β with respect to the Weyl chamber C. To see this it suffices to take λ to be the highest weight of σ with respect to the opposite Weyl Chamber $-C$. (Recall that $\Delta \equiv -\Delta$.) Then, a root $\beta \in \Delta$ is positive with respect to $-C$ if and only if it is negative with respect to C. V^{λ} is obviously 1–dimensional.

Given a representation ρ: $G \longrightarrow GL(V,C)$ there is a corresponding dual representation ρ': $G \longrightarrow GL(V^*,C)$ on the dual vector space V^* defined by $\rho'(\gamma)(v^*)(w) = v^*(\rho(\gamma^{-1})w)$ for $\gamma \in G$, $w \in V$ and $v^* \in V^*$. Fixing a basis on V and the corresponding dual basis on V^* in terms of matrices ρ' is given by $\rho'(\gamma) = (\rho^{-1}(\gamma))^t$, where A^t denotes the transpose of a matix A. The differential $\sigma' = d\rho'$ is given by $\sigma'(g)(v^*)(w) = -v^*(\sigma(g)w)$ for $g \in \mathfrak{g}$ and $\sigma = d\rho$. In terms of matrices as above we have $\sigma'(g) = -(\sigma(g))^t$. Let now ρ and σ be irreducible representations. To describe σ' in terms of weights write $W = V^*$ and; for each $\eta \in \Theta$ (the set of weights of σ), define $W_\eta \subset V^*$ to be the annihilator of $\Sigma_{\nu\in\Theta,\nu\neq\eta}V^{\nu}$. The formula $\sigma'(g)(v^*)(w) = -v^*(\sigma(g)w)$ then implies that $\sigma'(h)(w_\eta) = -\eta(h)w_\eta$ for $h \in \mathfrak{h}$ and $w_\eta \in W_\eta$. It follows easily that the set of weights Θ' of σ' is precisely $-\Theta$ and that for each $\eta \in \Theta$ we have $W^{-\eta} = W_\eta$. For $\varphi \in \Delta$, $\sigma'(e_\varphi)(W^{-\eta}) \subset W^{-\eta+\varphi}$, which is non–zero only if $\eta - \varphi \in \Theta$. It follows that we have

PROPOSITION 2

Let σ: $\mathfrak{g} \longrightarrow GL(V,C)$ be an irreducible representation of the complex semisimple Lie algebra $\mathfrak{g}$. Let $\lambda \in \Theta$ be the lowest weight of σ (cf. Remarks after Thm.1). Then, $-\lambda$ is the highest weight of the dual representation σ': $\mathfrak{g} \longrightarrow \mathfrak{gl}(V,C)$.

If $G_c \subset G$ (resp. $\mathfrak{g}_c \subset \mathfrak{g}$) is a compact real form of G (resp. $\mathfrak{g}$) and V is endowed with a Euclidean metric s so that G_c acts as unitary transformations, then $\rho|G_c$ acts as unitary transformations on (V,s). Choose the Cartan subalgebra $\mathfrak{h} \subset \mathfrak{g}$ of the form $\mathfrak{h} = \mathfrak{h}_c \otimes_R C$ for some real Cartan subalgebra $\mathfrak{h}_c \subset \mathfrak{g}_c$ so that $\mathfrak{h}_R^* = \sqrt{-1}\mathfrak{h}_c^*$ for $\mathfrak{h}_c^* = \mathrm{Hom}_R(\mathfrak{h}_c, R)$. In terms of an orthonormal basis of (V,s) $\rho'(\gamma) = \overline{\rho(\gamma)}$ for $\gamma \in G_c$ and $\sigma'(g) = -\sigma(g)$ for $g \in \mathfrak{g}_c$ as $\sigma(\mathfrak{g}_c)$ consists of skew–Hermitian matrices. For $h \in \mathfrak{h}_R = \sqrt{-1}\mathfrak{h}_c \subset \sqrt{-1}\mathfrak{g}_c$ we have therefore also $\sigma'(h) = -\sigma(h)$. Using s one can identify $W = V^*$ with the complex vector space $\overline{V}$. $\Theta' = -\Theta$ then follows from $\sigma'(h)(\overline{v^\theta}) = -\sigma(h)\overline{v^\theta}$ whenever $v^\eta \in V^\eta$.

For the classification of irreducible (finite–dimensional) representations of a complex semisimple Lie algebra $\mathfrak{g}$ we have

THEOREM 2
Given $\omega \in \mathfrak{h}_R^*$ there exists an irreducible representation $\sigma: \mathfrak{g} \to \mathfrak{gl}(V,C)$ (unique up to isomorphism) with highest weight ω if and only if $2\,\frac{<\omega,\alpha>}{<\alpha,\alpha>}$ is a nonnegative integer for any positive root α (or; equivalently, for any $\alpha \in S$.)

I.5 Some Results on Lie groups and Their Representations

Let G_c be a compact semisimple Lie group with Lie algebra $\mathfrak{g}_c$ and $\rho: G_c \to \mathfrak{gl}(V,C)$ be a complex representation. G_c acts on $\mathbb{P}(V)$ by projectivization. Write $\mathfrak{g}$ for the complexification of $\mathfrak{g}_c$. For every non–zero $v \in V$ denote by $[v]$ the projectivized element in $\mathbb{P}(V)$. Concerning orbits of irreducible representations of semisimple compact real Lie groups we have

THEOREM 1 (part of the Borel–Weil Theorem)
Let $\rho: G_c \to \mathrm{Aut}_C(V)$ be a complex representation, $\sigma: \mathfrak{g} \to \mathfrak{gl}(V,C)$ be the induced complex representation on $\mathfrak{g}$ and ω be the highest weight with respect to some choice of Cartan subalgebra $\mathfrak{h} \subset \mathfrak{g}$ and some choice of fundamental system of simple roots S. Then, for a non–zero weight vector v^ω with weight ω the orbit

of $[v^{\omega}]$ under G_c is complex–analytic. Conversely, if $0 \neq v \in V$ is such that $G_c[v]$ is complex–analytic, then v is a highest weight vector with respect to some Cartan subalgebra $\mathfrak{h}' \subset \mathfrak{g}$ and some fundamental system of simple roots S'. Consequently, there is only one complex G_c–orbit on $\mathbf{P}(V)$.

The proof of the complex–analyticity of $G_c[v^{\omega}]$ is exactly the same as in [Ch.6, (1.1), Prop.1]. It suffices to use the (*a priori*) weaker fact that for any $\alpha \in S$ such that $\omega - \alpha$ is a weight of σ we know that $\omega + \alpha$ is not a weight of σ. For the proof of the converse cf. Guillemin–Sternberg [G–ST, Prop.23.3, p.168].

Regarding complex semisimple Lie groups we have the following

THEOREM 2
Let G be a complex semisimple Lie group and K be a maximal compact real subgroup of G. Then, the injection $K \hookrightarrow G$ is a homotopy equivalence.

For the proof of Thm.2 cf. CHEVALLEY [CHE, Ch.6, §8–12].

II. SOME THEOREMS IN RIEMANNIAN GEOMETRY

(II.1) The de Rham Decomposition Theorem

Let (M,h) be a Riemannian manifold. (M,h) is said to be irreducible if any finite covering of (M,h) cannot be decomposed isometrically as a non–trivial product. When (M,h) is simply-connected and reducible it is clear that the tangent bundle T_M admits a non–trivial orthogonal decomposition into vector subbundles invariant under holonomy. (A vector subbundle V of T_M invariant under holonomy will be called a parallel vector subbundle.) Conversely, we have

THEOREM 1 (DE RHAM THEOREM)
Let (M,h) be a simply–connected complete Riemannian manifold. Then, there exists an orthogonal decomposition $T_M = \Sigma_{0 \leq i \leq p} T_i$ of the tangent bundle T_M into parallel vector subbundles such that the integral submanifolds of T_0 are flat and each T_i, $1 \leq i \leq p$, cannot be further decomposed into a non–trivial direct sum of parallel vector subbundles. Correspondingly (M,h) splits isometrically as a product $(M_0,g_0) \times (M_1,h_1) \times \ldots \times (M_p,h_p)$, where (M_0,g_0) is a flat Euclidean space and each (M_i,g_i), $1 \leq i \leq p$, is irreducible. Furthermore, the decomposition $T_M = \Sigma T_i$ and hence the isometric decomposition of (M,h) is unique up to permutation of factors. Except for permutation of isometric factors the isometry group $\mathrm{Aut}(M,h)$ splits correspondingly.

For a proof of Thm.1 cf. e.g. KOBAYASHI–NOMIZU [KN, Ch.IV, §6, Thm.(6.2), p.192].

(II.2) Some Theorems on Riemannian Locally Symmetric Manifolds

Let (X,g) be an m–dimensional Riemannian symmetric manifold, $G = \mathrm{Aut}_0(X,g)$ the identity component of its isometry group, $K \subset G$ the isotropy subgroup at some point. Thus, $X \cong G/K$ as a homogeneous manifold. We have

PROPOSITION 1
Let $H \subset SO(T_o(X)) \cong SO(m)$ be the holonomy group of (X,g) at $o = eK$. Then, H is contained in the isotropy subgroup $K \subset G$ at o of G.

Proof:

To prove the proposition we observe first of all that in the notations of [Ch.3, (1.1)] the transvections $\{\tau_t\}$ give rise to parallel transport along the geodesic $\gamma = \{\gamma(t); -\epsilon < t < \epsilon\}$ with $\gamma(o) = o$. To see this let $V \in T_o(X)$ be arbitrary and $\{V(t): -\epsilon < t < \epsilon\}$ be the corresponding vector field along the geodesic γ defined by $V(t) = (\tau_t)_* V \in T_{\gamma(t)}(X)$. Write $\dot{\gamma}(o) = \eta$ and $\nabla_\eta(V)(o) = W$. For the involution $\sigma = \sigma_o$ at o we have $\nabla_{\sigma\eta}(\sigma V) = \sigma W$ since σ is an isometry. Recall that $\sigma(m) = -m$ for $m \in \mathfrak{m} \cong T_o(X)$. From the definition $\tau_t = \sigma_{-\gamma(t/2)}\sigma$ one sees that $\sigma\tau_t\sigma = \tau_{-t}$, so that $\sigma(V(t)) = -V(-t)$, implying $\nabla_{-\eta}(-V)(o) = -W$, i.e. $\nabla_\eta(V)(o) = -W$. This gives $W = 0$. As any point on γ can be taken to be o this means $\nabla_\eta V \equiv 0$ on γ, i.e., the vector field V is parallel along γ.

To prove Prop.1, we show that parallel transport along a closed loop κ is given by a global isometry $\varphi \in G$. Suppose $\kappa = \{\kappa(t) : 0 \le t \le \lambda\}$ is a closed loop parametrized by arc–length such that $\kappa(0) = \kappa(\lambda) = o$. Let $\eta(t)$ denote the unit tangent vector $\dot{\kappa}(t) \in T_{\kappa(t)}(X)$. Let $\tau(t)$ be a transvection at $\kappa(t)$ corresponding to the tangent vector $\eta(t)$. Define by $\{\varphi(t): 0 \le t \le \lambda\}$ the one–parameter family of isometries $\{\varphi(t)\}$ satisfying $\frac{d\varphi}{dt}(\kappa(t)) = \tau(t)$. Then, $\varphi(t)(o) = \kappa(t)$ and $\varphi(\lambda) = 0$, so that $\varphi(\lambda) \in K$. From the fact that transvections induce parallel transport along geodesics it follows by integration that for any vector $V \in T_o(X)$, the vector field $V(t) = (\varphi(t))_*(V)$ is parallel on κ. As a consequence the element of the holonomy transformation given by parallel transport along κ is realized by the isometry $\varphi = \varphi(\lambda)$ at o, proving Prop.1. ▌

We have the following characterization of Riemannian locally symmetric manifolds using curvature conditions

PROPOSITION 2

Let (M,h) be a Riemannian manifold with the Riemannian connection ∇ and curvature tensor R. Then, (M,h) is Riemannian locally symmetric if and only if $\nabla R \equiv 0$.

The fact that $\nabla R \equiv 0$ on a Riemannian locally symmetric manifold (X,g) follows from the fact that ∇R is a tensor of odd total degree invariant under σ. To prove the converse, one sees that (M,h) is locally symmetric if and only if in a neighborhood of any point $x \in M$, the map σ_x defined by $\sigma_x(\exp_x(v)) = \exp_x(-v)$ in some neighborhood of x is an isometry. Here $\exp_x$ denotes the exponential map at x and $v \in T_x(M)$ is sufficiently near o. One has an expansion of the Riemannian metric in terms of normal geodesic coordinates. The expansion depends on the curvature tensor R and its covariant derivatives. A direct checking shows that σ_x is an isometry if $\nabla R \equiv 0$. For a proof of Prop.2 cf. KOBAYASHI–NOMIZU [KN, Vol.I, Ch.VI, §7, Thm.(7.4), p.261].

Regarding geodesic flats on a Riemannian symmetric manifold (X,g) we have

PROPOSITION 3

Let (X,g) be a Riemannian symmetric manifold of semisimple type. Let $x \in X$ and $K \subset G = \mathrm{Aut}_0(X,g)$ be the isotropy subgroup at x. Then, there exists a totally geodesic flat submanifold T such that $\cup_{\gamma \in K} \gamma T = X$.

Prop.3 in terms of Lie algebras has the following equivalent formulation.

PROPOSITION 4

Let (X,g) be a Riemannian symmetric manifold of semisimple type. Write $G = \mathrm{Aut}_0(X,g)$ and let $\mathfrak{g} = \mathfrak{k} + \mathfrak{m}$ be the Cartan decomposition of the Lie algebra $\mathfrak{g}$ of G. Let $\mathfrak{a} \subset \mathfrak{m}$ be a maximal abelian subspace in $\mathfrak{m}$ and write $A = \exp(\mathfrak{a}) \subset G$ be the corresponding abelian subgroup. Then, $G = KAK$.

Here a (maximal) abelian subspace $\mathfrak{a} \subset \mathfrak{m}$ means a (maximal) vector subspace such that $[\mathfrak{a},\mathfrak{a}] = 0$. For the correspondence between abelian subspaces of $\mathfrak{m}$ and totally geodesic submanifolds of (X,g) cf. HELGASON [HEL, Ch.V, Prop.(6.1), p.245]. For a proof of Prop.4 cf. [HEL, Ch.V, Thm.(6.7), p.249]. We note that any two maximal abelian subspaces $\mathfrak{a}, \mathfrak{a}' \subset \mathfrak{m}$ are conjugate under K (cf. [HEL, Ch.V, Lemma (6.3), p.247]) and that their common dimension as a vector space is called the rank of X (which is less than or equal to $\mathrm{rank}(\mathfrak{g})$, the common dimension of Cartan subalgebras of $\mathfrak{g}$).

III. CHARACTERISITIC PROJECTIVE SUBVARIETIES ASSOCIATED TO HERMITIAN SYMMETRIC MANIFOLDS

(III.1) Equivalent Definitions of Characteristic Vectors

Let Ω be an irreducible bounded symmetric domain equipped with the canonical metric g_0 and $\mathcal{S} \subset \mathbb{P}T(\Omega)$ be the characteristic bundle over Ω constructed as in [Ch.6, §1]. There are other equivalent ways of characterizing $\mathcal{S}$ in geometric terms. We prove

PROPOSITION 1

The following conditions on a unit vector $\alpha \in T_x(\Omega)$ are equivalent

(a) α is a characteristic vector ,

(b) α realizes the algebraic minimum of holomorphic sectional curvatures

(c) $\dim_{\mathbb{C}} \mathcal{N}_\alpha$ is maximum among non–zero vectors $\xi \in T_x(\Omega)$.

Here as in [Ch.6, (1.1)] $\mathcal{N}_\alpha$ denotes the zero eigenspace of the Hermitian bilinear form $H_\alpha(\mu,\nu) = R_{\alpha\bar{\alpha}\mu\bar{\nu}}$ for the curvature tensor R of (Ω,g_0). We will call a unit vector satisfying (b) a minimal vector.

For the proof of Prop.1 we need some facts on root systems associated to Hermitian symmetric manifolds. We use the terminology and notations of [Ch.5, §1]. Write $\Omega \cong G_0/K$ as usual. Fix a Cartan subalgebra $\mathfrak{h}$ of $\mathfrak{k}$ and let Δ_M^+ denote the set of positive non–compact roots as defined in [Ch5, (1.1)]. Let $\{\psi_1,\dots,\psi_r\} = \Psi \subset \Delta_M^+$ be a maximal strongly orthogonal set of roots constructed there by descending from the highest root $\mu = \psi_1$. Ψ corresponds to a distinguished (totally geodesic) polydisc $D \cong \Delta^r$ passing through the origin o such that $T_0(D) \cong \mathfrak{a}^+ = \Sigma_{\psi\in\Psi}\mathbb{C}e_\psi \subset \mathfrak{m}^+$. For $\psi \in \Psi$ let $[e_\psi,e_{-\psi}] = \sqrt{-1}H_\psi$, $H_\psi \in \sqrt{-1}\mathfrak{h}$, and $\mathfrak{h}(\Psi) \subset \mathfrak{h}$ be the (real) vector subspace generated by $\{\sqrt{-1}H_\psi\}$. Denote by W_K the subgroup of the Weyl group W of $G^{\mathbb{C}}$ which fixes the compact roots Δ_K and let $W_K(\Psi) \subset W_K$ be the subgroup of W_K fixing Ψ as a set. We need the Restricted Root Theorem of MOORE [MOO], which gives an explicit determination of the restriction $\rho(\varphi)$ of roots $\varphi \in \Delta$ to $\mathfrak{h}(\Psi)$ in terms of $\pm\Psi$ by combinatorial arguments.

Theorem 1 (Moore [Moo], The Restricted Root Theorem)

(A) Either (i) $\rho(\Delta) \cup \{0\} = \{\pm\frac{1}{2}\rho(\psi_i)\pm\frac{1}{2}\rho(\psi_j)\colon 1 \le i,j \le r\}$ or (ii) $\rho(\Delta) \cup \{0\} = \{\pm\frac{1}{2}\rho(\psi_i)\pm\frac{1}{2}\rho(\psi_j);\ \pm\frac{1}{2}\rho(\psi_i)\colon 1 \le i,j \le r\}$. Accordingly $\rho(\Delta_M^+) = \{\frac{1}{2}\rho(\psi_i)+\frac{1}{2}\rho(\psi_j)\colon 1 \le i,j \le r\}$ resp. $\rho(\Delta_M^+) = \{\frac{1}{2}\rho(\psi_i)+\frac{1}{2}\rho(\psi_j);\ \frac{1}{2}\rho(\psi_i)\colon 1 \le i,j \le r\}$.

(B) $W_K(\Psi)$ induces all possible permutations of the set Ψ. In particular, all roots in Ψ have the same length.

(B) follows from (A) by general theory of Weyl groups relative to maximal abelian subspaces $\mathfrak{a}$ of $\mathfrak{m}$ (cf. "Sophus Lie" [SL, Exp.23]).

Corollary 1

Let Ω be an irreducible bounded symmetric domain of rank r and $D \subset \Omega$ be a distinguished polydisc$\cong \Delta^r$. Let $\Phi \subset \mathrm{Aut}_0(\Omega)$ be the subgroup of automorphisms fixing D. Then, $\Phi|_D$ is the full group of automorphism $\mathrm{Aut}(D)$. Consequently, for any $\xi \in \mathfrak{m}^+ \cong T_0(\Omega)$ there exists $k \in \mathfrak{k}$ such that $ad(k)(\xi) = \Sigma_{1\le i\le r}\, a_ie_i$, $e_{\psi_i} = e_i$, such that $a_i \in \mathbb{R}$, $a_1 \ge a_2 \ge \dots \ge a_r \ge 0$. Moreover the vector $\Sigma_{1\le i\le r} a_ie_i$ $:= \eta$ is uniquely determined by ξ. We call η the normal form of ξ with respect to Ψ.

Proof:

By [(I.3), Prop.6] $W_K \subset W$ acts on $\mathfrak{t}^{\mathbb{C}}$ by inner automorphisms. It follows that all permutations of the factors of Δ^r are induced by elements of the identity component $\mathrm{Aut}_0(\Omega)$ of Ω. The rest of the proposition is immediate. ∎

Remarks

In case of the classical symmetric domains the normal forms of vectors $\xi \in \mathfrak{m}^+$ were explicitly described in [Ch.4, (2.3)–(2.5), Lemma 1 & (2.6), Lemma 2].

We proceed now to give a proof of Prop.1.

Proof of Prop. 1:

(a) $\Leftrightarrow$ (b) We may take $x \in \Omega$ to be the origin. Since both characteristic vectors and minimal vectors remain so under the adjoint action of K, to prove (a) ⇔ (b) we can assume without loss of generality that $\xi \in \mathfrak{a}^+$. For any $\psi \in \Psi$ denote by $e_\psi \in \mathfrak{m}^+$ a corresponding unit root vector. We claim that a unit vector $\xi \in \mathfrak{a}^+$ realizes the algebraic minimum of holomorphic sectional curvatures if and only if $\xi = e^{i\theta}e_\psi$ for some $\psi \in \Psi$ and for some real θ. For any $\xi = \Sigma\, a_\psi e_\psi \in \mathfrak{a}^+$, $R_{\xi\bar{\xi}\xi\bar{\xi}} = R^D_{\xi\bar{\xi}\xi\bar{\xi}} = \Sigma|a_\psi|^4 R(e_\psi,\overline{e_\psi};e_\psi,\overline{e_\psi})$. By Thm.1, $R(e_\psi,\overline{e_\psi};e_\psi,\overline{e_\psi}) = cB([e_\psi,e_{-\psi}];\overline{[e_\psi,e_{-\psi}]})$ (B denoting the Killing form on $\mathfrak{g}^{\mathbf{C}}$) is independent of $\psi \in \Psi$. Our claim follows readily. Consequently the set $\mathscr{M}_0 = \{[\alpha] \in \mathbb{P}(\mathfrak{m}^+)\colon \alpha$ is a minimal$\}$ constitutes a finite union of K–orbits of the form $K[e_\psi]$ for some $\psi \in \Psi$. By Cor.1 all e_ψ, $\psi \in \Psi$, are in the same K–orbit. It follows that $\mathscr{M}_0 = \mathcal{S}_0 = K[\mu]$, proving (a) ⇔ (b).

(b) ⇔ (c) It suffices to consider $0 \neq \alpha \in \mathfrak{a}^+ \subset \mathfrak{m}^+$. Write $\alpha = \Sigma_{\psi\in\Psi'}\alpha_\psi e_\psi$, where $\Psi' = \Psi'(\alpha) \subset \Psi$ and $\alpha_\psi \neq 0$ for $\psi \in \Psi'$. We are going to show

$$(*) \qquad \mathscr{N}_\alpha = \cap_{\psi\in\Psi'}\mathscr{N}_\psi := \mathscr{N}_{\Psi'}.$$

As e_ψ is a characteristic vector for any $\psi \in \Psi$, $\dim_{\mathbf{C}}\mathscr{N}_\psi$ is the same for any such ψ. Furthermore for Ψ'' a proper subset of Ψ' it is clear that $\mathscr{N}_{\Psi'}$ is a proper vector subspace of $\mathscr{N}_{\Psi''}$ since $e_\varphi \notin \mathscr{N}_{\Psi'}$ for $\varphi \in \Psi' - \Psi''$. It is immediate therefore that (*) implies that among unit vectors in $\mathfrak{a}^+$, $\dim_{\mathbf{C}}\mathscr{N}_\alpha$ is the largest if and only if $\Psi'(\alpha) = \{\psi\}$ for some $\psi \in \Psi$, proving (b) ⇔ (c). It remains to show (*). Let $\eta \in \mathfrak{m}^+$ be arbitrary. Reindex the unit vectors $\{e_{\psi'}\colon \psi' \in \Psi'\}$ as $\{e_i\}_{1\le i\le k}$, $1 \le k \le r$, so that $\alpha = \Sigma_{1\le i\le k}\alpha_i e_i$. We have

$$R_{\alpha\bar{\alpha}\eta\bar{\eta}} = \sum_{1\le i,j\le k}\alpha_i\bar{\alpha}_j R_{i\bar{j}\eta\bar{\eta}}. \tag{1}$$

On the other hand we have $R_{i\bar{i}j\bar{j}} = 0$ for $i \neq j$. Since (Ω,g) is of seminegative curvature in the dual Nakano sense (cf. [Ch.2, (3.1)]) it follows that $R_{i\bar{j}\eta\bar{\eta}} = 0$ for $i \neq j$. Consequently, we have by (1)

$$R_{\alpha\bar{\alpha}\eta\bar{\eta}} = \sum_{1\le i\le k}|\alpha_i|^2 R_{i\bar{i}\eta\bar{\eta}}. \tag{2}$$

Thus, $\eta \in \mathcal{N}_\alpha$ if and only if $R_{i\bar{i}\eta\bar{\eta}} = 0$ for $1 \leq i \leq k$, i.e., $\eta \in \bigcap_{\psi\in\Psi'}\mathcal{N}_\psi$, implying (*) and hence (b) $\Leftrightarrow$ (c). The proof of Prop.1 is completed. ∎

(III.2) Characteristic Projective Subvarieties as Symmetric Projective Submanifolds with Parallel Second Fundamental Forms

Let $\mathcal{S}_o \subset \mathbb{P}(\mathfrak{m}^+) \cong \mathbb{P}^{N-1}$ be the characteristic projective subvariety associated to the irreducible bounded symmetric domain Ω. The canonical metric g_o on Ω gives rise to a Fubini–Study metric h on $\mathbb{P}(\mathfrak{m}^+)$, whose Kähler form is given by Const.$\sqrt{-1}\partial\bar{\partial}\log\|\zeta\|^2$ for $\zeta \in \mathfrak{m}^+$. Here $\|.\|$ denotes the norm given by g_o. From now on we will normalize h so that $(\mathbb{P}(\mathfrak{m}^+),h)$ has constant holomorphic sectional curvature equal to 1. We are going to prove

THEOREM 1

$\mathcal{S}_o \subset \mathbb{P}(\mathfrak{m}^+)$ is a Kähler submanifold with parallel second fundamental form. Conversely, if $\tau\colon (V,h) \hookrightarrow (\mathbb{P}^k(1),h)$ is a full isometric embedding of a compact Kähler manifold into some projective space equipped with a Fubini–Study metric, then there exists some irreducible bounded symmetric domain Ω such that τ is equivalent to the inclusion $\iota\colon (\mathcal{S}_o(\Omega),g|\mathcal{S}_o(\Omega)) \hookrightarrow (\mathbb{P}(\mathfrak{m}^+),h)$.

Here by a full embedding into $\mathbb{P}^k$ we mean one in which the image does not lie in any proper projective subspace. We say that two isometric embeddings τ_i: $(V,s) \hookrightarrow (\mathbb{P}^k,g_i)$, $i = 1, 2$ are equivalent if and only if there exists a projective linear transformation Φ such that $\Phi^*(g_1) = cg_2$ for some constant c and such that $\tau_2 = \Phi\circ\tau_1$. The notation $(\mathbb{P}^k(1),g)$ indicates that g is (a Fubini–Study metric) of constant holomorphic sectional curvature 1.

In [ROS] ROS proved a pinching theorem characterizing isometric embeddings of compact Kähler manifolds $\tau\colon (V,s) \hookrightarrow (\mathbb{P}^k(1),g)$. He proved

THEOREM 2 (ROS [ROS])

Let $\tau\colon (V,s) \hookrightarrow (\mathbb{P}^k(1),g)$ be an isometric immersion of a Kähler manifold into some projective space. Then, τ has parallel second fundamental form if and only if the holomorphic sectional curvature $R^V_{\xi\bar{\xi}\xi\bar{\xi}}$ (for unit vectors ξ) of (V,s) is pinched by

$$\frac{1}{2} \leq R^V_{\xi\bar{\xi}\xi\bar{\xi}} \leq 1.$$

The condition $R^V_{\xi\bar{\xi}\xi\bar{\xi}} \leq 1$ is automatic because of the Gauss–Codazzi equation for Kähler manifolds. In [ROS] it was proved furthermore that if we actually have $\frac{1}{2} < R^V_{\xi\bar{\xi}\xi\bar{\xi}} \leq 1$, τ is a linear embedding of some projective space into $\mathbb{P}^k$. On the other hand, NAKAGAWA–TAKAGI [NT] had classified those τ with parallel second fundamental form. τ is necessarily an embedding and (V,s) must itself be a Hermitian symmetric manifold of compact type. Assuming that τ is full, it must be equivalent to one of 7 types which were enumerated in [NT], 2 of which are exceptional. Upon examining this list and the characteristic projective subvarieties $\mathcal{S}_o \subset \mathbb{P}(m^+)$ arising from irreducible bounded symmetric domains Ω (or, equivalently, from their compact duals X_c), one finds that there is a one–to–one correspondence between the list in [NT] and the list of $\mathcal{S}_o \subset \mathbb{P}(m^+)$. (In the exceptional cases some verification is needed.). Here we give a unified proof of the fact that $\mathcal{S}_o \subset \mathbb{P}(m^+)$ must have parallel second fundamental form by proving directly a pinching theorem on holomorphic sectional curvatures and invoking Thm.2 of Ros. The proof is very elementary and arises from variational inequalities on the curvature tensor.

PROPOSITION 1

Let $\mathcal{S}_o \subset \mathbb{P}(m^+)$ be a characteristic projective subvariety arising from some irreducible Hermitian symmetric manifold X_c of compact type. Equip $\mathbb{P}(m^+)$ with the Fubini–Study metric h as above. Then, for $R(\mathcal{S}_o)$ denoting the curvature tensor of $(\mathcal{S}_o, h|\mathcal{S}_o)$ and ξ denoting a unit tangent vector of type (1,0) on V, $(V, h|_V)$ satisfies the pinching condition

$$\frac{1}{2} \leq R_{\xi\bar{\xi}\xi\bar{\xi}}(\mathcal{S}_o) \leq 1.$$

Proof:

Let α be a characteristic vector of X_c at o. By Lemma 4 we have a decomposition $T_o(X_c) = C\alpha + \mathcal{H}_\alpha + \mathcal{N}_\alpha$ of $T_o(X_c)$ into eigenspaces of the Hermitian form $H_\alpha(\xi,\eta) = R_{\alpha\bar{\alpha}\xi\bar{\eta}}(X_c,g_c)$ corresponding to the eigenvalues $1, \frac{1}{2}$ and 0. Henceforth we write R for $R(X_c,g_c)$. By [Ch.7, (3.1), Prop.2] $T_{[\alpha]}(\mathcal{S}_o) = (\mathcal{H}_\alpha + C\alpha)/C\alpha \cong \mathcal{H}_\alpha$. Accordingly, we identify the normal space of the embedding $\mathcal{S}_o \subset \mathbf{P}(m^+)$ at $[\alpha]$ with $\mathcal{N}_\alpha$. Near $[\alpha]$, there is a holomorphic projection π from $\mathcal{S}_o$ to $\mathbf{P}(C\alpha + \mathcal{H}_\alpha)$ arising from the orthogonal projection $m^+ \longrightarrow C\alpha + \mathcal{H}_\alpha$. Let $\xi \in \mathcal{H}_\alpha$ be of unit length. Let $C_\xi = \{[\alpha + \theta\xi + s\theta^2\zeta_\theta]: \zeta_\theta \in \mathcal{N}_\alpha, \|\zeta_o\| = 1, |\theta| < \epsilon\}$ for some $\epsilon > 0$, with ζ_θ holomorphic in θ, be a local holomorphic curve on $\mathcal{S}_o$. We may assume s real and nonnegative. For any $\mu \in T_{[\alpha]}(C_\xi)$ and $R(C_\xi)$ denoting the curvature tensor of $(C_\xi,h|C_\xi)$. The Gauss–Codazzi equation for Kähler manifolds gives

$$R_{\mu\bar{\mu}\mu\bar{\mu}}(C_\xi) \le R_{\mu\bar{\mu}\mu\bar{\mu}}(\mathcal{S}_o)$$

To prove the proposition it suffices to show that $R_{\mu\bar{\mu}\mu\bar{\mu}}(C_\xi) \ge \frac{1}{2}$ for μ of unit length. Recall that in the definition of C_ξ, $\|\zeta_o\| = 1$. From the Gauss–Codazzi equations it follows readily that

$$R_{\mu\bar{\mu}\mu\bar{\mu}}(C_\xi) = 1 - 2s^2, \tag{1}$$

To prove $R_{\mu\bar{\mu}\mu\bar{\mu}}(C_\xi) \ge \frac{1}{2}$ it is equivalent to show that $s \le \frac{1}{2}$. Write ζ for ζ_o. Define $\alpha_\theta = \alpha + \theta\xi + s\theta^2\zeta_\theta = \alpha + \theta\xi + s\theta^2\zeta + O(|\theta|^3)$. Consider $\theta = t$ real. Since $[\alpha_\theta] \in \mathcal{S}_o$ we have by [(III.1), Prop.1],

$$R(\alpha_t,\bar{\alpha}_t;\alpha_t,\bar{\alpha}_t) = \|\alpha_t\|^4 = 1 + 2t^2 + (1 + 2s^2)\,t^4 + O(t^6). \tag{2}$$

On the other hand expanding the left hand side we have

$$\begin{aligned} &R(\alpha_t,\bar{\alpha}_t;\alpha_t,\bar{\alpha}_t) \\ &= 1 + 4R_{\alpha\bar{\alpha}\xi\bar{\xi}}t^2 + (4\mathit{Re}R_{\alpha\bar{\xi}\xi\bar{\xi}})t^3 + (R_{\xi\bar{\xi}\xi\bar{\xi}} + 4s\mathit{Re}R_{\alpha\bar{\xi}\zeta\bar{\xi}})t^4 + O(t^5) \\ &= 1 + 2t^2 + (4\mathit{Re}R_{\alpha\bar{\xi}\xi\bar{\xi}})t^3 + (R_{\xi\bar{\xi}\xi\bar{\xi}} + 4s\mathit{Re}R_{\alpha\bar{\xi}\zeta\bar{\xi}})t^4 + O(t^5) \end{aligned} \tag{3}$$

Here we use the facts that

(a) $R_{\alpha\bar{\zeta}i\bar{j}} = 0$ since (X_c, g_c) is of semipositive curvature in the dual sense of Nakano (cf. proof of [Ch.3, (1.3), Prop.1]);

(b) $R_{\alpha\bar{i}\alpha\bar{j}} = 0$ for $(i,j) \neq (1,1)$.

To see (b) without loss of generality we may assume that $\{e_i\}$ is a basis of $T_0(X)$ consisting of unit root vectors, such that $\alpha = e_1 = e_\mu$ corresponds to a dominant root μ. Write $e_i = e_\varphi$ and $e_j = e_{\varphi'}$ for some $\varphi, \varphi' \in \Delta_M^+$. Then (b) follows immediately from the curvature formula [Ch.3, (1.3), Prop.1] and [(I.3), Prop.3], which give $R_{\alpha\bar{i}\alpha\bar{j}} = 0$ unless both $\mu - \varphi$ and $\mu - \varphi'$ are roots, in which case we have $R_{\alpha\bar{i}\alpha\bar{j}} = N_{\mu,-\varphi}N_{\varphi',-\mu}B(e_{\mu-\varphi}, \overline{e_{\varphi'-\mu}})$, except when $\mu - \varphi = \varphi' - \mu$, i.e., $2\mu = \varphi + \varphi'$. The latter case is impossible since μ dominates φ, φ'. Comparing (2) and (3) we obtain

$$Re R_{\alpha\bar{\xi}\xi\bar{\xi}} = 0; \tag{4}$$

$$R_{\xi\bar{\xi}\xi\bar{\xi}} + 4s Re R_{\alpha\bar{\xi}\zeta\bar{\xi}} = 1 + 2s^2. \tag{5}$$

We remark that if s is replaced by any real number r we have the inequality

$$R_{\xi\bar{\xi}\xi\bar{\xi}} + 4r Re R_{\alpha\bar{\xi}\zeta\bar{\xi}} \le 1 + 2r^2 \tag{6}$$

since for $\beta_t = \alpha + t\xi + rt^2\zeta$, we have $R(\beta_t,\bar{\beta}_t;\beta_t,\bar{\beta}_t) \le (1 + t^2 + r^2t^4)^2$, as the maximal holomorphic sectional curvature is 1. We claim that $|R_{\alpha\bar{\xi}\zeta\bar{\xi}}| \le \frac{1}{2}$. Since (X_c, g_c) is of semipositive holomorphic bisectional curvature we have $\partial_t^2 R(\alpha+t\xi,\overline{\alpha+t\xi};\zeta+t\xi,\overline{\zeta+t\xi})|_{t=0} \le 0$, which gives $4Re R_{\alpha\bar{\xi}\zeta\bar{\xi}} + 4R_{\alpha\bar{\alpha}\xi\bar{\xi}} = 4Re R_{\alpha\bar{\xi}\zeta\bar{\xi}} + 2 \ge 0$. Replacing ξ by $e^{i\varphi}\xi$ for some real φ so that $Re(e^{i\varphi}R_{\alpha\bar{\xi}\zeta\bar{\xi}}) = -|R_{\alpha\bar{\xi}\zeta\bar{\xi}}|$ we get the desired inequality. Rewrite (6) as

$$2r^2 + 4r Re R_{\alpha\bar{\xi}\zeta\bar{\xi}} + (1 - R_{\xi\bar{\xi}\xi\bar{\xi}}) \ge 0 \tag{7}$$

with equality at $r = s$. Thus the discriminant $\delta = 0$ and we have

$$s = -Re R_{\alpha\bar{\xi}\zeta\bar{\xi}} \le \frac{1}{2}, \tag{8}$$

as desired, proving the pinching condition $\frac{1}{2} \le R_{\mu\bar{\mu}\mu\bar{\mu}}(C_\xi) \le R^V_{\mu\bar{\mu}\mu\bar{\mu}} \le 1$. ∎

REMARKS

The proof of Prop.1 actually gives a very precise formula for the curvature tensor R^V_μ of $V = \mathcal{S}_0$. From $\delta = 0$, (1) and (8) we have

$$2(ReR_{\alpha\bar{\xi}\zeta\bar{\xi}})^2 - (1 - R_{\xi\bar{\xi}\xi\bar{\xi}}) = 0, \quad \text{and hence}$$
$$R_{\mu\bar{\mu}\mu\bar{\mu}}(C_\xi) = 1 - 2s^2 = 1 - 2(ReR_{\alpha\bar{\xi}\zeta\bar{\xi}})^2 = R_{\xi\bar{\xi}\xi\bar{\xi}}.$$

We argue that actually $R_{\mu\bar{\mu}\mu\bar{\mu}}(\mathcal{S}_0) = R_{\xi\bar{\xi}\xi\bar{\xi}}$ by an improvement of the proof as follows. There always exists a local holomorphic curve C'_ξ on $\mathcal{S}_0$ tangent to C_ξ at $[\alpha]$ such that in fact $R_{\mu\bar{\mu}\mu\bar{\mu}}(C'_\xi) = R_{\mu\bar{\mu}\mu\bar{\mu}}(\mathcal{S}_0)$, described now by the graph of $\theta \longrightarrow [\alpha + \theta\xi_\theta + s'\theta^2\zeta'_\theta] = [\alpha'_\theta]$, with s' real, as in the definition of C_ξ, except that $\xi_\theta \in \mathcal{H}_\alpha$, $\xi_0 = \xi$ is not necessarily constant. Expanding $R(\alpha'_t,\bar{\alpha}'_t;\alpha'_t,\bar{\alpha}'_t)$ as in (3) the whole argument goes through to get the equation (5) with s replaced by s'. Here one uses the facts that $R(\alpha,\bar{\alpha};\xi_t,\bar{\xi}_t) = \frac{1}{2}$ for any t (as $\xi_t \in \mathcal{H}_\alpha$) and that $R(\alpha,\bar{\xi}_t;\xi_t,\bar{\xi}_t) = 0$ by (4), as $\xi \in \mathcal{H}_\alpha$ is arbitrary. Consequently, we obtain by polarization that $R(\mathcal{S}_0)([\alpha]) \equiv R|\mathcal{H}_\alpha$, if one makes the identification $T_{[\alpha]}(\mathcal{S}_0) \cong (\mathbf{C}\mathcal{H}_\alpha + \mathbf{C}\alpha)/\mathbf{C}\alpha \cong \mathcal{H}_\alpha$.

(III.3) Enumeration of the Characteristic Projective Subvarieties

We enumerate in this section the set of characteristic projective subvarieties $\mathcal{S}_0 \subset \mathbb{P}(\mathfrak{m}^+)$ corresponding to the 6 types of irreducible bounded symmetric domains Ω.

($D^I_{p,q}$), p, q ≥ 1 Identifying as usual $\mathfrak{m}^+$ with $M(p,q;\mathbf{C})$, $\mathcal{S}_0$ is given by $\{[X]$: X is of rank 1$\}$. Any such X can be written in the form $X = uv^t$, where u and v are non–zero column vectors of order p and q resp. Writing $u = [u_1,\dots,u_p]$ and $v = [v_1,\dots,v_q]$, we have $X_{ij} = u_iv_j$. The mapping $\sigma\colon \mathbb{P}^{p-1} \times \mathbb{P}^{q-1} \longrightarrow \mathbb{P}(\mathfrak{m}^+)$ defined by

$$\sigma([u_1,\dots,u_p]\,;\,[v_1,\dots,v_q]) = [X] \text{ with } X_{ij} = u_iv_j$$

is equivalent to the Segre embedding of $\mathbb{P}^{p-1} \times \mathbb{P}^{q-1}$ into $\mathbb{P}^{pq-1} \cong \mathbb{P}(\mathfrak{m}^+)$. (When p (resp. q = 1), we get the identity map on projective space $\mathbb{P}^{p-1}$ (resp. $\mathbb{P}^{q-1}$.) The semisimple part K_s of $K = S(U(p)\times U(q))$ is $SU(p)\times SU(q)$. From [Ch.4, (2.2), Lemma 1] we have $\mathcal{S}_0(D^I_{p,q}) \cong (SU(p)\times SU(q))/(U(p-1)\times U(q-1)) \cong SU(p)/U(p-1) \times SU(q)/U(q-1) \cong \mathbb{P}^{p-1} \times \mathbb{P}^{q-1}$.

(D^{II}_n), n ≥ 2 In the same notations as above we have $\mathcal{S}_0 = \{[X] : X \in M_a(n,n;C)$ is of rank 2$\}$. Identify an anti–symmetric matrix $X = (x_{ij})$ with $\xi = \Sigma\, x_{ij}\, e^i \wedge e^j \in \Lambda^2 V$, where $\{e^i\}$ is an orthonormal basis of $V \cong C^n$. An element $\xi \in \Lambda^2 V$ is said to be decomposable if $\xi = \eta \wedge \eta'$ for some $\eta, \eta' \in V$. Under this identification $\mathcal{S}_0$ is identified with the space of $[\alpha]$, where α is a decomposable element of $\Lambda^2 V$. Thus, $\mathcal{S}_0 \subset \mathbb{P}(m^+)$ is identified with the Grassmannian of 2–planes of $V \cong C^n$. The semisimple part K_s of $K \cong U(n)$ is $SU(n)$. From [Ch.4 (2.4), Lemma 1] $\mathcal{S}_0(D^{II}_n) \cong SU(n)/S(U(2)\times U(n-2)) \cong G(2,n)$.

(D^{III}_n), n ≥ 2 In the same notations as above we have $\mathcal{S}_0 = \{[X] : X \in M_s(n,n;C)$ is of rank 1$\}$. Any such X can be written in the form of uu^t, where u is a non–zero column vector of order n. The mapping $v\colon \mathbb{P}^{n-1} \longrightarrow \mathbb{P}(m^+)$ defined by

$$v([u_1,\ldots u_n]) = [X] \quad \text{with} \quad X_{ij} = u_i u_j$$

is equivalent to the Veronese embedding of $\mathbb{P}^{n-1}$ into $\mathbb{P}^{\frac{n^2+n}{2}-1} \cong \mathbb{P}(m^+)$, which is the second canonical embedding of the projective space $\mathbb{P}^{n-1}$. The semisimple part K_s of $K \cong U(n)$ is $SU(n)$. From [Ch.4, (2.3) Lemma 1] we have $\mathcal{S}_0(D^{III}_n) \cong SU(n)/U(n-1) \cong \mathbb{P}^{n-1}$.

(D^{IV}_n), n ≥ 3 $\mathcal{S}_0$ is simply the set $\{[z_1,\ldots,z_n]\colon \Sigma z_i^2 = 0\} \cong Q^{n-2}$. The semisimple part K_s of $K = SO(n)\times SO(2)$ is $SO(n)$. From [Ch.4, (2.5), Lemma 2] we have $\mathcal{S}_0(D^{IV}_n) \cong SO(n)/SO(n-2)\times SO(2)) \cong Q^{n-2}$.

(D^V) We have $D^V \cong E_6/(\mathrm{Spin}(10)\times S^1)$, of complex dimension 16. The semisimple part K_s of $K = \mathrm{Spin}(10)\times S^1$, where Spin(10) is a simply-connected 2-fold covering of SO(10). From the classification of compact Kähler submanifolds of $\mathbb{P}^k$ with parallel second fundamental form we must have $\mathcal{S}_o(D^V) \cong SO(10)/U(5)$, which is the Hermitian symmetric manifold $G^{II}(5,5) \subset G(5,5)$ of compact type.

(D^{VI}) We have $D^{VI} \cong E_7/(E_6\times S^1)$, of complex dimension 27. The semisimple part K_s of $K = E_6\times S^1$ is E_6. From the classification of compact Kähler submanifolds of $\mathbb{P}^k$ with parallel second fundamental form we must have $\mathcal{S}_o(D^V)$ $\cong E_6/(\mathrm{Spin}(10)\times S^1)$, which is the first exceptional Hermitian symmetric manifold of compact type.

Since the second fundamental form of $\iota: (\mathcal{S}_o,h|\mathcal{S}_o) \hookrightarrow (\mathbb{P}(m^+),h)$ is parallel, it follows that the curvature tensor $R(\mathcal{S}_o)$ is parallel, so that $(\mathcal{S}_o,h|\mathcal{S}_o)$ is necessarily symmetric. From the pinching condition on holomorphic sectional curvatures in Prop.1 it follows that $\mathcal{S}_o$ must be of rank 1 or 2. $\mathcal{S}_o(\Omega)$ is of rank 1 exactly when $\Omega = D^I_{p,q}$ with $p = 1$ or $q = 1$, or when $\Omega = D^{III}_n$. Except in the case of D^{III}_n, the inclusion $\iota: (\mathcal{S}_o(\Omega),h|\mathcal{S}_o(\Omega)) \hookrightarrow (\mathbb{P}(m^+),h)$ gives the first canonical embedding. The list of $\iota: (\mathcal{S}_o,h|\mathcal{S}_o) \hookrightarrow (\mathbb{P}(m^+),h)$ exhausts precisely the first and second canonical embeddings of $\mathbb{P}^n$, $n \geq 1$, and the first canonical embedding on a (not necessarily irreducible) Hermitian symmetric manifold of compact type and of rank 2. $\mathcal{S}_o(\Omega)$ is irreducible except in the case of $D^I_{p,q}$, $p, q > 1$, where $\mathcal{S}_o \cong \mathbb{P}^{p-1} \times \mathbb{P}^{q-1}$ and in the case of D^{IV}_4, where $\mathcal{S}_o \cong Q^2 \cong \mathbb{P}^1 \times \mathbb{P}^1$. The list agrees with that in NAKAGAWA–TAKAGI [NT]. In particular, we have proved [(III.2), Thm.1].

(III.4) Higher Characteristic Bundles

We use the notations as in (III.1). Let $\Omega \subset\subset \mathbb{C}^N$, $\Omega \cong G_0/K$, be the

Harish–Chandra realization of an irreducible bounded symmetric domain and $X = \Omega/\Gamma$ be the quotient of Ω by a torsion–free discrete group of automorphisms. We defined the characteristic bundle $\mathcal{S}(\Omega) \longrightarrow \Omega$ on Ω so that the fiber $\mathcal{S}_x(\Omega)$ over $x \in \Omega$ is given by $K_x[\alpha]$ for $K_x \subset G$ the isotropy subgroup and $[\alpha] \in \mathbb{P}T_x(X)$ the projectivization of some dominant weight vector $\alpha \in T_x(X)$ of the isotropy K_x–representation on $T_x(\Omega)$. We may define $\mathcal{S}(X)$ as $\mathcal{S}(\Omega)/\Gamma$. From now on assume that Ω is of rank $r \geq 2$. For $1 \leq k \leq r-1$ we are going to define a k–th characteristic bundle $\mathcal{S}_k(\Omega)$ in such a way that $\mathcal{S}(\Omega) = \mathcal{S}_1(\Omega)$. Take $x = o$ and $\xi \in m^+ \cong T_o(\Omega)$. As in (III.1) write $\Psi = \{\psi_1,...,\psi_r\}$. Write also e_i for e_{ψ_i}. By [(III.1), Cor.1] there exists $k \in \mathfrak{k}$ such that $ad(k)(\xi) = \Sigma_{1\leq i\leq r} a_i e_i$ with a_i real, $0 \leq a_i \leq 1$, such that $a_1 \geq a_2 \geq ... \geq a_r$. Moreover the r–tuple $(a_1,...,a_r)$ is uniquely determined by ξ. Suppose $a_i > 0$ for $1 \leq i \leq p$ and $a_{p+1} = a_{p+2} = ... = 0$. We call $p = r(\xi)$ the rank of ξ. A similar definition can be applied to any point $x \in \Omega$. With this terminology a characteristic vector ξ at x is then equivalently a unit vector of rank 1. We define

DEFINITION 1

Let $1 \leq k \leq r(\Omega)$ and let $\mathcal{S}_{k,x}$ denote $\{[\xi]: \xi \in T_x(\Omega)$ and $1 \leq r(\xi) \leq k\}$. We call $\mathcal{S}_{k,x}(\Omega) \subset \mathbb{P}T_x(X)$ the k–th characteristic projective subvariety at $x \in \Omega$. The union $\mathcal{S}_k(\Omega) := \cup_{x\in\Omega} \mathcal{S}_{k,x} \subset \mathbb{P}T(\Omega)$ is called the k–th characteristic bundle over Ω. The quotient $\mathcal{S}_k(\Omega)/\Gamma$ (noting that $\mathcal{S}_k(\Omega)$ is invariant under the standard action of Γ), written $\mathcal{S}_k(X)$, is called the k–th characteristic bundle over X.

As a generalization to [Ch.6, (1.2), Prop.2] we prove

PROPOSITION 1

The k–th characteristic bundle $\mathcal{S}_k(\Omega) \longrightarrow \Omega$ is holomorphic. Moreover, in terms of the Harish–Chandra embedding $\Omega \hookrightarrow \mathbb{C}^N$, $\mathcal{S}_k(\Omega)$ is parallel on Ω in the Euclidean sense, i.e., identifying $\mathbb{P}T(\Omega)$ with $\Omega \times \mathbb{P}^{N-1}$ using the Euclidean coordinates we have the identification $\mathcal{S}_k(\Omega) \cong \Omega \times \mathcal{S}_{k,o}$.

Proof:

It suffices to show that $\mathcal{S}_{k,o} \subset \mathbb{P}(m^+)$ is a complex–analytic subvariety. The rest of the proof is the same as that given in [Ch.6, (1.2), Prop.2]. For $\xi \in \mathfrak{m}^+$ of rank k, with normal form $\eta = \Sigma_{1\le i\le k} a_i e_i$ with respect to Ψ, we have by the proof of [(III,1), Prop.1, (b) $\Leftrightarrow$ (c)] that $\dim_{\mathbb{C}} \mathcal{N}_\xi = \dim_{\mathbb{C}} \mathcal{N}_\eta = \dim_{\mathbb{C}}(\cap_{1\le i\le k} \mathcal{N}_{e_i}) := n_k(\Omega)$, which depends only on the rank $k = r(\xi)$ of ξ. $n_1(\Omega) = n(\Omega)$ is the null dimension of Ω. We call $n_k = n_k(\Omega)$ the k–th null dimension of Ω. Consider the vector bundle $V = \Lambda^{n_k} T(\Omega)$. The Bergman metric g on Ω induces a canonical metric h on V in such a way that (V,h) is of seminegative curvature in the dual sense of Nakano (and hence in the sense of Griffiths). Denote by Θ the curvature of (V,h). Then for a non–zero vector $\xi \in \mathfrak{m}^+$, $\Theta_{v\bar{v}\xi\bar{\xi}} = 0$ for some non–zero vector $v \in V_o$ if and only if ξ is of rank $\le$ k. In this case v is in $\Lambda^{n_k} \mathcal{N}_\xi \subset \Lambda^{n_k} T_o(\Omega) = V_o$. Let Q be the Hermitian bilinear form on $\mathfrak{m}^+ \otimes V_o$ defined by $Q(\xi \otimes \bar{v}; \xi' \otimes \bar{v}') = \Theta_{v'\bar{v}\xi\bar{\xi}'}$ and extended by Hermitian bilinearity. Q is negative semi–definite. Let $W \subset \mathfrak{m}^+ \otimes V_o$ be the null subspace of Q. W is a complex vector subspace of $\mathfrak{m}^+ \otimes V_o$. The space of decomposable elements $\eta \otimes \bar{u} \in \mathfrak{m}^+ \otimes V_o$ constitutes a complex subvariety Σ. By the description of the zeros of curvature (in the sense of Griffiths) above it follows that $[\xi] \in \mathcal{S}_{k,o}$ if and only if there exists $v \in V_o$, $v \neq 0$, such that $\xi \otimes \bar{v} \in W \cap \Sigma$. It follows readily that $\mathcal{S}_{k,o} \subset \mathbb{P}(\mathfrak{m}^+)$ is complex–analytic, as desired. ▌

In general, the k–th characteristic bundles $\mathcal{S}_k(\Omega)$ are not smooth for $k \ge 2$. There is a stratification of $\mathcal{S}_k(\Omega)$ into the union $(\mathcal{S}_k(\Omega) - \mathcal{S}_{k-1}(\Omega)) \cup \ldots (\mathcal{S}_2(\Omega) - \mathcal{S}_1(\Omega)) \cup \mathcal{S}_1(\Omega)$ of k non–singular complex–analytic bundle over Ω.

Using Prop.1 one can also obtain integral formulas on $\mathcal{S}_k(X)$ for $1 \le k \le r(\Omega) - 1$ to prove metric rigidity theorems. The singularities of $\mathcal{S}_k(\Omega)$ for $k \ge 2$ do not pose any problems since one can always integrate over the smooth part and justify the integration by parts by lifting to a non–singular model. We did not need this so far. In the dual case of Hermitian symmetric manifolds of compact type an integral formula on some dual k–th characteristic bundle $\mathcal{S}_k^*(\Omega)$ turns out to be useful (cf. [Ch.7, (2.2), Remarks]).

IV. A DUAL GENERALIZED FRANKEL CONJECTURE FOR COMPACT KÄHLER MANIFOLDS OF SEMINEGATIVE BISECTIONAL CURVATURE

(IV.1) Background

Throughout this monograph we have been studying Kähler/Hermitian metrics of seminegative or semipositive curvature on locally Hermitian symmetric manifolds of semisimple type by exploiting the particular structure of zeros of curvatures of underlying canonical metrics. One may go one step further to study Kähler/Hermitian manifolds X of semipositive curvature in general. We restrict our attention to the case of compact manifolds X. As explained in [Ch.7, (2.1)] on irreducible Hermitian symmetric manifolds of compact type X_c there are many Hermitian metrics h of semipositive curvature. They were however shown to be canonical if assumed Kähler. In general, one can construct on any compact homogeneous manifolds M metrics of semipositive curvature. Nonetheless, the canonical metrics on M are not of semipositive curvature unless X is symmetric. For compact Kähler manifolds X of semipositive curvature we proved more recently

THEOREM 1 (MOK [MO4])
Let (X,h) be an n-dimensional compact Kähler manifold of semipositive holomorphic bisectional curvature and let $(\tilde{X},\tilde{h})$ be its universal covering space. Then, there exist nonnegative integers k, p; $N_1, \dots, N_q$ and irreducible Hermitian symmetric manifolds of compact type $(M_1,g_1), \dots, (M_p,g_p)$ of rank ≥ 2 such that (X,h) is isometrically biholomorphic to

$$(\mathbb{C}^k,g_0) \times (\mathbb{P}^{N_1},\theta_1) \times \dots (\mathbb{P}^{N_q},\theta_q) \times (M_1,g_1) \times \dots \times (M_p,g_p),$$

where θ_i, $1 \leq i \leq q$, is a Kähler metric on the complex projective space $\mathbb{P}^{N_i}$ of semipositive holomorphic bisectional curvature.

In [FRA,1961] FRANKEL conjectured that any compact Kähler manifold of positive bisectional curvature is biholomorphic to a projective space. This was resolved in the affirmative by MORI [MO,1979] and SIU–YAU [SY1,1980]. Thm.1 confirms in particular what could be called the generalized Frankel conjecture which asserts that a compact Kähler manifold of semipositive bisectional curvature is biholomorphic to a Hermitian locally symmetric manifold. In analogy with Thm.1, one may pose a dual Frankel conjecture at least for compact Kähler manifolds of seminegative bisectional curvature. In this case, however, there are many compact Kähler manifolds X of strictly negative bisectional curvature not biholomorphic to a quotient of the Euclidean ball B^n. First of all, MOSTOW–SIU [MO–S] constructed a countable infinite set of compact Kähler surfaces of strictly negative Riemannian sectional curvature not covered by the ball B^2. On the other hand, because of the curvature–decreasing property of Kähler submanifolds one can always take X to be submanifolds of some compact quotients of the ball. They are in general not of negative Riemannian sectional curvature. Thus, in any dual version of the generalized Frankel conjecture one should single out a class of manifolds which include such examples.

To start with let me state the classification theorem in Riemannian Geometry of compact Riemannian manifolds (M,g) of seminegative Riemannian sectional curvature. There is a notion of rank for such manifolds defined in terms of Jacobi fields as follows: Let v be a unit vector at $x \in M$ and $\gamma_v = \{\gamma_v(t): -\infty < t < \infty\}$, $\gamma(o) = x$, be the unique geodesic on M parametrized by arc–length determined by v. We define rank(v) to be the maximal dimension of parallel Jacobi fields $J^p(v)$ (including $\dot{\gamma}(t)$) along γ. The rank of (M,g) is defined to be the minimum of rank(v) as v ranges over the unit sphere bundle S(M) of M (cf. BALLMANN-GROMOV–SCHROEDER [BGS, App.1]). For a Riemannian locally symmetric manifold of non–compact type this definition of rank agrees with the usual definition. We say that a Riemannian manifold (M,g) is irreducible if any finite covering of (M,g) is irreducible in the sense of de Rham. Then, we have

THEOREM 2 (BALLMANN [BAL] & BURNS–SPATZIER [BS])
Let (M,g) be an irreducible compact Riemannian manifold of seminegative sectional curvature. Then, either rank(M,g) = 1 or (M,g) is isometric to an irreducible Riemannian locally symmetric manifold of non–compact type.

Thm.2 was actually proved under the weaker hypothesis that (M,g) is complete and of finite volume, and that the sectional curavatures of (M,g) is pinched by $-C \leq$ Sectional Curvatures ≤ 0 for some constant C.

There is something in common between Thms. 1 and 2 above. They are both proved using Berger's characterization ([BER1]) of Riemannian locally symmetric manifolds in terms of the holonomy group (cf. also SIMON [SIM]). In case of Thm.2 of BALLMANN [BAL] and BURNS–SPATZIER [BS] they constructed integrals of the geodesic flow. i.e., non–constant functions f on the unit sphere bundle $S(M)$ invariant under the geodesic flow, so that the level sets of f are invariant under holonomy. In case of Thm.1 we constructed a space $\mathscr{C}$ of rational curves on X such that some subset $\mathcal{S} \subset \mathbb{P}T_X$ associated to $\mathscr{C}$ is invariant under holonomy. For an irreducible Hermitian symmetric manifold of compact type the $\mathcal{S}$ thus constructed actually agrees with $\mathcal{S}(X_c)$, as defined in [Ch.7, (2.2)].

(IV.2) Formulation of a Dual Generalized Frankel Conjecture

Our formulation of a dual generalized Frankel conjecture will also in some sense be a Kähler analogue of Thm.2. In place of the geometric condition of rank one should have some notion which depends only on the complex structure of X. To start with we define the notion of almost negative holomorphic vector bundles.

Let $L \longrightarrow X$ be a holomorphic line bundle over a compact complex manifold X. We say that L is ample if for some positive integer p the complex vector space of holomorphic sections $\Gamma(X,L^p)$ has no base points and defines canonically a holomorphic embedding $\sigma_p: X \longrightarrow \mathbb{P}^N$ (well–defined up to a projective transformation), $N = \dim_{\mathbb{C}}\Gamma(X,L^p) - 1$, of X into a projective space. By the Kodaira Embedding Theorem (KODAIRA–MORROW [KM, Ch.3, §7,8]) this is the case if and only if L carries a Hermitian metric of positive curvature.

DEFINITION 1

Let $L \longrightarrow X$ be a holomorphic line bundle over an n–dimensional compact complex manifold X. We say that L is almost ample if for some positive integer p the complex vector space $\Gamma(X,L^p)$ is non–trivial and defines a meromorphic mapping $\sigma_p: X \longrightarrow \mathbb{P}^N$, $N = \dim_{\mathbb{C}}\Gamma(X,L^p) - 1$, such that σ_p is holomorphic and of rank n at some point $x \in X$. We say that L is almost negative if the dual bundle $L^* \longrightarrow X$ is almost ample.

From standard algebro–geometric arguments involving vanishing theorems and dimension counting we have

PROPOSITION 1

The following conditions are equivalent for a holomorphic line bundle L over a compact manifold X,

(1) $L \longrightarrow X$ is almost ample

(2) For a sufficiently large positive integer p, $\Gamma(X,L^p)$ (is non–trivial and) defines a birational embedding $\sigma_p: X \longrightarrow \mathbf{P}^N$ into some projective space.

(3) There exists a positive number c such that for p sufficiently large

$$\dim_{\mathbf{C}} \Gamma(X,L^p) \geq cp^n.$$

(2) $\Rightarrow$ (1) is obvious. For the proof of (1) $\Rightarrow$ (2) cf. UENO [UE, Ch.II, Prop.5.7, p.56]. We remark that by the argument in Siegel's Theorem (for the field of meromorphic functions) of counting Taylor coefficients it is always true that $\dim_{\mathbf{C}}\Gamma(X,L^p) \leq \mathrm{Const}.p^n$ for any holomorphic line bundle L over a compact manifold X (cf. e.g. ADAMS-GRIFFITHS [AG]). The same argument yields a proof of (1) $\Leftrightarrow$ (3).

For any holomorphic line bundle $L \longrightarrow X$ over a compact complex manifold X and a positive integer p, we denote by $\mathcal{B}_0^p(L)$ the base point set of $\Gamma(X,L^p)$. We denote by $\mathcal{B}^p(L)$ the union of $\mathcal{B}_0^p(L)$ and the subset of $X - \mathcal{B}_0^p(L)$ on which $\sigma_p|(X-\mathcal{B}_0^p(L))$ fails to be of maximal rank. It is easy to see that $\mathcal{B}^p(L)$ is a complex–analytic subvariety of X. We call $\mathcal{B}^p(L) \subset X$ the bad set associated with L^p. Define $\mathcal{B}_0(L) = \cap_{p>0}\mathcal{B}_0^p(L)$ and $\mathcal{B}(L) = \cap_{p>0}\mathcal{B}^p(L)$. We call $\mathcal{B}_0(L)$ the base point set of the system $\{L^p\}_{p>0}$, and $\mathcal{B}(L)$ the bad set associated to the system $\{L^p\}_{p>0}$.

For holomorphic vector bundles $V \longrightarrow X$ over a compact complex manifold X we define

DEFINITION 2

Let $V \longrightarrow X$ be a holomorphic vector bundle over a compact complex manifold X. Denote by $\pi\colon \mathbb{P}(V) \longrightarrow X$ the projectivization of V and by $L_V \longrightarrow \mathbb{P}(V)$ the tautological line bundle over V. Let $\mathcal{B} := \mathcal{B}(L_V^*) \subset \mathbb{P}(V)$ be the bad set associated to the system $\{L_V^{-p}\}_{p>0}$. We say that V is almost negative if and only if L_V is almost negative over $\mathbb{P}(V)$ and $\pi(\mathcal{B}) \neq X$. We say that V is almost ample if and only if the dual bundle V^* is almost negative.

Here the tautological bundle $L_V \longrightarrow \mathbb{P}(V)$, when restricted to each projective fibre of $\mathbb{P}(V) \longrightarrow V$, is isomorphic to the dual of the hyperplane section line bundle. It is for this reason that we use V^* instead of V in defining almost ample vector bundles.

For holomorphic line bundles $L \longrightarrow X$ as a generalization of the Kodaira Embedding Theorem we have as solutions to the Grauert–Riemenschneider Conjecture [GR] the following

THEOREM 1 (SIU [SIU4] & DEMAILLY [DE])

Let $(L,h) \longrightarrow X$ be a Hermitian holomorphic line bundle over an n–dimensional compact complex manifold X. Suppose the curvature (1,1)–form $\Theta(h)$ of h is positive semi–definite on X and positive definite at some point. Then, there exists a positive constant c such that $\dim_{\mathbb{C}}\Gamma(X,L^p) \geq cp^n$ for $p > 0$ sufficiently large. In other words, L is almost ample.

We are now ready to formulate a dual generalized Frankel conjecture.

DUAL GENERALIZED FRANKEL CONJECTURE

Let (X,h) be an irreducible compact Kähler manifold of seminegative holomorphic bisectional curvature. Denote by Θ the curvature form of (X,h). Then, one of the following (mutually exclusive) possibilities occur.

(1) The holomorphic tangent bundle $T(X)$ is almost negative;

(2) (X,h) is biholomorphic to an irreducible compact quotient of a bounded symmetric domain of rank ≥ 2.

(3) For any tangent vector $\alpha \in T_x(X)$ there exists a non–zero $\beta \in T_x(X)$ such that $\Theta_{\alpha\bar{\alpha}\beta\bar{\beta}} = 0$.

We proceed to explain that the three possibilities are indeed mutually exclusive and to give non–trivial examples of (3). For the latter consider M a abelian variety, equipped with the locally Euclidean metric h_0. Let $X \subset M$ be a complex submanifold and define $h = h_0|_X$. We assert that if $\dim_{\mathbf{C}} X > \dim_{\mathbf{C}} M - \dim_{\mathbf{C}} X$, then X satisfies condition (3). (Note that a generic submanifold X is irreducible.) As (M,h_0) is locally Euclidean (X,h) is Kähler and of seminegative bisectional curvature. Moreover for $x \in X$ and $\alpha, \beta \in T_x(X)$ we have $\Theta_{\alpha\bar{\alpha}\beta\bar{\beta}} = \|\sigma(\alpha,\beta)\|^2$ for the second fundamental form $\sigma\colon T_x(X) \times T_x(X) \longrightarrow N_x$, with N denoting the (1,0)–component of the complexified normal bundle of X in M, by the Gauss–Codazzi equation. Consequently, $\Theta_{\alpha\bar{\alpha}\beta\bar{\beta}} = 0$ if and only if $\sigma(\alpha,\beta) = 0$. The mapping $\sigma_\alpha\colon T_x(X) \longrightarrow N_x$ defined by $\sigma_\alpha(\beta) = \sigma(\alpha,\beta)$ is complex linear in β. When $\dim_{\mathbf{C}} T_x(X) = \dim_{\mathbf{C}} X > \dim_{\mathbf{C}} M - \dim_{\mathbf{C}} X = \dim_{\mathbf{C}} N_x$, σ_α must have a non–trivial kernel for any $\alpha \in T_x(X)$, i.e., $\Theta_{\alpha\bar{\alpha}\beta\bar{\beta}} = 0$ for $\beta \in \mathrm{Ker}(\sigma_\alpha) \neq 0$, proving our assertion.

It remains to show that the possibilities (1), (2) and (3) are indeed mutually exclusive. For simplicity of notations write T for T_X and L for L_T. By [Ch.2, (4.1), Prop.2], (T,h) is of seminegative curvature in the sense of Griffiths if and only if $c_1(L_T,\hat{h})$ is negative semi–definite. By the solutions to the Grauert–Riemenschneider Conjecture (Thm.1) it follows that $L \longrightarrow \mathbb{P}(T)$ is almost negative if and only if $\int_{\mathbb{P}(T)} [-c_1(L_T,\hat{h})]^{2n-1} > 0$. Again by [Ch.2, (4.2), Prop.1] we have $[-c_1(L,\hat{h})]^{2n-1} \equiv 0$ if and only if (3) is valid. Thus, (1) and (3) are mutually exclusive. For (X,g) a compact quotient of a bounded symmetric domain of rank $r \geq 2$, endowed with canonical metrics, we know from the description of the zeros of bisectional curvatures in [(III.4), Prop.1] that $-c_1(L,\hat{g})([\xi]) > 0$ for any tangent vector $\xi \in T$ of rank r. Thus, $-c_1(L,\hat{g})([\xi]) > 0$ for a dense open subset of $\mathbb{P}(T)$, so that by Thm.1 $L \longrightarrow \mathbb{P}(T)$ is almost negative, showing that (2) and (3) are mutually exclusive. To contrast (1) and (2) in the Hermitian locally symmetric case $X = \Omega/\Gamma$ with Ω irreducible and of rank ≥ 2 consider the $(r-1)$–th characteristic bundle $\mathcal{S}_{r-1}$, of dimension s, as defined in [(III.4), Def.1]. By [(III.4), Prop.1] $Z(\mathbb{P}(T),L^*) = \mathcal{S}_{r-1}$. Moreover $\int_{\mathcal{S}_{r-1}} [-c_1(L,\hat{g})]^s \equiv 0$. When (X,h) is Hermitian locally symmetric but Ω is reducible one has a similar integral formula obtained by replacing $\mathcal{S}_{r-1}$ by subbundles of $\mathbb{P}(T)$ arising from the local direct factors of X. To show that (1) and (2) are mutually exclusive it remains to show (using notations in this paragraph)

PROPOSITION 2

Let (X,h) be a compact Hermitian manifold of seminegative curvature. Suppose $S \subset \mathbb{P}(T)$ is a complex–analytic subvariety such that $c_1(L,\hat{h})|_S$ has at least one zero eigenvalue at every smooth point of S. Then, $S \subset \mathcal{B}(L^*)$.

Proof:

Let $\xi \in S$. Suppose for some positive integer p there exists $\sigma_0, \sigma_1, \ldots, \sigma_{2n-1} \in \Gamma(\mathbb{P}(T),L^{-p})$, $n = \dim_{\mathbb{C}} X$, such that $\sigma_0([\xi]) \neq 0$ and such that for $f_i = \sigma_i/\sigma_0$, $(df_1 \wedge \ldots \wedge df_{2n-1})([\xi]) \neq 0$. Define a Hermitian metric θ on L by

$$\|\nu\|_\theta^{2p} = \|\nu\|_{\hat{g}}^{2p} + \left[\sum_{0\le i\le 2n-1} |\sigma_i(\nu^p)|^2\right]$$

for $\nu \in L$, where we use the natural pairing between L^{-p} and L^p. An application of the Lagrange identity (cf. [Ch.4, (1.2), proof of Prop.1] shows that near $[\xi] \in \mathbb{P}(T)$, θ^p is the sum of $\hat{g}$ and a Hermitian metric of strictly negative curvature, so that $-c_1(L,\theta)([\xi]) > 0$. Consequently we have $\int_S [-c_1(L,\theta)]^s > 0$ while by assumption $\int_S [-c_1(L,\hat{h})]^s = 0$, where integration is understoood to be performed over the regular part. This gives a contradiction to Stokes' Theorem and establishes Prop.2. The justification of Stokes' Theorem for the possibly singular S can be performed by lifting S and L to a non–singular model $\tilde{S}$ of S, noting that the corresponding liftings of $c_1(L,\hat{h})$ and $c_1(L,\theta)$ to $\tilde{S}$ are closed (1,1)–forms as both of these forms are actually defined on the smooth manifold $\mathbb{P}(T)$. ∎

From Prop.2 we deduce that actually the possibilities (1), (2) and (3) in the Dual Generalized Frankel Conjecture are mutually exclusive. Let now (X,g) be a quotient of an irreducible bounded symmetric domain of rank ≥ 2. In the notations above denote by $Z(\mathbb{P}(T),L)$ the subset of $\mathbb{P}(T)$ on which the curvature (1,1)–form $\Theta(\hat{g})$ has some zero eigenvalues. Prop.2 implies that $\mathcal{B}(L^*) \supset Z(X,L)$ for the bad set $\mathcal{B} \subset \mathbb{P}(T)$ associated to the system $\{L^{-p}\}_{p>0}$. As a motivation for the formulation of the Dual Generalized Frankel Conjecture we have

PROPOSITION 3

Let $X = \Omega/\Gamma$ be a compact quotient of an irreducible bounded symmetric domain Ω of rank ≥ 2 and of dimension n, we have actually $\mathcal{B}(L^*) = Z(\mathbb{P}(T),L^*) = \mathcal{S}_{r-1}$ for $r = \text{rank}(X)$. In other words, for any $[\eta] \in \mathbb{P}(T) - \mathcal{S}_{r-1}$, there exists a positive integer $p > 0$ and $\sigma_i \in \Gamma(\mathbb{P}(T),L^{-p})$, $0 \leq i \leq 2n-1$ such that $\sigma_0([\eta]) \neq 0$ and $[\sigma_0,\ldots,\sigma_{2n-1}]$ defines a local embedding at $[\eta]$.

Prop.3 is a motivation for formulating the conjecture because the subvariety $\mathcal{S}_k(\Omega)$ and hence $\mathcal{S}_k = \mathcal{S}_k(X)$, for $1 \leq k \leq r-1$, are invariant under holonomy. To see this let $D_r \cong \Delta^r$ be a distinguished polydisc in Ω as defined in the Polydisc Theorem [Ch.5, (1.1), Thm.1)]. Recall from the definition of $\mathcal{S}_k(\Omega)$ that there exists a polydisc $D_k \cong \Delta^k \hookrightarrow \Delta^r \cong D_r$, $T_0(D_k) \cong \mathfrak{a}_k^+ \subset \mathfrak{a}^+ \cong T_0(D_r)$ such that $\mathcal{S}_k(\Omega)$ is the G_0–orbit of $\mathbb{P}(\mathfrak{a}_k^+)$ in $\mathbb{P}T(\Omega)$. On the other hand the holonomy group H_x at $x \in \Omega$ is contained in the isotropy subgroup $K_x \subset G_0$ at x (cf. (I.1)). It follows that $\mathcal{S}_k(\Omega)$ is invariant under holonomy, as asserted.

We continue with the proof of Prop.3.

<u>Sketch of proof of Prop.3</u>:

A section $\sigma \in \Gamma(\mathbb{P}(T),L^{-p})$ can be identified with a holomorphic section $\tau \in \Gamma(X,S^pT^*)$. The locally homogeneous Hermitian vector bundle S^2T admits a decomposition as $V \oplus W$, where $V \subset S^2T$ is the highest weight summand given by

$$V = \Sigma_{[\alpha]\in\mathcal{S}_0} C(\alpha\circ\alpha).$$

Here $\alpha\circ\alpha$ denotes the symmetric square of α. Correspondingly there is a direct sum decomposition $S^2T^* = V' \oplus W'$, where $V' \subset S^2T^*$ is the highest weight subspace generated by dual characteristic vectors. From the natural pairing between S^2T and S^2T^* it can be deduced that V and W' are annihilators of each other. For X of rank 1 W and W′ are trivial, while for rank(X) ≥ 2 W and W′ are irreducible, as can be read from CALABI–VESENTINI [CV] and BOREL [BO4]. Consider first of all the case when rank(X) = 2. In this case it can be shown that W is of strictly negative curvature in the sense of Griffiths while V is properly seminegative curvature in the sense of [Ch.10, (2.1), Def.1]. To produce sections in $\Gamma(\mathbb{P}(T),L^{-p}) \cong \Gamma(X,S^pT^*)$ it suffices to produce holomorphic sections of symmetric tensor powers of the positive vector bundle W′. Any section $\sigma \in \Gamma(\mathbb{P}(T),L^{-p})$ thus produced has to vanish at any $[\alpha] \in \mathcal{S}$ because W′ and V are annihilators of each other. On the other hand, by the L^2–estimates of $\bar\partial$ applied to the positive vector bundle W′ it is easy to deduce that for $[\eta] \in \mathbb{P}(T) - \mathcal{S}$ one can find holomorphic sections $\sigma_i \in \Gamma(\mathbb{P}(T),L^{-p})$ with the desired properties at $[\eta]$ for p sufficiently large.

In the case of arbitrary rank ≥ 2 one considers similarly direct sum decompositions of S^rT^* to extract direct summands of strictly positive curvature. We omit the details. ▌

So far we have been considering the bad set $\mathcal{B}(L^*)$ consisting of the base point set $\mathcal{B}_0(L^*)$ and the branching locus. It is interesting to note that we have actually the stronger

PROPOSITION 4

Let $X = \Omega/\Gamma$ be a compact quotient of an irreducible bounded symmetric domain Ω of rank ≥ 2. Then, $\mathcal{B}_0(L^*) = \mathcal{B}(L^*) = Z(\mathbb{P}(T),L^*) = \mathcal{S}_{r-1}$ for $r = \text{rank}(X)$. In other words, for any integer $p > 0$ and any $\sigma \in \Gamma(\mathbb{P}(T),L^{-p})$, σ vanishes identically on $\mathcal{S}_{r-1}$.

Proof:

Again we consider first of all the case when rank(X) = 2. In this case $r - 1 = 1$ and $\mathcal{S}_1$ is simply the characteristic bundle $\mathcal{S}$. Let $p > 0$ and $\sigma \in \Gamma(\mathbb{P}(T),L^{-p})$. Then, $\theta = (\hat{g}^p + \sigma \otimes \bar{\sigma})^{1/p}$ defines a Hermitian metric of seminegative curvature on L over $\mathbb{P}(T)$, as explained in the proof of Prop.1. By the second proof of the

Hermitian metric rigidity theorem in the seminegative case given in [Ch.6, (3.1) and (3.2)] using Moore's Ergodicity Theorem we conclude that for $\theta = e^u\hat{g}$, u is a constant on $\mathcal{S}$. It follows readily that $\|\sigma\|_{\hat{g}^p}$ is a constant c on $\mathcal{S}$. If $c \neq 0$ we would have a non–vanishing holomorphic section in $\Gamma(\mathcal{S},L^{-p})$, showing that L^p is holomorphically trivial. This is impossible since $\int_{\mathcal{S}} [-c_1(L,\hat{g})] \wedge \nu^{s-1} > 0$ for the Kähler form $\nu = -c_1(L,\hat{g}) + \pi^*\omega$ (cf. [Ch.6, (2.1), Lemma 1], π: $\mathbb{P}(T) \to X$ the base projection and ω the Kähler form of (X,g).

To complete the proof of Prop.3 we need to generalize to the case of higher rank. To start with we assume the following statement

(*) For X locally irreducible and for $1 \leq i \leq r - 2$ we have $\dim_{\mathbb{C}}\mathcal{S}_k \leq \dim_{\mathbb{C}}\mathcal{S}_{k+1} - 2$.

As before let $p > 0$, $\sigma \in \Gamma(\mathbb{P}(T),L^{-p})$ and define $\theta = (\hat{g}^p + \sigma \otimes \overline{\sigma})^{1/p}$, $\theta = e^u\hat{g}$ on $\mathbb{P}(T)$. Consider the filtration $\mathcal{S} = \mathcal{S}_1 \subset \mathcal{S}_2 \subset \ldots \subset \mathcal{S}_{r-1}$. By the proof in the rank–2 case we know that $\sigma|\mathcal{S} \equiv 0$. We are going to show that $\sigma|\mathcal{S}_{r-1} \equiv 0$ by induction. Consider the following integral formula, which is a generalization of [Ch.6, (3.1), Prop.1].

$$\int_{\mathcal{S}_k} (-c_1(L,\theta)) \wedge (-c_1(L,\hat{g}))^{2n-2q(k)-1} \wedge \nu^{q(k)-1}$$
$$= \int_{\mathcal{S}_k} (-c_1(L,\hat{g}))^{2n-2q(k)} \wedge \nu^{q(k)-1} = 0,$$

where $q(k) = n_k(X)$ is the k–th null–invariant of X. As in the dimension formula of [Ch.6, (1.2), Prop.4] it follows easily that $\dim_{\mathbb{C}}\mathcal{S}_k = 2n - q(k) - 1$. From the integral formula we conclude that

$$(-c_1(L,\theta)) \wedge (-c_1(L,\hat{g}))^{2n-2q(k)-1} \equiv 0 \quad \text{on } \mathcal{S}_k.$$

Let $[\xi] \in \mathcal{S}_k - \mathcal{S}_{k-1}$. Let $[\tilde{\xi}] \in \mathcal{S}_k(\Omega) - \mathcal{S}_{k-1}(\Omega)$ be a lifting of $[\xi]$ to $\mathbb{P}T(\Omega)$. Write $\Omega = G_0/K$ as usual we consider the natural G_0 action on $\mathbb{P}T(\Omega)$. Define $\mathcal{S}[\xi]$ as $G_0[\tilde{\xi}]/\Gamma$. As in the [Ch.6, (3.2)] by an argument using Moore's Ergodicity

Theorem we deduce that u is constant on $\mathcal{S}[\xi]$. For $k > 1$, $\mathcal{S}_k(\Omega) - \mathcal{S}_{k-1}(\Omega)$ is however not homogeneous under G_0.

We now deduce by induction that $\sigma|\mathcal{S}_{r-1} \equiv 0$ starting with $\sigma|\mathcal{S} \equiv 0$, $\mathcal{S} = \mathcal{S}_1$. Suppose $\sigma|\mathcal{S}_k \equiv 0$ for some k, $1 \le k \le r-2$. We argue first of all that σ must vanish somewhere on $\mathcal{S}_{k+1} - \mathcal{S}_k$ by using (*). Suppose otherwise. Then, σ^{-1} is holomorphic on $\mathcal{S}_{k+1} - \mathcal{S}_k$. Since $\dim_{\mathbf{C}}\mathcal{S}_{k+1} - \dim_{\mathbf{C}}\mathcal{S}_k \ge 2$ by (*) we conclude that σ^{-1} extends holomorphically to $\mathcal{S}_{k+1}$. As $\sigma.\sigma^{-1} \equiv 1$ it follows that σ is non–vanishing on $\mathcal{S}_k$, contradicting with our inductive hypothesis. We have therefore showed that $\sigma[\xi] = 0$ and hence $u([\xi]) = 1$ for some $[\xi] \in \mathcal{S}_{k+1} - \mathcal{S}_k$. Thus $\sigma|\mathcal{S}[\xi] \equiv 0$ by the preceding paragraph. As the zero set of σ is complex–analytic and $S[\xi] = G_0[\xi]/\Gamma$ it follows that σ vanishes on $(G^{\mathbf{C}}[\tilde{\xi}]\cap\mathbf{P}T(\Omega))/\Gamma$. Write $\mathcal{S}_i(X_c)$ for the i–th characteristic bundle on the compact dual X_c. From the fact that $\mathrm{Aut}(\mathbf{P}^r)$ acts transitively on $\mathcal{S}_{k+1}(\mathbf{P}^r) - \mathcal{S}_k(\mathbf{P}^r)$ for $r = \mathrm{rank}(X)$ one can easily deduce that $G^{\mathbf{C}}$ acts transitively on $\mathcal{S}_{k+1} - \mathcal{S}_k$. It follows that $G^{\mathbf{C}}[\tilde{\xi}] \cap \mathbf{P}T(\Omega) = \mathcal{S}_{k+1} - \mathcal{S}_k$ and that σ vanishes identically on $\mathcal{S}_{k+1} - \mathcal{S}_k$ and hence on $\mathcal{S}_{k+1}$. Assuming (*) the proof that $\sigma|\mathcal{S}_{r-1} \equiv 0$ now follows by induction.

The simplest proof of (*) is by a case–by–case verification. By the dimension formula $\dim_{\mathbf{C}}\mathcal{S}_k = 2n - n_k(X) - 1$ it suffices to show that $n_{k+1}(X) \le n_k(X) - 2$. In case of classical symmetric domains of type (I) – (III) this follows from the description of zeros of bisectional curvatures in [Ch.4, (3.3)] in terms of matrices. D_n^{IV} and D^V are rank–2 symmetric spaces. It remains to verify (*) for $D^{VI} \cong E_7/(E_6 \times S^1)$, which is of rank 3. In the notations of [(III.1), proof of Prop.1] write $\Psi = \{\psi_1,\psi_2,\psi_3\}$ and write $\mathcal{N}_{\Psi'}$ as $\mathcal{N}_1$ and $\mathcal{N}_2$ for $\Psi' = \{\psi_1\}$ resp. $\{\psi_1,\psi_2\}$. Their dimensions are $n_1(X) = \dim_{\mathbf{C}}\mathcal{N}_1 = 10$ and $n_2(X) = \dim_{\mathbf{C}}\mathcal{N}_2 = 3$, as can be read from ZHONG [ZHO], for example. ▌

BIBLIOGRAPHY

[AD–G] ADAMS, J. & GRIFFITHS, P. *Topics in Algebraic and Analytic Geometry*, Math. Notes, No. 13, Princeton Univ. Press, Princeton & Univ. of Tokyo Press, Tokyo, 1974.

[AG] ANDREOTTI, A. & GRAUERT, H. Algebraische Körper von automorphen Funktionen, Nachr. Akad. Wiss. Göttingen, Math.–phy. Klasse (1961), 39–48.

[AMRT] ASH, A., MUMFORD, D., RAPOPORT, M. & TAI, Y.–S. *Smooth Compactification of Locally Symmetric Varieties*, Lie Groups: History, Frontiers and Applications, Vol. 4, Math. Sci. Press, Brookline, 1975.

[BAL] BALLMANN, W., Nonpositively curved manifolds of higher rank, Ann. of Math. **122** (1985), 597–609.

[BB] BAILY, W. L., Jr. & BOREL, A. Compactification of arithmetic quotients of bounded symmetric domains, Ann. of Math. **84** (1966), 442–528.

[BER1] BERGER, M. Sur les groupes d'holonomie homogènes des variétés a connexion affine et des variétés riemanniennes, Bull. Soc. Math. France **83** (1955), 279–330.

[BER2] BERGER, M. Sur les variétés d'Einstein compactes, C.R. IIIe Réunion Math., Expression Latine, Namur (1965), 35–55.

[BGS] BALLMANN, W., GROMOV, M. & SCHROEDER, V. *Manifolds of Nonpositive Curvature*, Progress in Math., Vol. 61, Birkhäuser, Boston–Basel–Stuttgart, 1985.

[BI] BISHOP, E. Mappings of partially analytic spaces, Amer. J. Math. **83**, (1961), 209–242.

[BM] BANDO, S. & MABUCHI, T. Uniqueness of Einstein Kähler metrics modulo connected group actions, *Algebraic Geometry*, Sendai, Advanced Studies in Pure Math., Kinokuniya, Tokyo & North–Holland, Amsterdam, New York, Oxford, 1985.

[BO1] BOREL, A. Density properties for central subgroups of semi–simple groups without compact components, Ann. of Math. **72** (1960), 179–188.

[BO2] BOREL, A. Compact Clifford–Klein forms of symmetric spaces, Topology **2** (1963), 111–122.

[BO3] BOREL, A. Pseudo–concavité et groupes arithmétiques, in *Essays on Topology and Related Topics*, articles dedicated to G. de Rham, ed. by A. Haefliger and R. Narasimhan, Springer–Verlag, Berlin–Heidelberg–New York, 1970, p.70–84.

[BO4] BOREL, A. On the curvature tensor of the Hermitian symmetric manifolds, Ann. of Math. **71** (1970), 508–521.

[Bo5] Borel, A. Some metric properties of arithmetic quotients of symmetric spaces and an extension theorem, J. Diff. Geom. **6** (1972), 543–560.

[BS] Burns & Spatzier Manifolds of nonpositive curvature and their buildings, Inst. Hautes Études Sci. Publ. Math. **65** (1987), 5–34.

[BT] Bedford, E. & Taylor, B. A. A new capacity for p.s.h. functions, Acta Math. **149** (1982), 1–39.

[CCL] Chen, E.–H., Cheng, S.–Y. & Lu, Q.–K. On the Schwarz lemma for complete Kähler manifolds, Scientia Sinica **22** (1979), 1238–1247.

[CE] Cheeger J. & Ebin, D. G. *Comparison Theorems in Riemannian Geometry*, North–Holland, Amsterdam, 1975.

[Che] Chevalley, C. *Theory of Lie Groups*, Princeton Univ. Press, Princeton, 1946.

[Chee] Cheeger, C. Some examples of manifolds of nonnegative curvature, J. Diff. Geom. **8** (1973), 623–628.

[CM] Cao, H.–D. & Mok, N. Holomorphic immersion between complex hyperbolic space forms. Preprint.

[Cor] Corlette, K. Flat G–bundles with canonical metrics, J. Diff. Geom. **28** (1988), 623–628.

[CV] Calabi, E. & Vesentini, E. On compact locally symmetric Kähler manifolds, Ann. of Math **71** (1960), 472–507.

[CY] Chen, Z.–H. & Yang, H.–C. On the Schwarz lemma for complete Hermitian manifolds, Proc. Intern. Conf. Several Complex Variables, Hangzhou, ed. by Kohn–Lu–Remmert–Siu, Birkhäuser, Boston, 1984, p.99–116.

[De] Demailly, J.–P. Champs magnétiques et inégalités de Morse pour la $\bar{\partial}$ cohomologie, Ann. Inst. Fourier, **35** (1985) 189–225.

[Dru] Drucker, D. *Exceptional Lie Algebras and the Structure of Hermitian Symmetric Spaces*, Memoirs A.M.S. **208**, Providence, 1978.

[Ebe] Eberlein, P. Rigidity of lattices of non–positive curvature, Ergod. Th. & Dyn. Sys. **3** (1983), 47–85.

[El] El Mir, H. Sur le prolongement des courants positifs fermés, Acta Math. **153** (1984), 1–45.

[Eh] Ehresmann, C. Sur la notion d'espace complet en géométrie différentielle, C. R. Acad. Sci. Paris **202** (1936), 2033.

[ES] Eells, J. & Sampson, H. Harmonic maps of Riemannian manifolds, Amer. J. Math. **86** (1964), 109–160.

[FE] FEDER, S. Immersions and imbeddings in complex projective spaces, Topology **4** (1965), 143–158.

[FG] FISCHER, G. & GRAUERT, H. Lokale–triviale Familien kompakter komplexer Mannigfaltigkeiten, Nach. Akad. Wissen. Göttingen Math.–phy. Kl. II (1965), 89–94.

[FI] FISCHER, G. *Complex Analytic Geometry*, Lecture Notes Math., Vol. 538, Springer–Verlag, Berlin–Heidelberg–New York, 1976.

[FN] FORNÆSS, E. & NARASIMHAN, R. The Levi problem on complex spaces with singularities, Math. Ann. **248** (1980), 47–72.

[FRA] FRANKEL, L. Manifolds with positive curvature, Pacific J. Math. **11** (1961), 165–174.

[GH] GRIFFITHS, P. & HARRIS, J. *Principles of Algebraic Geometry*, Pure & Appl. Math., Wiley–Interscience Publishers, New York, 1978.

[GR] GRAUERT, H. & REMMERT R., Theory of Stein Spaces, Grundlehren der mathematische Wissenschaften, Springer–Verlag, Berlin–Heidelberg–New York, 1979.

[GRA1] GRAUERT, H. On Levi's problem and the imbedding of real–analytic manifolds, Ann. of Math. **68** (1958), 460–472.

[GRA2] GRAUERT, H. Über Modifikationen und exzeptionelle analytische Menge, Math. Ann. **146** (1962), 331–368.

[GRI1] GRIFFITHS, P. Hermitian differential geometry, Chern classes and positive vector bundles, in *Global Analysis*, articles dedicated to K. Kodaira, Univ. of Tokyo Press, Tokyo & Princeton Univ. Press, Princeton, 1969, p.185–251.

[GRI2] GRIFFITHS, P. Two theorems on extension of holomorphic mappings, Invent. Math. **41** (1977), 33–43.

[GRO] GROTHENDIECK, A. Sur la classification des fibrés holomorphes sur la sphère de Riemann, Amer. J. Math. **79** (1957), 121–138.

[GROM] GROMOV, M. Lectures in Collège de France, Paris, 1981.

[GS] GRIFFITHS, P. & SCHMID, W. Locally homogeneous complex manifolds, Acta Math. **123** (1969), 253–302.

[G–ST] GUILLEMIN V. & STERNBERG, S. *Symplectic Techniques in Physics*, Cambridge Univ. Press, Cambridge 1984.

[HA] HARISH–CHANDRA Representation of semisimple Lie groups, VI, Amer. J. Math. **78** (1956), 564–628.

[HEL] HELGASON, S. *Differential Geometry, Lie Groups and Symmetric Spaces*, Academic Press, New York, 1978.

[HÖR] HÖRMANDER, L. *An Introduction to Complex Analysis in Several Variables*, North–Holland, Amsterdam–London, 1973.

[HP] HARVEY, R. & POLKING, J. Extending analytic objects, Comm. Pure & Appl. Math., **28** (1975), 701–727.

[HUA] HUA, L.–K. *Harmonic Analysis of Functions of Several Complex Variables in Classical Domains*, English translation, Amer. Math. Soc., Providence, 1963.

[JA] JACOBSON, N. *Lie Algebras, Interscience Tracts,* No. 10, J. Wiley & Sons, New York, 1962.

[JY1] JOST, J. & YAU, S.–T. Harmonic mappings and Kähler manifolds, Math. Ann. **262** (1983), 145–166.

[JY2] JOST, J. & YAU, S.–T. The strong rigidity of locally symmetric complex manifolds of rank one and finite volume, Math. Ann. **275** (1986), 291–304.

[JY3] JOST, J. & YAU, S.–T. On the rigidity of certain discrete groups of algebraic varieties, Math. Ann. **278** (1987) 481–496.

[KM] KODAIRA, K. & MORROW, J. *Complex Manifolds*, Holt, Reinhart & Winston, New York, 1971.

[KN] KOBAYASHI, S. & NOMIZU, N. *Foundations of Differential Geometry* I, II, Interscience, New York, 1963, 1969.

[KO] KOBAYASHI, S. On compact Kähler manifolds with positive Ricci curvature, Ann. of Math. **74** (1961). 570–574.

[KOE] KOECHER, M. Zur Theorie der Modulfunktionen n–ten Grades I. Math. Zeit. **59** (1954), 399–416.

[KOH] KOHN, J.–J. Harmonic integrals on strongly pseudo–convex manifolds I, Ann. of Math. **78** (1963), 206–213.

[KOS] KOSTANT, B. On the conjugacy of real Cartan subalgebras, Proc. Nat. Acad. Sci. **41** (1955), 967–970.

[KW] KORANYI, A. & WOLF, J. A. Realizations of Hermitian symmetric spaces as generalized half–planes, Ann. of Math. **81** (1965), 265–288.

[LA] LAZARSFELD, R. Some applications of the theory of positive vector bundles, *Complete Intersections* (Acireale, 1983), Lecture Notes in Math., Vol. 1092, Springer–Verlag, Berlin–Heidelberg–New York, 1984, p.29–61.

[LE] LELONG, P. Fonctions entières (n variables) et fonctions plurisousharmoniques d'ordre fini dans $\mathbf{C}^n$, J. Analyse Math. **12** (1964), 365–407.

[LIV] LIVNE, R. A. On certain covers of the universal elliptic curve, Ph.D. thesis, Harvard Univ., 1981.

[LM] LAI, K.–F. & MOK, N. On a vanishing theorem on irreducible quotients of finite volume of polydiscs, Séminaire d'Analyse Lelong–Dolbeault–Skoda, Lecture Notes in Math., Vol. 1198, Springer–Verlag, Berlin–Heidelberg–New York, 1986, p.163–171.

[MA1] MATSUSHIMA, Y. Sur la structure du groupe d'homéomorphismes analytiques d'une certaine variété kählérienne, Nagoya Math. J. **11** (1957), 145–150.

[MA2] MATSUSHIMA, Y. On Betti numbers of compact locally symmetric Riemannian manifolds, Osaka Math. J. **14** (1962), 1–20.

[MAR] MARGULIS, G. A. Discrete groups of motion of manifolds of nonpositive curvature, A.M.S. Transl. (2) **109** (1977), 33–45.

[MA–S] MATSUSHIMA, Y. & SHIMURA, G. On the cohomology groups attached to certain vector–valued differential forms on products of the upper half–plane, Ann. of Math. **78** (1963), 417–449.

[MO] MORI, S. Projective manifolds with ample tangent bundles, Ann. of Math. **110** (1979), 593–606.

[MOK1] MOK, N. The holomorphic or anti–holomorphic character of harmonic maps into irreducible quotients of polydiscs, Math. Ann. **272** (1985), 197–216.

[MOK2] MOK, N. Metric rigidity theorems on locally symmetric Hermitian spaces, Proc. Natl. Acad. Sci. U.S.A. **83** (1986), 2288–2290.

[MOK3] MOK, N. Uniqueness theorems of Hermitian metrics of seminegative curvature on quotients of bounded symmetric domains, Ann. of Math. **125** (1987), 105–152.

[MOK4] MOK, N. Uniqueness theorems of Kähler metrics of semipositive bisectional curvature on compact Hermitian symmetric spaces, Math. Ann. **276** (1987), 177–204.

[MOK5] MOK, N. The uniformization theorem for compact Kähler manifolds of nonnegative holomorphic bisectional curvature, J. Diff. Geom. **27** (1988), 179–214.

[MOK6] MOK, N. Strong rigidity of irreducible quotients of polydiscs of finite volume, Math. Ann. **282** (1988), 555–477.

[MOK7] MOK, N. Local biholomorphisms between Hermitian locally symmetric spaces of non–compact type. Preprint.

[MOO] MOORE, C. C. Compactifications of symmetric spaces. II (the Cartan domains), Amer. J. Math. **86** (1964), 258–378.

[MOS1] MOSTOW, G. D. *Strong Rigidity of Locally Symmetric Spaces*, Ann. Math. Studies **78**, Princeton Univ. Press, 1973.

[MOS2] MOSTOW, G. D. On a remarkable class of polyhedra in complex hyperbolic space, Pacific J. Math. **86** (1980), 171–276.

[MS] MYERS S. B. & STEENROD, N. The group of isometries of a Riemannian manifold. Ann. of Math. **40** (1939), 400–416.

[MO–S] MOSTOW, D. & SIU, Y.–T. A compact Kähler surface of negative curvature not covered by the ball, Ann. of Math. **112** (1980), 321–360.

[MT] MOK, N. & TSAI, I.–H. Rigidity of convex realizations of irreducible bounded symmetric domains of rank ≥ 2. In preparation.

[MUM1] MUMFORD, D. Hirzebruch's proportionality theorem in the non–compact case, Invent. Math. **42** (1977), 239–272.

[MUM2] MUMFORD, D. An algebraic surface with K ample, $(K^2) = 9$, $p_g = q = 0$, Amer. J. Math. **10** (1979), 233–244.

[NA1] NARASIMHAN, R. Holomorphically complete complex spaces, Amer. J. Math. **82** (1961), 917–934.

[NA2] NARASIMHAN, R. On the homology group of Stein spaces, Inv. Math. **2** (1967), 377–385.

[NA3] NARASIMHAN, R. *Several Complex Variables*, Chicago Lectures in Math., Chicago Univ. Press, Chicago–London, 1971.

[NA–T] NADEL, A. & TSUJI, H. Compactification of complete Kähler manifolds of negative curvature, J. Diff. Geom. **28** (1988), 503–512.

[NN] NEWLANDER A. & NIRENBERG, L. Complex analytic coordinates in almost complex manifolds, Ann. of Math. **65** (1957), 391–404.

[NT] NAKAGAWA, H. & TAKAGI, R. On locally symmetric Kähler manifolds in a complex projective space, J. Math. Soc. Japan **28** (1976), 638–667.

[PRA] PRASAD, G. Strong rigidity of Q–rank 1 lattices. Invent. Math. **21** (1973), 255–286.

[RA] RAGHUNATHAN, M. S. *Discrete Subgroups of Lie Groups*, Springer–Verlag, Berlin–Heidelberg–New York, 1972.

[ROS] ROS, A. A characterization of seven compact Kähler submanifolds by holomorphic pinching, Ann. of Math. **121** (1985), 377–382.

[ROY] ROYDEN, The Ahlfors–Schwarz lemma in several complex variables, Math. Helv. **55** (1980), 547–558.

[SA1] SATAKE, I. On compactifications of the quotient spaces for arithmetically defined discontinuous groups, Ann. of Math. **72** (1960), 555–580.

[SA2] SATAKE, I. *Algebraic Structures of Symmetric Spaces*, Iwanami Shoten, Publishers, & Princeton Univ. Press, Princeton, 1980.

[SER1] SERRE, J.–P. Représentations linéaires et espaces homogènes kählériens des groupes de Lie compacts, Séminaire Bourbaki **100** (1954), Paris.

[SER2] SERRE, J.–P. *Complex Semisimple Lie Algebras*, Springer–Verlag, Berlin–Heidelberg–New York, 1987.

[SHI] SHIMIZU, H. On discontinuous groups operating on the product of upper half–planes, Ann. of Math. **77** (1963), 33–71.

[SI] SIBONY, N. Quelques problèmes de prolongement de courants en analyse complexe, Duke Math. J. **52** (1985), 157–197.

[SIM] SIMON, J. On the transitivity of holonomy systems, Ann. of Math. **76** (1962), 213–234.

[SIU1] SIU, Y.–T. Analyticity of sets associated to Lelong numbers and the extension of closed positive currents, Invent. Math. **27** (1974), 53–156.

[SIU2] SIU, Y.–T. The complex analyticity of harmonic maps and the strong rigidity of compact Kähler manifolds, Ann. of Math. **112** (1980), 73–111.

[SIU3] SIU, Y.–T. Strong rigidity of compact quotients of exceptional bounded symmetric domains, Duke Math. J. **48** (1981), 857–871.

[SIU4] SIU, Y.–T. A vanishing theorem for semipositive line bundles on non–Kähler manifolds, J. Diff. Geom. **19** (1984), 431–452.

[SIU5] SIU, Y.–T. Strong rigidity for Kähler manifolds and the construction of bounded holomorphic functions, in *Discrete Groups in Geometry and Analysis*, papers in honor of G. D. Mostow, ed. by R. Howe, Birkhäuser, Boston–Basel–Stuttgart, 1987, p.124–151.

[SKO1] SKODA, H. Applications des techniques L^2 à la theorie des idéaux d'un algèbre de fonctions holomorphes avec poids, Ann. Sci. Éc. Norm. Sup. (4) **5** (1972), 548–580.

[SKO2] SKODA, H. Prolongement des courants positifs fermés de masses finies, Invent. Math. **66** (1982), 361–376.

[SL] SEMINAIRE SOPHUS LIE "Théorie des algèbres de Lie, Topologie des groupes de Lie", 1954/1955, École Norm. Sup., Paris.

[STO] STOLZENBERG, G. *Volumes, Limits and Extension of Analytic Varieties*, Lecture Notes in Math., Vol. 19, Springer–Verlag, Berlin–New York, 1966.

[SY] SIU, Y.–T. & YAU, S.–T. Compactifications of negatively curved complete Kähler manifolds of finite volume, in *Seminar on Differential Geometry*, ed. by Yau., Ann. of Math. Studies, Vol. 102, Princeton Univ. Press, Princeton, 1982, p.363–380.

[THI] THIE, P. The Lelong number in a point of a complex analytic set, Math. Ann. **172** (1967), 269–312.

[TO] TO, W.–K. Hermitian metrics of seminegative curvature on quotients of bounded symmetric domains, Invent. Math. **95** (1989), 559–578.

[TSA] TSAI, I.–H. Rigidity of holomorphic maps between Hermitian symmetric spaces of compact type. Preprint.

[UE] UENO, K. *Classification Theory of Algebraic Varieties and Compact Complex Spaces*, Lecture Notes Math., Vol. 439, Springer–Verlag, Berlin–Heidelberg–New York, 1975.

[VA] VARADARAJAN, V. S. *Lie Groups, Lie Algebras and Their Representations*, Prentice–Hall, Engelwood Cliffs, 1974.

[WOL1] WOLF, J. A. *Spaces of Constant Curvature*, McGraw–Hill, New York, 1967.

[WOL2] WOLF, J. A. Fine structure of Hermitian symmetric spaces, in *Geometry of Symmetric Spaces*, ed. by Boothby–Weiss, Marcel–Dekker, New York, 1972, p.271–357.

[YAN] YANG, P. C. On Kähler manifolds with negative holomorphic bisectional curvature, Duke Math. J. **43** (1976), 871–874.

[YAU] YAU, S.–T. A general Schwarz lemma for Kähler manifolds, Amer. J. Math. **100** (1978), 197–203.

[ZIM] ZIMMER, R. J. *Ergodic Theory and Semisimple Groups*, Monographs in Mathematics, Birkhäuser, Boston–Basel–Stuttgart, 1984.

[ZHO] ZHONG, J.–Q. The degree of strong nondegeneracy of the bisectional curvature of exceptional bounded symmetric domains, Proc. Intern. Conf. Several Complex Variables, Hangzhou, ed. by Kohn–Lu–Remmert–Siu, Birkhäuser, Boston, 1984, p.127–139.

INDEX

F

G

H

I

S

T

V

W